Working with Map Projections

A Guide to Their Selection

Working with Map Projections

A Guide to Their Selection

Fritz C. Kessler and Sarah E. Battersby

CRC Press is an imprint of the
Taylor & Francis Group, an **informa** business

CRC Press
Taylor & Francis Group
6000 Broken Sound Parkway NW, Suite 300
Boca Raton, FL 33487-2742

Printed on acid-free paper

International Standard Book Number-13: 978-1-138-30498-7 (Paperback)

Visit the Taylor & Francis Web site at
http://www.taylorandfrancis.com

and the CRC Press Web site at
http://www.crcpress.com

This book is dedicated to John Snyder and Waldo Tobler, two giants in the map projection field who did much to inspire the present work.

Contents

Preface

As a map maker, you have collected the appropriate data. But you are now faced with the task of needing to symbolize that data. You feel confident in most cartographic processes such as choosing a visual variable or an appropriate symbolization method. Yet, knowing which map projection to choose to represent your data is something few map makers truly understand or give serious consideration to when making a map. A projection fundamentally impacts three aspects of the map making process.

First, some symbolization methods require that specific kinds of measurements take place. For example, isarithmic maps are created using spatial interpolation methods. Some of these interpolation methods use distance measurements to estimate data values at unknown locations, which are then used to create surfaces. If the distance measurements are not accurate, then the isarithmic surface representation will not be accurate. Thus, the mapped surface does not accurately reflect the original data values and the underlying phenomenon. In short, if the isarithmic symbolization method requires measurements to be made, then the projection needs to be selected where those measurements are accurate.

Second, distortion is an inevitable consequence of the projection process. Distortion can impact the overall shape and size of the mapped area. In turn, this distortion can also impact the appearance of the symbols used to represent the data. For example, proportional symbol maps use the same symbol to represent data collected from individual locations. The symbols at each location are sized in relation to the underlying data values. Since distortion can alter the shape and size of the mapped area, the location of the symbols can also change, which impacts the overall apparent distribution and the map reader's ability to extract meaningful patterns in the data.

Third, the map's overall design and aesthetic can be greatly impacted by the projection choice. Some of these design and aesthetic choices include the projection's overall shape, the required visual tasks projections, map scale and levels of projection distortion. Projections have overall shapes (e.g., circular, rectangular, and oval) that need to be considered as a map design variable. For example, a projection that is circular in shape placed into a rectangular page space leaves gaps in the corners which may not be aesthetically pleasing. The map maker also needs to be aware of the visual tasks required of the map reader. For instance, using interrupted projections is not a good strategy for proportional flow lines as the flow lines cross over the interruptions causing difficulties interpreting the pattern of flow lines. In addition, global-scale maps suffer disproportionate levels of distortion near the map's periphery. To avoid placing the symbols near the map's periphery where distortion is highest, the projection's center should be wisely chosen. Choosing an appropriate projection center is especially true for dot maps where dot clusters and their patterns falling near the periphery can be greatly distorted to the point of being unrecognizable.

The crux of this book is to offer guidance to the map maker on selecting a map projection. Accepting the default projection or its parameters should not be an option. Neither should the map maker use a projection just because that projection is commonly used. Doing so will reduce the effectiveness in representing the data's pattern and diminish the ability of your map reader to carry out any expected visual tasks. This book is designed to get you to pause and think about the projection, its impact on symbolization methods, and how to work with a projection to maximize the overall map information and its design.

Acknowledgments

Fritz: One morning early in the fall semester of 2017, I met with Sarah on the campus of Penn State University when she was in town to give a talk in the Geography Department on her spatial binning research. It was during this meeting when I asked her if she would be interested in co-authoring a book with me on a topic about map projections, a topic about which both of us are passionate. I gave a basic overview of the scope and purpose of the book idea I had circulating in my head. After I finished my presentation, she immediately agreed to work on this collaborative work. To say the least, I was excited. Throughout the writing process, her substantial expertise and insights on projections have greatly added to the overall quality of this book. It has been an honor to work with her on this project. So, thank you Sarah for teaming up with me again to work on map projections!

Sarah: I've known Fritz for many years—long enough that I can't even remember when or where we met. I imagine it was at an International Cartography Association Commission on Map Projections workshop or one of the many map projection-related sessions we end up in together at conferences like AutoCarto and the American Association of Geographers annual meeting. Since we met and discovered our mutual interest in map projections and how they are used and abused, we've collaborated on a few projects and papers. When Fritz asked me to join him in working on a whole book dedicated to the topic of map projections, there was no question in my mind that this was something I absolutely wanted to collaborate with him on. So, thank you Fritz for many years of great collaborations. I learn something new in every project we've worked on (and have a lot of fun doing it).

Our collaborative effort would not have been possible without the support from others who share our passion for map projections. We are indebted to Daniel "daan" Strebe, Bojan Šavrič, and Mike Finn for their ongoing support and commentary in all things projection-related. Their added perspectives helped to refine our own thoughts on map projections and writing this book. Special additional thanks to daan for help with his Geocart software when we needed to bend it to our will. We also extend thanks to Elaine Guidero who generously provided her editorial assistance to our writing.

We would also like to acknowledge the importance of the International Cartographic Association's Commission (ICA) on Map Projections (http://ica-proj.kartografija.hr/home.en.html). As you will read in this book, the field of map projections does not have a single academic or professional home. The topic is scattered among many different related fields like surveying and geodesy. However, the ICA hosts a centralized home for the dissemination of map projection topics. The commission is currently headed by the gracious efforts of Dr. Miljenko Lapaine of the University of Zagreb, Hungry.

Authors

Fritz C. Kessler, PhD, is an associate teaching professor in the geography department at Penn State University, where he teaches a course on datums, map projections, and grid systems. Dr. Kessler earned a PhD in geography at the University of Kansas. He is a member of the International Cartographic Association Commission on Map Projections and is a past president of the North American Cartographic and Information Society.

Sarah E. Battersby, PhD, is a research scientist at Tableau Software. Dr. Battersby earned a PhD in geography (focusing on cognitive cartography) at the University of California at Santa Barbara. She is a member of the International Cartographic Association Commission on Map Projections and is a past president of the Cartography and Geographic Information Society.

Part I

Projection Basics, Cartographic Symbolization, Projection Influences on People's Mental Maps, and Selecting Projections

This book is divided into two parts. Part I provides foundational knowledge about map projections, how projections influence people's world mental maps, considerations necessary when selecting a projection, and the cartographic symbolization process involved when making a map. Collectively, these topics are beneficial background preparation for Part II of the book, where we focus on a detailed discussion of working with and selecting projections for different map symbolization methods.

To begin, we will briefly discuss the four guiding themes from Part I that are important for the reader to review as preparation for examining the material included in Part II.

1. **General Background on Map Projections.** (See Chapter 2 for an overview of projections, what they are, and how they are created.)

   The projection is an inherently mathematical process that takes latitude and longitude values on Earth's spherical surface and projects them to a map. The projection process can preserve certain spatial relationships that are found on Earth's surface such as areas, angles, distances, and directions, but it cannot preserve all of these spatial relationships at the same time. Not all projections preserve a specific spatial relationship property, and those that do have limitations in the way the property is preserved. The inability for any given projection to preserve all of Earth's spatial relationships is due to the change in dimensions from the sphere (Earth, 3-dimensional) to a plane (map, 2-dimensional). This loss of dimension can be described as distortion. Every projection includes distortion and the amount, type, and location of distortion can impact the map's ability to fulfill its purpose. Thus, you, the map maker, must be familiar with projections, their properties, and their distortion patterns. This knowledge is essential to the process of selecting an appropriate projection that will to support the map's purpose and lead to an easy to interpret visualization.

2. **Overview of the Cartographic Process.** (See Chapter 3 for a discussion of how phenomena on Earth's surface are abstracted and symbolized on a map.)

   Earth's surface is complex and full of details, and a map can only present a simplification and abstraction of that complexity. No map can preserve and display all of Earth's complexities and details. There within lies the struggle. How can we design maps that capture the essential data from a phenomenon and symbolize it so that it appropriately represents the phenomenon as it exists on Earth's surface? To overcome this struggle, the map maker begins by conceptualizing how a specific phenomenon on Earth's surface is distributed, for instance, does it exist everywhere or only in isolated instances (i.e., continuous or discrete), and how does the phenomenon change across space (i.e., abruptly or smoothly). Next, the map maker collects the necessary data to represent the phenomenon as points, lines, or areas. The collected data is described as having a specific measurement level (e.g., qualitative or quantitative). This assignment is needed to select a visual variable that is appropriate for the data measurement level. Specifically, the visual variable helps the map reader intuitively perceive the graphic marks placed on the map to represent the collected data. Next, the map maker selects an appropriate symbolization method (e.g., choropleth, proportional symbol, dot method, isarithmic, dasymetric, or cartogram) that incorporates all the above considerations while representing the phenomenon's distribution on Earth's surface. Finally, the map maker identifies the projection that will best highlight the spatial patterns to reflect the original phenomenon being mapped.

3. **Map Projection's Influence on People's Mental Maps.** (See Chapter 4 that examines the influence that projections have on people's mental maps.)

   Everyone has a mental atlas, so to speak. That atlas contains mental maps that reflect the perceived size and shape of the world and its many landmasses. For most people, those maps are constructed according to what they have seen on map, and the projections that the map reader is most familiar with will likely serve as the mental atlas's framework. As we will discuss in Chapter 2, every projection contains distortion. Thus, the maps that are instrumental in helping construct a person's mental atlas are distorted and could influence the way the world is remembered and interpreted. This chapter highlights research on how projection may influence mental maps, and how those mental maps are compared to reality.

4. **Selecting a Map Projection.** (See Chapter 5 for an examination of the decisions and trade-offs involved when selecting a projection.)

   Although there are claims to the contrary, there really is no single *best* projection. Any projection distorts Earth and everything upon it. If you agree with this statement, then you must accept that no one projection is best suited for a given map purpose. In order to select a projection, the map maker must be willing to enter into a give and take relationship; some projection characteristics and parameters will be favorable to support a specific map's purpose while others will need to be sacrificed.

   When selecting a projection, there are several questions the map maker must ask. Some of these questions focus on the need for the projection to preserve a specific property so the map reader can use the map for a given purpose. Other questions examine the geographic extent of the landmass and data with respect to the distribution of distortion across the map's surface. Additional questions focus on the aesthetics (e.g., how the overall shape of the projection fits the display space and should the poles be represented as lines or points) and interpretability (e.g., does the inclusion of the graticule provide the map reader with a geographic context when interpreting the map). These questions are discussed at length in this chapter explaining how each impacts the appearance and function of the map.

   Until the advent of computers, there were only a few written guidelines available to recommend a projection. Even so, those guidelines were not very helpful as they expected those using the guideline to have a certain level of projection knowledge. Today, automated projection selection guidelines allow map makers with limited projection knowledge to interactively work through decisions needed to select a projection. The logic of several automated projection selection guidelines will be examined in this chapter.

Readers who are unfamiliar with any of these topics are encouraged to read through Part I before examining Part II. Readers with a sufficient background in the topics covered in Part I can skip ahead and read chapters of interest in Part II. We note that the chapters in Part II were not intended to be read in a particular order; thus, they can be read individually as they present information relevant to your particular map type of interest.

1

Introduction

For most of our professional careers, we (Fritz and Sarah) have worked with, taught, and researched various facets of map projections. We thoroughly enjoy the subject. However, we have come to realize that many people who work with geospatial data need guidance when working with projections on a mapping project. Specifically, we find that many people need help selecting a projection, and that this is a process which is fraught with challenges. Beşdok et al. (2012, p. 666) offer that "map projections are among the most difficult topics" while Muehrcke and Muehrcke (1978, p. 457) go a step further and state that projections are the "most bewildering aspect of map appreciation."

If you are reading this book, then you most likely work with projections and have found the subject challenging or even bewildering. Projections don't have to be either; in fact, working with projections can be fun and a rewarding part of making maps. We hope that reading this book will help demystify the challenges you may face when working with projections. To do this, we present five themes that describe the common challenges that people face when working with projections. These themes helped organize our thoughts when writing this book. First, projections are inherently mathematical. Second, projection terminology is filled with confusing jargon. Third, the projection literature is scattered. Fourth, the projection is often overlooked as a variable in the map design process. Fifth, selecting a projection is not a clear-cut process. We will briefly discuss each of these themes here, and we have also woven them into the discussion throughout this book.

Mathematical Complexity

Projections are inherently mathematically based, and this can make the topic intimidating to many. In fact, John Snyder (1993, p. 276) once stated "working with projections still strikes fear in the hearts of many trained cartographers and geographers because of the mathematical aspects." For those readers who are familiar with mathematics (primarily calculus and differential equations) books are available that may enlighten your mathematical understanding of projections. However, we suspect that for most readers learning about the mathematics of projections is scary and viewed as not necessary. To some extent, we agree. Someone who works with geospatial data doesn't necessarily need to know a projection equation as the computations are built into mapping software and operate "behind the scenes," so to speak. However, we feel that some understanding of the mathematics behind projections is important and, as examples, we include a few mathematical equations in our book. To provide a bit of mathematical grounding on our projection discussion, we present two mathematical threads that are relevant to people who work with projections: ellipsoidal versus sphere Earth models and projection parameters.

Thread 1: Ellipsoid vs. Sphere Earth Models

The comparison of ellipsoidal and spherical Earth models is overlooked in most elementary discussions of projections. Nonetheless, there is an important distinction between the spherical and ellipsoidal forms of projection equations. Earth's exact shape is complex, and to project this shape, the complexity needs to be simplified. The simplest way to think of the Earth's shape is to model it as a sphere. A sphere has a constant radius and, as such, the mathematical equations that are used to calculate the x and y plotting coordinates resulting from projection are relatively simple. A consequence of this sphere model, however, is the diminished accuracy of the x and y plotting coordinates. A more accurate model of Earth's shape is an ellipsoid. Due to rotational and gravitational forces, Earth's shape bulges at the equatorial regions and is flattened in the polar areas. To better approximate Earth's shape, an ellipsoid can be used. With an ellipsoid, a unique radius value for its equatorial and polar axes are specified which approximates overall Earth's shape better than a sphere. Since the ellipsoid is a closer approximation of Earth's shape, the inherent accuracy of the projection equations' calculation of the x and y plotting coordinates is greater; however, the equations are much more complex. To illustrate this difference, we'll start by examining Equations 1.1 and 1.2, where we present equations used to calculate the x and y plotting coordinates for a cylindrical equal area projection on a *sphere*.

$$x = R(\lambda - \lambda_0)\cos\phi s \tag{1.1}$$

$$y = R\sin\phi / \cos\phi s \tag{1.2}$$

Here, λ represents longitude values, φ represents latitude values, λ_0 is the central meridian, φs is the latitude assigned as the standard line, and R is the radius of the sphere that corresponds to the final map scale (note that these are common symbols used to represent latitude and longitude and will be used as such throughout this book). For now, it is not important to understand the specifics of what these parameters mean and how they impact a projection. We will discuss each in more detail in the chapters to come.

A model that corresponds more closely to Earth's true shape uses an ellipsoid and yields a more accurate calculation of the x and y plotting coordinates; this, in turn, creates a more accurate map. This added accuracy requires more complex equations that take into consideration Earth's nonspherical shape (Equations 1.3–1.6).

$$k_0 = \cos\phi s / (1 - e^2 \sin^2 \phi s)^{½} \quad (1.3)$$

$$q = \left(1 - e^2\right) [\sin\phi / (1 - e^2 \sin^2 \phi) - \left[1/(2e)\right] \ln[(1 - e\sin\phi / 1 + e\sin\phi)] \quad (1.4)$$

$$x = ak_0(\lambda - \lambda_0) \quad (1.5)$$

$$y = aq / (2k_0) \quad (1.6)$$

In Equations 1.3–1.6, φ, λ, λ_0 and φs is the same as seen in Equations 1.1 and 1.2. Equations 1.3–1.6 include the additional variables k_0 and q that compute x and y, a which defines the semi-major axis of the reference ellipsoid, and e which represents the eccentricity of the ellipsoid, or the amount of departure from a circle that characterizes the ellipsoid. The ellipsoidal forms of projection equations are used for navigational charts or topographic maps where accurate distances or directions are calculated. Highly accurate measurements are not associated with most thematic maps and thus, a spherical Earth model can be used. Overall, it is important to realize that the map purpose dictates the appropriate Earth model to use and in turn the projection equations that are used ensure those accurate measurements can be determined using a map. Throughout this book, we will present examples where calculations are carried out using spherical and ellipsoidal projection equations so that the differences in results can be realized and compared.

Thread 2: Projection Parameters

The second thread focuses on understanding the various parameters associated with a given projection equation. As seen in Equations 1.1–1.6, there are different parameters associated with the calculation of x and y plotting coordinates. These and other parameters are associated with a projection that helps control, for example, the projection's center and the distribution of the projection's distortion pattern. You have already seen three common projection parameters: central meridian (λ_0), central latitude (φ_0), and standard line (φ_s). The first two parameters define the projection's center. The third parameter controls the location of the line of no distortion (called a standard line) which, in many cases, coincides with a specific line of latitude and is referred to as a standard parallel. Note that not all projection equations include each of these parameters. In fact, some projections have additional parameters that offer greater control over a projection's appearance and distortion pattern. You do not need to study every projection equation to be able to take advantage of the available parameters. Rather, a general familiarity of projection equations and the possible parameters is all that you need to work with projections. Understanding which projection parameters are worth knowing will be discussed throughout this book.

Map Projection Terminology

Like most fields of study, the projection field is filled with specialized concepts and terminology. By itself, this fact shouldn't cause concerns. Yet, to the casual reader, the myriad of terms and their usage poses some challenges. Some projection concepts have different words that have the same meaning and have been used interchangeably throughout the literature. For example, *azimuthal* and *zenithal* both refer to a projection property that preserves directions from the center of the map to any other point on the map. *Zenithal* was more commonly used in the early part of the 20th century and to some extent has disappeared from the literature while *azimuthal* has persisted across time. *Conformal* and *orthomorphic* both refer to a projection property that preserves angular relationships. *Conformal* is more commonly used today but *orthomorphic*, popular in the early 20th century, still appears in the literature. In another instance, the same word can have two meanings. *Azimuthal* can refer to either the projection property that preserves directions on a projection *or* a term used to classify projections that has the property.

Another challenge to working with projections is that some projection terms have often been applied in an inconsistent manner. The way in which a projection's center is described is a good example of this inconsistency. *Aspect* describes the appearance of the graticule (Maling, 1992) and relates to the projection's center. One term describing a projection's aspect is *normal*. Historically, the normal aspect of a projection is usually associated with the arrangement of the graticule that is "simplest" in both mathematical derivation and appearance. To most readers of the projection literature, simplest is a matter of perspective. Figures 1.1A and 1.1B illustrate the normal aspect of the stereographic azimuthal conformal planar and Gall stereographic cylindrical projections, respectively. Figures 1.1C and 1.1D show the transverse aspect of the same two projections, respectively. The two projections in the top row exhibit the simpler graticule arrangement with concentric circles and straight lines. A more visually complex graticule arrangement is shown by the same projections in the bottom row using various types of curves. While it is easy to see that the top row uses a simpler graticule arrangement, this side-by-side comparison is neither convenient nor intuitive to many cartographers and map readers. A more intuitive method, and the one we adopted for this book, is to refer to the projection aspect according to

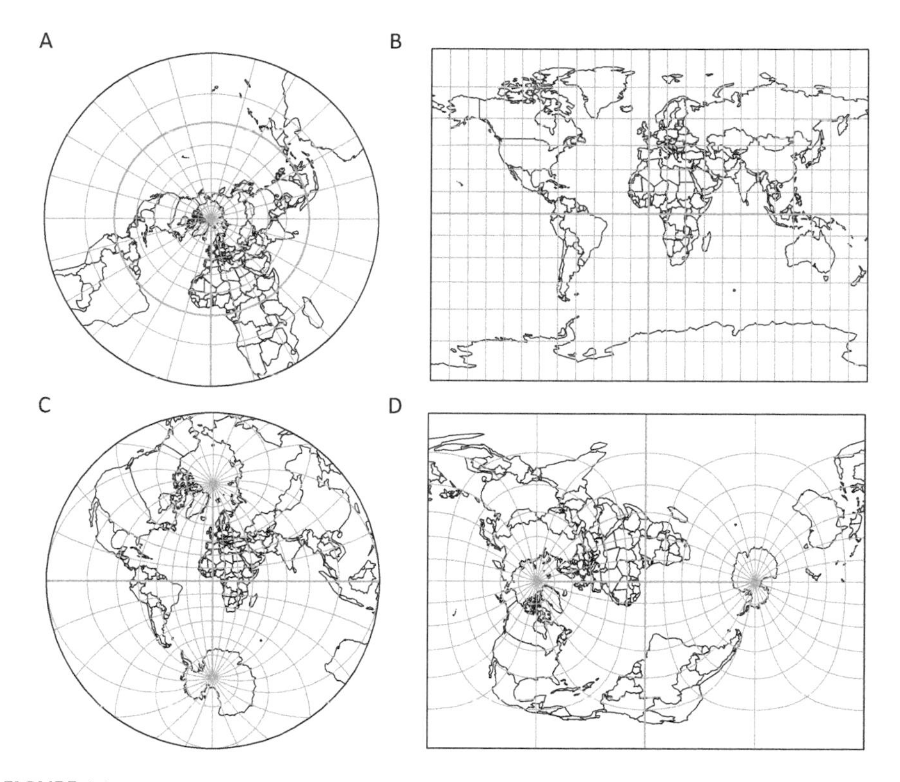

FIGURE 1.1
The normal aspect of the stereographic azimuthal conformal planar (A) and Gall stereographic compromise cylindrical (B) projections. The transverse aspect of the stereographic (C) and Gall stereographic (D) projections.

location-based terms such as polar, oblique, and equatorial-centered at a pole (Figure 1.2A), somewhere between a pole and the equator (Figure 1.2B), and along the equator (Figure 1.2C), respectively. Wherever possible, we adopt this approach throughout the book by using terms that we feel will be more intuitive to a broader audience.

Throughout the book, we have avoided the use of the more complex projection terms, favoring those that are widely used in contemporary projection literature. There are a few specific terms that we use and will define here. To begin, we distinguish *cartographic* and *geographic* scale. Cartographic scale refers to map scale. Large-scale maps show small geographic extents with considerable detail while small-scale maps show greater geographic extents with less detail. Geographic scale refers to the geographic extent that is shown on the map: *Global-scale*, *continental-scale*, and *local-scale* refers to maps that show the entire globe, continent, or a local geographic extent, respectively. We also distinguish between *projection* and *map*. A projection is the mathematical process where Earth's latitude and longitude is projected onto a plane surface or a map. Thus, a projection is a process while the map

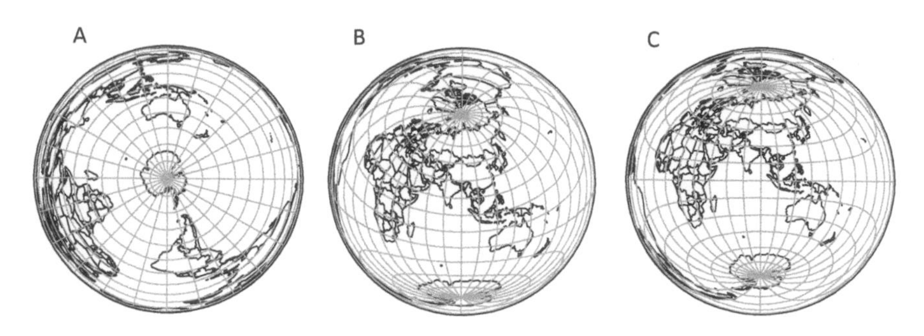

FIGURE 1.2
The polar (A), oblique (B), and equatorial (C) aspect on the Lambert azimuthal equal area planar projection.

results from that process. The terms *area* and *size* also appear throughout the chapters. *Area* refers to a quantifiable measurement such as square-miles, while *size* is a relative descriptor (one country appears larger relative to another country).

Some projection terms are cryptic to read, for instance, *isoanamorphism*, *aphylactic*, and *geodesic*. *Isoanamorphism* defines lines of equal distortion drawn on a projection's surface, *aphylactic* refers to a projection that does not possess a specific property, and *geodesic* describes the shortest path between two points on an ellipsoidal surface. There are glossaries specific to coordinate systems that can help you interpret cryptic coordinate system terms. We recommend three references here that will help you navigate the terminology. The first is the *Glossary of Mapping Sciences*. Thiswork is a collaborative effort between the American Society of Civil Engineers (ASCE), American Congress on Surveying and Mapping (ACSM), and American Society for Photogrammetry and Remote Sensing (ASPRS).. This *Glossary* (ASCE, ACSM, and ASPRS, 1994) contains approximately 10,000 terms from geodesy, mapping, remote sensing, and surveying. The *Glossary of Mapping, Charting, and Geodetic Terms*, published by the United States Defense Mapping Agency (1973), provides detailed information on maps, charts, and associated terms. This glossary is designed to assist readers in understanding complex terms associated with coordinate systems, geodesy, and the broader mapping field.* The *Geodetic Glossary*, published by the National Geodetic Survey (1986), covers many of the same terms as the other two glossaries but focuses on surveying and geodesy terms.† If you are interested in learning more about projection terminology and how researchers have wrestled with

* An online version of the third edition of the *Glossary of Mapping, Charting, and Geodetic Terms* can be found at https://catalog.hathitrust.org/Record/001272002.

† The *Geodetic Glossary* is available at https://www.ngs.noaa.gov/CORS-Proxy/Glossary/xml/NGS_Glossary.xml.

organizing this terminology and naming projections we recommend Maling (1968) and Lee (1944).

Map projection names are another confusing facet of the projection field. There are over 400 projections that have been developed and more, undoubtedly, are on their way. The literature is not consistent on how projections are named. We have adopted the following format in naming a projection in this book. We refer to a projection by its common name (Mercator), property (conformal), and class (cylindrical). The full name of this projection is the Mercator conformal cylindrical with the developer's name capitalized. We feel that this naming convention should provide you with a useful account of the projection, its property, and its class while avoiding confusion with similar projections. For example, the web Mercator compromise cylindrical is not the same projection as the Mercator conformal cylindrical. Thus, we have adopted a naming convention with the first instance of the projection being mentioned in this book (e.g., Mercator conformal cylindrical). Thereafter, we switch to the common name (e.g., Mercator). A complete listing of all projections referenced in this book, including the projection distortion patterns, are included in the Appendix.

Scattered Map Projection Literature

Most fields of study have at least one or two "homes" where that subject matter is published. This is not the case with projections. Research on projections is published in geography, cartography, geodesy, surveying, civil engineering, petroleum engineering, physical geology, and others. In short, the projection field doesn't have a single "home" and because of this, the topic has matured in different fields and ways across time. Unlike many other fields of study, projections also have multiple professional and academic publication outlets. If you want to find information on projections you will need to scour through disparate and sometimes obscure journals and texts published in different fields and languages.

There are dozens of well-researched and well-written books on projections. Unfortunately, many are focused on deriving and reporting mathematical equations used to calculate the x and y plotting coordinates. Other projection books simply provide a cursory overview of projection basics such as the projection properties, classes, and graticule appearances. Some projection books are limited to showing galleries of projections' graticule arrangement. Thus, piecing together projection knowledge takes time and energy and you may not be able to easily find answers to the specific questions you are asking. The bibliography at the end of this book provides a listing of projection-related readings that we feel are good, comprehensive references to help you in understanding how to work with projections.

Map Projections as a Map Design Variable

The cartography literature has well-established guidelines for many decisions that are associated with making a map. For example, selecting a color scheme, symbolization method, and visual variable are all well discussed in cartography books, articles, and online resources. Specifically, choosing a correct symbolization method for a dataset generally involves a relatively limited number of options. Projections, on the other hand, are very different. Over 400 projections have been developed. Most projections have at least one parameter that you can manipulate in some manner. For example, establishing the projection's center involves selecting a single pairing of latitude and longitude values. There are an infinite number of latitude and longitude values on Earth's surface, which creates an infinite number of possible projection *centers*. Where the projection is centered alters how the geographic area to be mapped appears on the map, the arrangement of the graticule, and the appearance of the landmasses on the map's periphery. Collectively, choosing a single projection center may not be a clear-cut decision. These and other decisions impact the map's design and the map reader's interpretation of the map content.

Projections can have a dramatic effect on the overall appeal and communicative effect of the map design in three ways. First, projections control the appearance of the graticule arrangement and the shape of the landmasses. The appearance is caused by the inherent distortion that is part of the projection process and relates to the projection's center and the map scale. For instance, assume we wished to create a map of Australia. Figure 1.3A shows a small-scale map of the world centered about the South Pole and Figure 1.3B is a map that is zoomed in on Australia. Both maps use the Lambert azimuthal equal area planar projection. The small-scale map in Figure 1.3A exhibits a considerable amount of distortion, especially near the map's periphery where the landmasses and graticule become barely recognizable. In addition, Australia looks "stretched out" on this map in an east-to-west fashion. Figure 1.3B appropriately centers Australia in the map. As such, distortion at the map's center is reduced and Australia looks similar to its appearance on a globe.

Second, projections come in a variety of overall shapes, which can add a pleasing aesthetic to the map. Figures 1.4A and 1.4B show the Robinson compromise pseudocylindrical and plate carrée equidistant cylindrical projections, respectively. These projections were the subject of a user preference survey of world map projections and their shapes and characteristics were

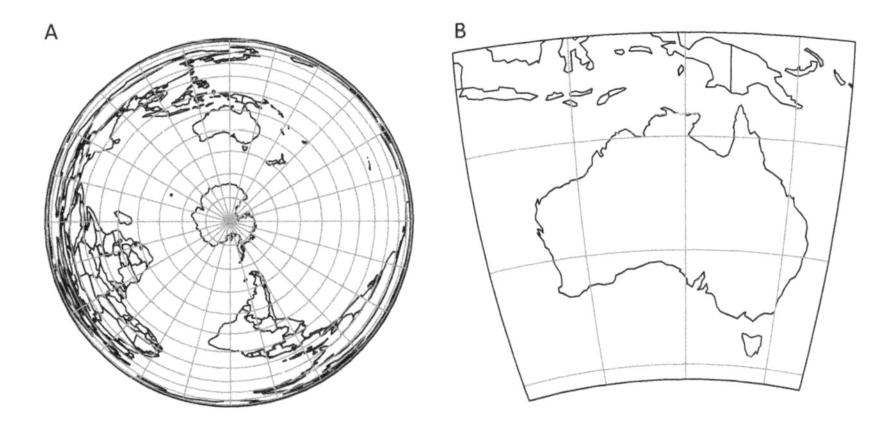

FIGURE 1.3
A small-scale map centered on the South Pole (A) and a large-scale map centered on Australia (B). Both maps are cast on the Lambert azimuthal equal area planar projection.

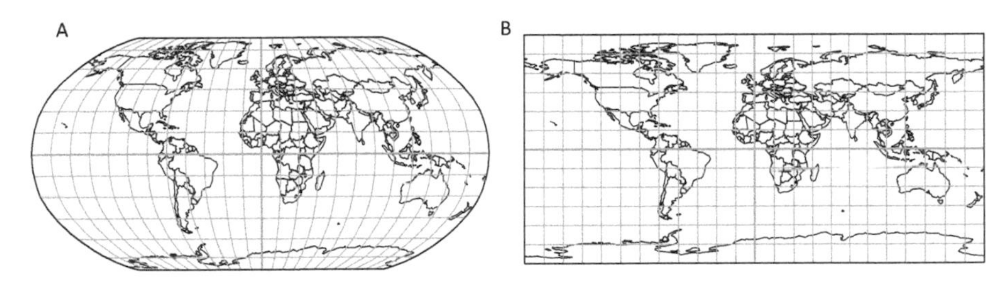

FIGURE 1.4
The Robinson compromise pseudocylindrical (A) and plate carrée equidistant cylindrical (B) projections.

found to be some of the most preferred (Šavrič et al., 2015). Participants in this study preferred projections with straight rather than curved parallels and meridians with smooth rather than complex curves. In addition, the survey results indicated a preference for projections that represent poles as lines compared to projections that show poles with sharp edges. In another study, Werner (1993) examined map readers' preferences for equatorial-centered world map projections. The Robinson and pseudocylindrical projections, in general, were also ranked high by the participants in his study.

Aside from user preferences of overall projection shapes, some shapes fit a specific space (e.g., computer screen) better than others. Figures 1.5A and 1.5B show that the equirectangular equidistant cylindrical projection fits a rectangular space better than the sinusoidal equal area pseudocylindrical. Due to its curved meridians, the sinusoidal leaves "gaps" between the projection's bounding meridian and the rectangular frameline.

Third, some projections are just more interesting to view. Figures 1.6A and 1.6B show the fisheye azimuthal and apple equal area quaziazimuthal projections, respectively. The fisheye projection is characterized by the variable

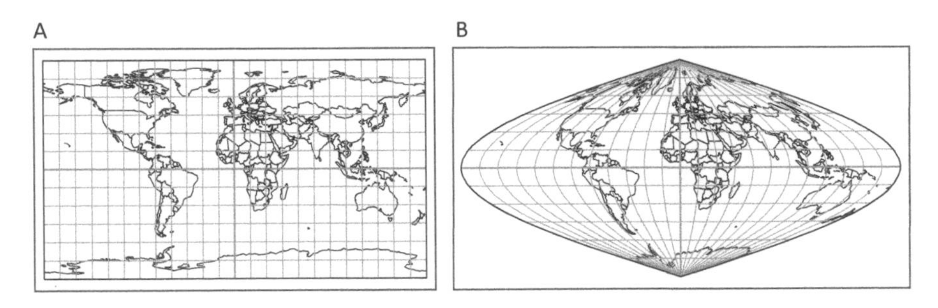

FIGURE 1.5
The equirectangular equidistant cylindrical (A) and sinusoidal equal area pseudocylindrical (B) projections.

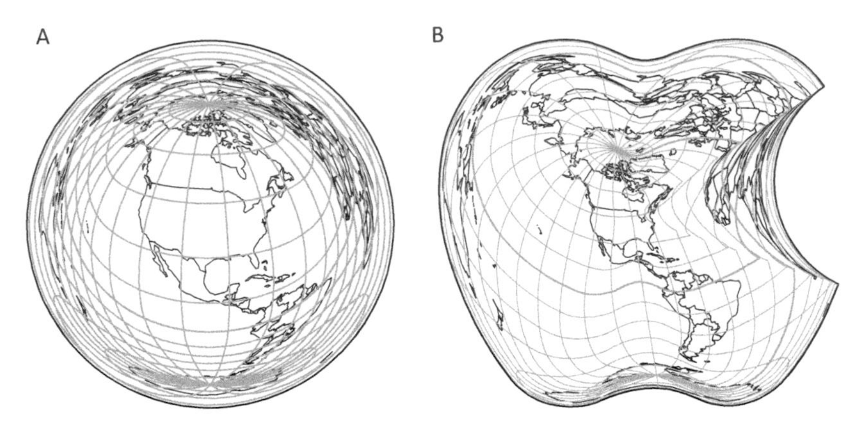

FIGURE 1.6
The fisheye azimuthal (A) and apple (B) projections.

scale that exists across the map's surface. The center of this map in Figure 1.6A is set at a larger scale. As one moves from the map's center outward, the map scale tapers and becomes smaller toward the periphery. Thus, the landmasses at the map's center are brought to the focus of the reader's attention for an eye-catching design. Other projections have been developed that take on familiar shapes. For example, the apple projection (Figure 1.6B) shows the world as an "apple" with a bite taken out of it. Other shapes include projecting the world into a heart, square, oval, or star. Some of these more creative designs have ended up being used as backdrops for company logos. For example, the Berghaus star projection (Figure 1.7) is used by the American Association of Geographers as their logo.

Very little has been written about the role that projections play in map design. One significant article on this topic is by Hsu (1972), in which she presents a good discussion of the importance the projection plays in map design. Hsu (1972, p. 152), reflecting on this importance, suggests that the projection provides the "structural framework for a map and which characterizes

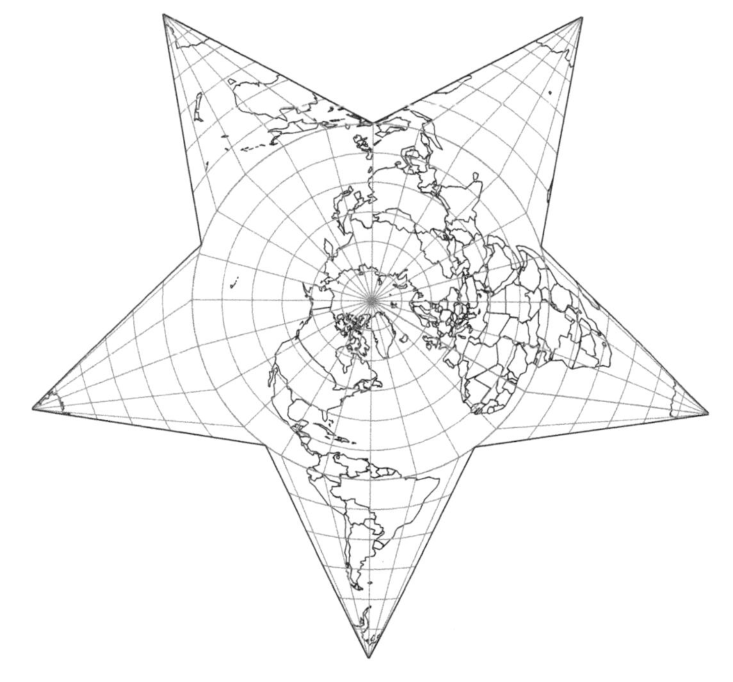

FIGURE 1.7
The Berghaus star projection logo used by the American Association of Geographers.

its scale relationships must also determine the effectiveness of the map as a communication medium." She continues that an "expertly chosen map projection helps to focus and amplify the geographic message of a map" while a poorly chosen "projection can easily destroy the map" (1972, p. 152). An undercurrent that we include throughout this book is that there are many facets of projections that you should consider as design variables when making maps.

The Problem of Selecting a Map Projection

Selecting a suitable map projection for a mapping project is probably the most important and challenging aspect of working with projections. However, it is our experience that the projection literature doesn't do a good job of providing guidance on this important task. General recommendations on how to select a projection do exist. For example, a map of the equatorial region should be based on a cylindric projection; or thematic maps require equal area projections. While there is some truth to these recommendations, most individuals would be left adrift as to which specific cylindrical or equal area projection out of the hundreds that exist would be appropriate for the map. In large part, the motivation for us to write this book came from the realization that the task that most people face when working with geospatial data is how to select an appropriate map projection and that the literature was ill-equipped to assist with this task.

In summary, the fundamental problem this book tackles focuses on the fact that projections are inherently mathematical, the field includes a lot of jargon, the projection literature is scattered about many disciplines, projections have not been thoroughly treated as a variable in the map design process, and selecting a projection is not necessarily a straight-forward process. Given the large number of projections that have been developed throughout history, the number of projection parameters that are available in an equation, and the possible requirements from the map purpose, selecting a *single best* projection is not possible. Every projection has pros and cons. Since no projection is a completely accurate representation of Earth, you must balance the trade-offs involved in selecting an appropriate projection that meets the needs of the specific mapping project coupled with the overall map design goals. For most map purposes, you won't be able to find a single projection that does everything equally well. Compromises will be necessary. This book gives our perspectives on what you should consider as trade-offs when working with geospatial data and projections so that the projection is well-suited to the map purpose.

This Book's Structure

This book is organized into two parts. The first part contains four chapters (2–5). Chapter 2 gives an overview of what a map projection is, what is involved with projecting the Earth to a map, and the distortion patterns that are the inevitable consequence of the projection process. Included in this chapter is a discussion on the various parameters that are associated with a projection and how those parameters impact the overall appearance of the graticule and landmasses on the map. Chapter 3 provides a primer on cartographic symbolization, which is the organizing theme for the second part of this book. The type of geospatial data that is being mapped should be a primary consideration when working with and ultimately choosing a projection. For example, if you are creating a dot map, then you need to be aware of how projections preserve area. This preservation of area is important in order to maintain the apparent dot density distribution across the map. Chapter 4 examines map projections and their influence on data. Among the topics included in this chapter is a discussion on how people interpret (correctly and incorrectly) the projected data. It is important to realize that what people see on a map helps to form their mental map of Earth's landmasses.

Also important is the influence the projection has on the map reader's ability to correct or incorrectly interpret the mapped distribution of a dataset. Chapter 5 lays out our general ideas on how to select a map projection. No projection is the "best" for a given map purpose as no projection is a 100% accurate model of Earth. One of our ideas is that the selection process should be viewed as more of a compromise between the needs of the map purpose and what a map projection can offer. Another consideration on selecting projections we discuss is that sometimes the map maker doesn't have many choices in selecting a projection (e.g., some organizations have a standard projection that is used).

The second part of the book is divided into six chapters (6–11) and an appendix. Each chapter focuses on one of the main cartographic symbolization methods, and the chapters are intended to be accessible whether read in sequence or one at a time, depending on the method you are interested in at the time. Specifically, Chapters 6–9 report our ideas on how to select a projection for maps that use the choropleth, isarithmic, dot, and proportional symbol symbolization, respectively. Each chapter provides a specific overview of the symbolization method, the main visual analysis tasks associated with reading and interpreting data represented by that method, and how projection characteristics and parameters can be adjusted to create a better map. Each chapter uses real data and walks the reader through the complicated decisions that are involved with selecting a projection. Chapter 10 focuses on special types of maps such as heatmaps, spatial bin maps, and special map use tasks such as measuring distances and directions on maps. This chapter details the projection considerations that are needed to make these maps more effective at communicating their information. Chapter 11 discusses map projection resources that are available through the web. The appendix includes a listing of all projections mentioned in this book. Along with the projection name and class, a series of images that illustrate the areal, angular, scale, and overall distortion patterns are included.

2

A Gentle Introduction to Map Projections

Earth's shape can be considered spherical. Maps, on the other hand, are flat. There within lies a problem, because these shapes are topologically different; spheres are 3-dimensional (3-D) while maps are 2-dimensional (2-D). The map projection provides a mathematical solution to the process of projecting a spherical surface to a plane; however, this comes with a consequence of unavoidable distortion from the projection process. A quick examination of a world map (especially near the map's periphery) will show that some landmasses appear, for example, stretched or compressed more than they do on a globe. This is particularly noticeable when examining multiple maps of the world and comparing how they stretch or compress landmasses. Fortunately, there are specific types of projections, and ways to adjust the parameters that define the projections, that will allow for control over some of this stretching and compression that distorts the map. Controlling distortion is often desirable to the map maker as certain map purposes require specific patterns of distortion to be minimized or eliminated (e.g., controlling the distortion of distances on a map designed to evaluate how far away different world cities are from a single major airport). In this chapter, we present a basic overview of the projection process, the inevitable distortion that results, and the different projection parameters that can be modified to control distortion.

In this chapter, we discuss issues surrounding:

- Latitude and longitude, which are coordinate values that function to locate features that exist on Earth's surface.
- The different models (ellipsoid and sphere) that have been developed to approximate that shape and size. These models impact the accuracy of measurements carried on a map.
- The projection process where latitude and longitude values are mathematically projected from Earth's 2-D curved surface to a plane (a map).
- The role that mathematics plays in the projection process.
- The projection properties (equal area, conformal, equidistant, azimuthal, and compromise) and their utility in meeting the needs of a particular map purpose.
- Two projection characteristics that can impact map design: class and aspect. A projection class can be thought of as the overall "shape" of the projection. Common shapes include rectangles, ovals, circles,

and cones. The projection aspect defines the location of the projection's center which results in a geographic region of interest (e.g., Amazon rain forest) centered in the map.

- Distortion as an unavoidable consequence of the projection process but can be mitigated by choosing appropriate projection parameters such as the location of a standard line along which there is no distortion or a property value where one type of distortion is absent.
- The implications of distortion on the map, including its visual impacts on landmasses, how to control it, and how to illustrate the resulting patterns that exist across a projection's surface.

Earth's Latitude and Longitude

Latitude and longitude function as an address system which enables features on Earth's surface to be located. Latitude defines a point location north or south of the equator which is arbitrarily assigned an origin of 0°. Latitude values are expressed in degrees, minutes, and seconds—a system referred to as sexagesimal (or base 60) and range from 90° N to 90° S—the locations which define the North and South Poles, respectively (Figure 2.1A). Connecting identical latitude values together traces an infinite number of circles upon the spherical surface where each circle is parallel to the equator (Figure 2.2A)—for instance, the point marked B is on the equator, which connects all points at 0° latitude. Any specific circle is referred to as a parallel of latitude. The equator is referred to as a great circle as the plane that contains the tracing of the equator passes through Earth's center. The radius of the great circle is the same as the radius of the Earth. All other lines of latitude are described as small circles as the planes that contain the tracing of these lines do not pass through Earth's center.

Assume we wish to determine the latitude value of point A shown in Figure 2.1A. Determining point A's latitude begins by passing an imaginary plane through the Earth that is perpendicular to the equator, intersects Earth's center, contains point A, and intersects both poles. Note that point A is located on the tracing of this plane that passes through both poles. A second plane, coincident with the equator, contains B and Earth's center, and intersects the previous plane at a right angle. Point A's latitude value is the angle between point A and point B. If we assume Earth's shape to be spherical, then the vertex of this angle is at Earth's center. The Greek letter ϕ (phi) is often used to symbolize latitude.

Longitude locates a point east or west of the Prime Meridian, which is arbitrarily assigned an origin of 0° (Figure 2.1B). Longitude values also use the same sexagesimal system as latitude. Longitude values range from

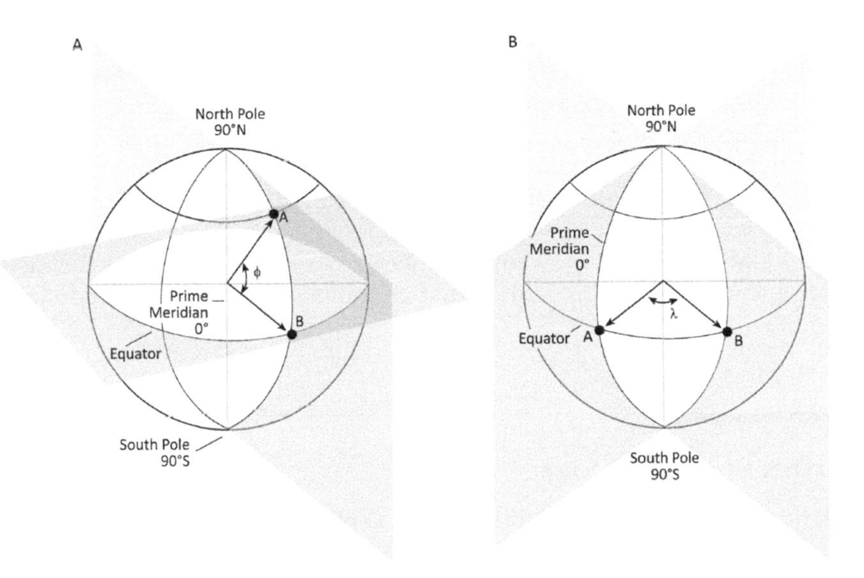

FIGURE 2.1
Determining a latitude value (A) and longitude value (B) on Earth's spherical surface.

0° to 180° either side of the Prime Meridian, with the 180° meridian partially coinciding with the International Dateline. Connecting identical longitude values together creates an infinite number of lines that connect the North and South Poles (Figure 2.2B). Connecting identical longitude values together traces a circle upon the spherical surface and the plane that contains any given circle is referred to as a meridian of longitude (Figure 2.2B). All meridians are great circles as each meridian plane intersects Earth's center.

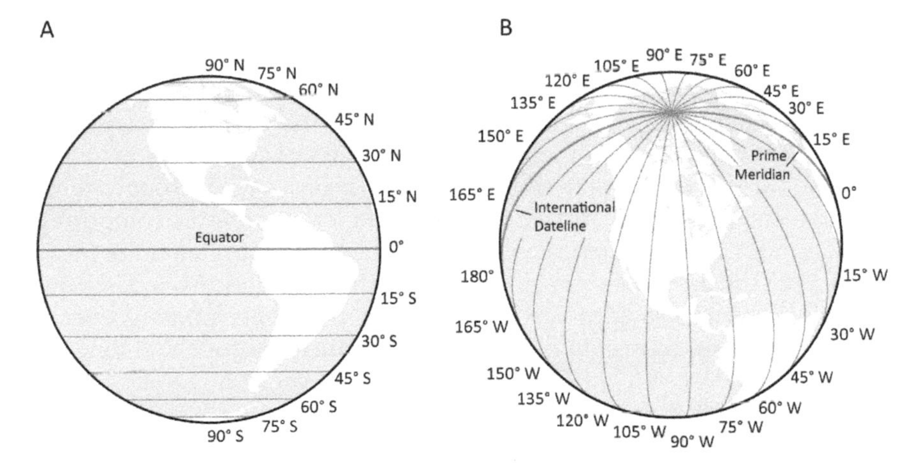

FIGURE 2.2
Numerical range of latitude (A) and longitude (B) on Earth's spherical surface.

Assume we wish to know the longitude value of point A. Determining point A's longitude begins with two planes. While each plane passes through both poles, one plane coincides with the Prime Meridian and contains point B while the second plane coincides with point A (Figure 2.1B). Both planes pass through Earth's center. Point A's longitude value is the vertex of the angle formed between these two planes. Again, if we assume Earth's shape to be spherical, then the vertex of the angle is at Earth's center. The Greek letter λ (lambda) is often used to symbolize longitude. Point A's location on Earth's surface would be defined by a unique latitude and longitude value.

Earth's Shape and Size

Earth is commonly thought of as being a sphere having an nominal equatorial radius of 6,371,000 meters. While Earth can be treated as a sphere, its exact shape is more complex.[1] Earth is a dynamic planet experiencing a variety of forces that act to define its shape, such as crustal motion, differing rock densities, gravitational attraction from other planets, and crustal rebound (isostasy) from continental glaciation. Generally speaking, a *geoid* describes Earth's shape as a smoothly undulating surface that is influenced by the aforementioned forces. Figure 2.3 shows a gravity model of Earth recorded by the National Aeronautical Space Administration's GRACE (Gravity Recovery and Climate Experiment) satellite beginning in 2002. On this model, the geoidal heights range from about −100 meters (shades of blue and purple) to roughly +80 meters (shades of red and orange). Blue and purple shades represent locations on Earth's surface that have less mass and are closer to its center while red and orange shades indicate locations that have greater mass and are further from its center. We do not necessarily see the undulations of the geoid on Earth's surface, but its surface defines the vertical control or heights of Earth's surface with respect to mean sea level. Unless you are mapping heights (either topographic or bathymetric), the geoid isn't necessarily a component of your mapping project. Since Earth is dynamic, new geoid definitions are produced on a somewhat regular basis (e.g., through satellite measurement). For instance, the National Geodetic Survey is computing a new datum for North America called the National Spatial Reference System (NSRS) 2022 (https://www.ngs.noaa.gov/datums/newdatums/index.shtml).

While the geoid provides vertical control to define elevation, it does not provide horizontal control to define exactly where 0° latitude and longitude coordinates are located on Earth's surface.[2] The geoid is a very complex surface and using it to provide horizontal control would be computationally complex. A simpler model of Earth's geoid shape that can be used to define locations is an ellipsoid. We know that Earth rotates on its polar axis and this rotation causes Earth's equatorial region to bulge outward

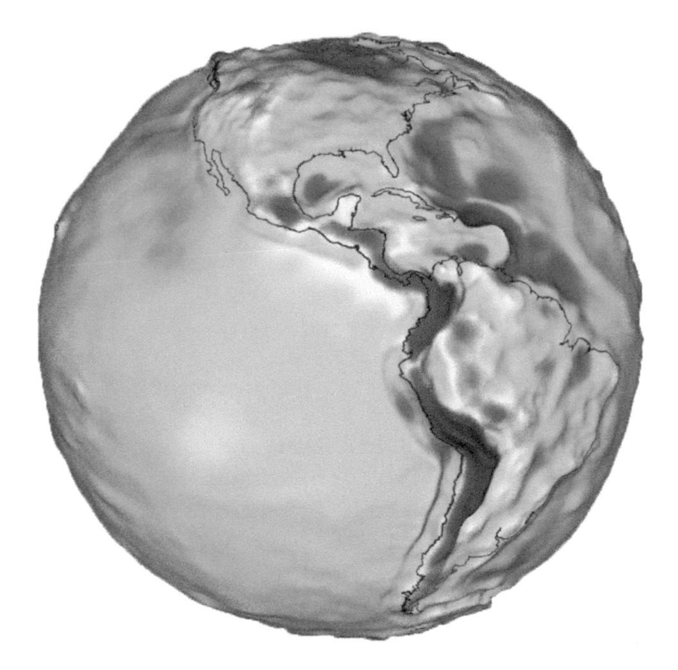

FIGURE 2.3
The GRACE (Gravity Recovery and Climate Experiment) Earth gravitational model 2002. (Reprinted from NASA's Visualization Explorer, in Mapping Earth's Gravity, 2013. Retrieved December 28, 2018, from https://nasaviz.gsfc.nasa.gov/11234.)

through centrifugal forces. At the same time, the rotation compresses the poles, creating an oblate ellipsoid.[3] This oblateness can be defined using two parameters—the semi-major and semi-minor axes (Figure 2.4), which are assigned letters a and b, respectively. By changing these parameters, the degree of polar flattening (symbolized as f) is controlled. This flattening can then be set to match Earth's shape departure from a circle. Armed with values of a and b, calculating a reference ellipsoid's flattening is carried out by inserting these values into Equation 2.1.

$$f = (a - b) / a \tag{2.1}$$

Since Earth is a dynamic planet (e.g., crustal motion) and gravitational measurement techniques have evolved over time (e.g., orbiting satellites) various reference ellipsoids[4] have been developed to model Earth's shape. Reference ellipsoids developed during the 1800s attempted to model or "fit" the reference ellipsoid's parameters to mirror Earth's curvature at a specific geographic region. Beginning in the 1950s satellite measurements made it possible to define reference ellipsoids that modeled Earth's overall shape. Figure 2.5A shows the Clarke 1880 reference ellipsoid and how it models Earth's curvature at Cape Town, South Africa and Vladivostok, Russia. The Clarke 1880 reference ellipsoid was developed to fit Earth's

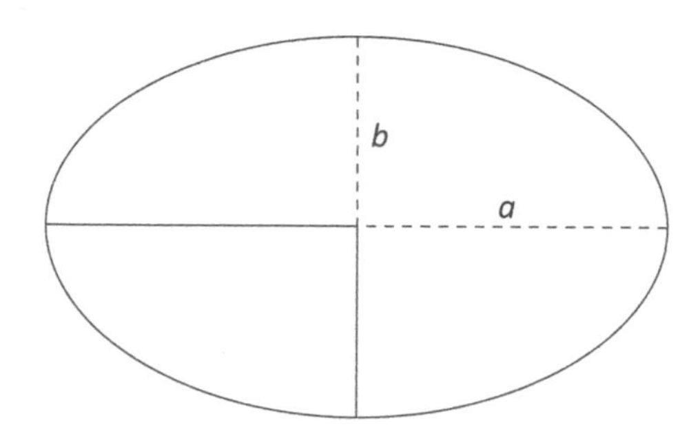

FIGURE 2.4
An ellipsoid defined by the semi-major axis (a) and semi-minor axis (b).

curvature over South Africa. Table 2.1 lists the parameters that define the Clarke 1880 reference ellipsoid. Note that the difference in the lengths of *a* and *b* for this reference ellipsoid is 21,734.178 meters. Figure 2.5B shows the World Geodetic System 1984 (WGS84) reference ellipsoid, which was designed to better approximate the entirety of Earth's shape rather than any specific geographic location. Here, WGS84 is a better average fit for *both* locations. The difference in the lengths of *a* and *b* for this reference ellipsoid is 21,384.686 meters.

A close examination of the lengths of *a* and *b* values in Table 2.1 suggests that the Clarke 1880 reference ellipsoid is wider and shorter than the WGS84 reference ellipsoid. In addition, a comparison of the inverse flattening values suggests that the Clarke 1880 reference ellipsoid deviates more from a circle than the WGS84. Other ellipsoids have slightly different values for *a*, *b*, and inverse flattening[5] values close to 300. Obviously, the flattening apparent in Figure 2.5 is greatly exaggerated to show the flattening effect.

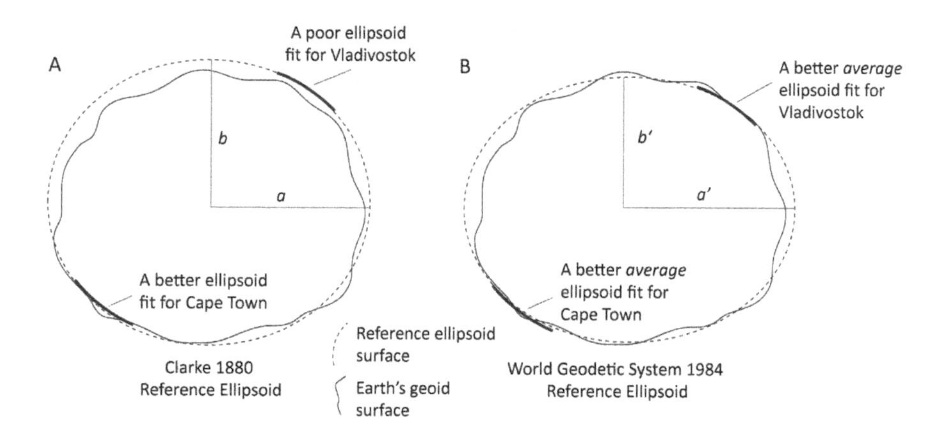

FIGURE 2.5
Illustrating how the Clarke 1880 reference ellipsoid fits Earth's curvature for Cape Town, South Africa (A) but does not fit Vladivostok, Russia. The World Geodetic System of 1984 reference ellipsoid gives a better average fit to Earth's curvature for both cities (B).

TABLE 2.1

A Comparison of the Parameters for the Clarke 1880 and World Geodetic Survey 1984 Reference Ellipsoids

Reference Ellipsoid Name	*a* (meters)	*b* (meters)	*f*	1/*f*
Clarke 1880	6,378,249.145	6,356,514.967	0.0034075	293.4663
World Geodetic System 1984	6,378,137.0	6,356,752.3142	0.0033528	298.2572

If you looked down on Earth from outer space, you would not be able to see this approximate 21,000 meter length difference between these or any other reference ellipsoids. However, this difference can have a considerable impact on the ability to carry out accurate measurements (e.g., distances) on a map. We will use the following example to illustrate the impact that different reference ellipsoids and a sphere Earth model have on distance measurements. Table 2.2 shows Cape Town, South Africa and Vladivostok, Russia and reports their coordinate locations. Table 2.2 reports the measured distances between these cities using two reference ellipsoids and a sphere Earth model. The two reference ellipsoids are the Clarke 1866 and WGS84 (Figure 2.6). The Clarke 1866 reference ellipsoid was developed to fit Earth's curvature of North America and does not coincide with the curvature of the geographic area between these two cities. On the other hand, the WGS84 reference ellipsoid was developed to model Earth's overall shape and produces a more correct measured distance between the two cities. Table 2.2 also shows that treating Earth as a sphere does not take into consideration its correct shape and results in the largest measured distance. Accurately measuring, for example, distances requires knowledge of choosing the appropriate

TABLE 2.2

Comparing the Distances between Cape Town, South Africa and Vladivostok, Russia as Measured on the WGS 1984 and Clarke 1866 Reference Ellipsoid and an Assumed Sphere Surface

Cities' Latitude and Longitude	
Cape Town, South Africa	Vladivostok, Russia
Latitude: 33°55′31″S	*Latitude: 43°08′ N*
Longitude: 18°25′26″E	*Longitude: 131°54′ E*
Distance Comparison	
Clarke 1866:	14,283,583.16 meters
WGS 1984:	14,283,685.22 meters[a]
Sphere:	14,290,906.9 meters

[a] Earth distance measurements were obtained using the National Geodetic Agency's Geodetic Toolkit Inverse (www.ngs.noaa.gov/cgi-bin/Inv_Fwd/inverse2.prl).

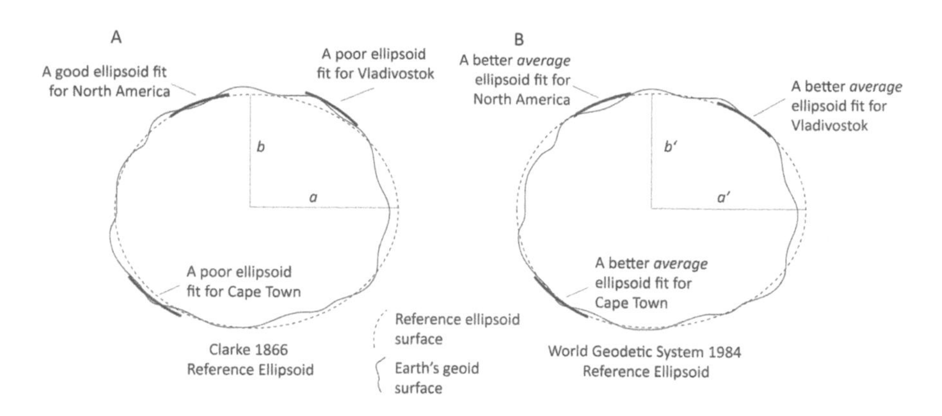

FIGURE 2.6
Illustrating how the Clarke 1866 reference ellipsoid fits Earth's curvature for North America but does not fit Cape Town, South Africa or Vladivostok, Russia (A). The World Geodetic System of 1984 reference ellipsoid gives a better average fit to Earth's curvature for all locations (B).

reference ellipsoid that fits the geographic area under consideration (if only mapping a localized area) or requires use of a reference ellipsoid that was developed to fit Earth as a whole.

The Map Projection Process

As we described above, the projection process takes latitude and longitude values from Earth's spherical surface and mathematically converts them to a plane surface. Earlier we discussed that distortion was a consequence of the projection process. In Figure 2.7A we show a view of Earth as if we were looking down on the Northern Hemisphere from outer space. Note that four points A, B, C, and D are highlighted in Figure 2.7A. Figure 2.7B displays the same points on a conic projection (note the "fan" appearance of the map's overall shape relating to a cone). Points A and B in Figure 2.7A are represented as points in Figure 2.7B. However, point C in Figure 2.7A is represented as two points in Figure 2.7B. On Earth's surface, recall that the 180° meridian of longitude runs from the North Pole to the South Pole and partially coincides with the International Dateline. In Figure 2.7B, we see that the projection process creates two lines out of the single 180° meridian shown in Figure 2.7A. Point D was located at the North Pole in Figure 2.7A but is now shown as a line in Figure 2.7B. Collectively speaking, these dimensional changes (e.g., point to a line) demonstrate that the projection process can introduce distortion on the map. In fact, no map is a completely faithful representation of Earth's surface. As we discuss later in the book, projections can be selected so that this distortion can be controlled according to, for example, the geographic area mapped.

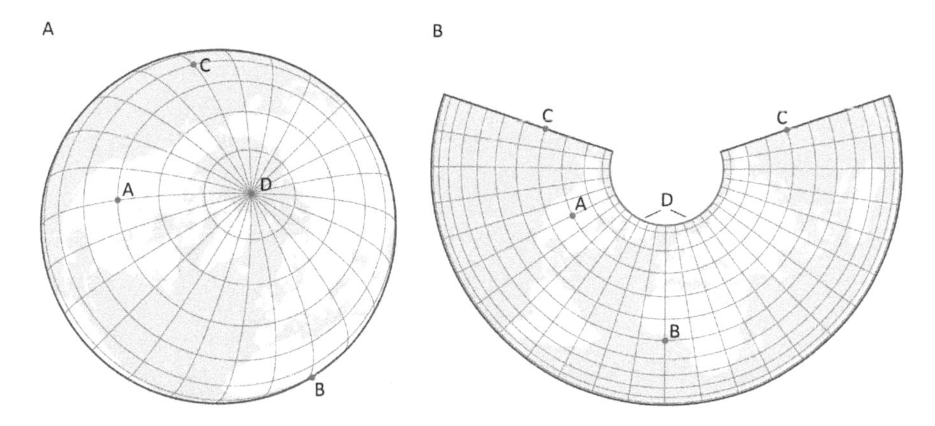

FIGURE 2.7
Four points (A, B, C, and D) appearing on Earth's surface (A) and the impact of the projection process on those four points (B).

Projection Equations

Projecting Earth's curved surface (e.g., a point defined by a latitude and longitude value) involves a mathematical process that projects latitude and longitude values onto a plane surface (Figure 2.8). This process is carried out through a set of equations. For example, assume you want to project a point on Earth's surface to a plane and the point in question is defined by the latitude and longitude values of 40° N and 60° E, respectively, as shown in Figure 2.8. The goal is to project these coordinate values expressed in degrees to a corresponding set of *x* and *y* Cartesian coordinates. Projections usually involve at least two mathematical equations: one defining the *x* value and another defining the *y* coordinate value. Equations 2.2 and 2.3 describe one of the simplest projections, the plate carrée equidistant cylindrical projection:

$$x = R * (\lambda - \lambda_o) \tag{2.2}$$

$$y = R * \phi \tag{2.3}$$

where λ is the longitude value, ϕ is the latitude value, λ_o is the value of the central meridian (the longitude of the projection's east–west center), and R is the radius of the reference globe. The *x* and *y* coordinate values are commonly referred to as plotting coordinates, specify the longitude and latitude values in a Cartesian coordinate system, and are expressed in radians.

To compute the *x* and *y* values using Equations 2.2 and 2.3, four steps are necessary. First, the longitude value for the central meridian must be selected. Any longitude value is valid, but if 0° is chosen, the central meridian will coincide with the Prime Meridian. Second, a value of *R* must be specified, which ultimately establishes the final scale of the map. Although

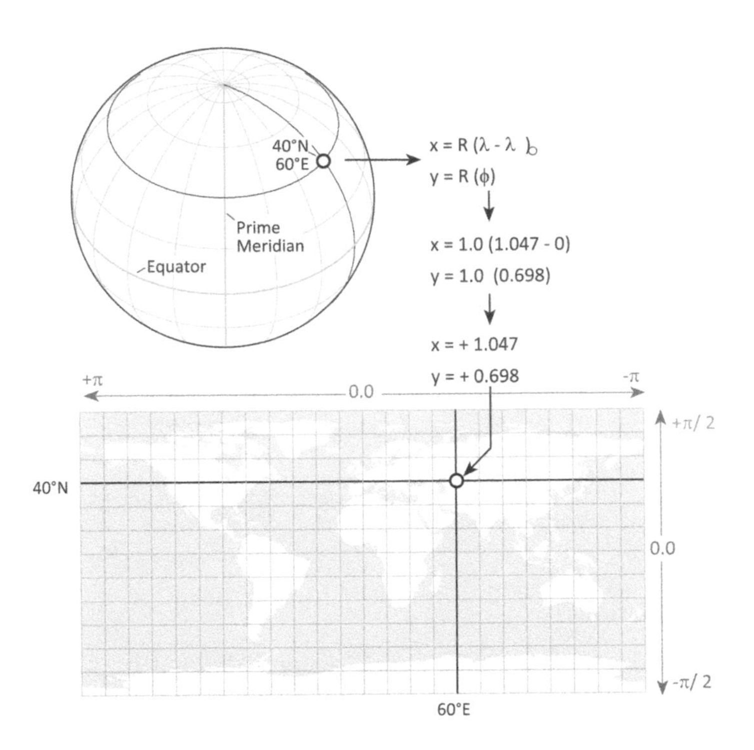

FIGURE 2.8
An example of taking a point on Earth's curved surface defined by a latitude and longitude pair and projecting that pair to an x and y plotting coordinate pair through equations. The equations shown produce the plate carrée equidistant cylindrical projection. Note that during the projection process, the latitude and longitude values have been converted from degrees (40° N and 60° E) into radians.

you can specify values of R, we simplify the computations that follow and assume R to be 1.0.

Third, all latitude and longitude degree values must be converted into radians. This conversion is necessary when using projection software or computer programming languages as the drawing environment is usually set to plot in radians. Converting degree values to radians involves multiplying the degree measurement by the constant $\pi/180$, where π is approximately 3.1415. For instance, 90° in radians equals $90° \times (\pi/180) = 1.5707$ or $\pi/2$, while 180° equals πradians. Since longitude values on Earth range from –180° to +180° and all x values are projected to longitude degree values, then x values range from $-\pi$ to $+\pi$. On the other hand, since latitude values on Earth range from –90° to +90°, and all y values are projected latitude values, then y values range from $-\pi/2$ to $+\pi/2$. For instance, 60° E converts to 1.047 radians and 40° N converts to 0.698 radians.

The fourth step in computing x and y values involves inserting longitude and latitude radian values into Equations 2.4 and 2.5. Using our values of 1.047 and 0.698, we have:

$$x = 1.0 * (1.047 - 0) \tag{2.4}$$

$$y = 1.0 * 0.698 \tag{2.5}$$

Which produces the following values for x and y:

$x = +1.047$
$y = +0.698$

If all remaining point locations on Earth's curved surface defined by latitude and longitude were computed (e.g., every 15°) and plotted in the Cartesian coordinate system, the plate carrée projection shown on the bottom of Figure 2.8 would be generated. The network of lines appearing on the map in Figure 2.8 represents latitude and longitude and is referred to as the graticule. The graticule is often shown on maps and serves a useful role in helping map readers determine the latitude and longitude value of a point location.

We can expand on the simple pair of equations introduced above by adding trigonometric functions which create different projections. For example, Equations 2.6 and 2.7 includes the sine of the latitude applied to the y values:

$$x = R * (\lambda - \lambda_o) \tag{2.6}$$

$$y = R * \mathrm{Sin}\,\phi \tag{2.7}$$

This mathematical transformation produces the Lambert equal area cylindrical developed in 1772 by Johann H. Lambert (Figure 2.9). This graticule arrangement is similar to the plate carrée projection Figure 2.8 where the meridians are equally spaced, but the spacing of the parallels decreases as their distance from the equator increases, showing the impact of taking the sine of the latitude.

As an alternative, Equations 2.8 and 2.9 produce the sinusoidal pseudocylindrical which results from introducing the cosine function.

$$x = R * (\lambda - \lambda_o) * \mathrm{Cos}\,\phi \tag{2.8}$$

$$y = R * \phi \tag{2.9}$$

Figure 2.10 shows the sinusoidal which shares some of the visual characteristics of the graticule with the plate carrée projection, namely that the lines of latitude are parallel and, in this case, equally spaced. However, applying the cosine function results in plotting curved meridians causing them to converge at the North and South Poles which tends to compress and stretch landmasses in the polar areas.

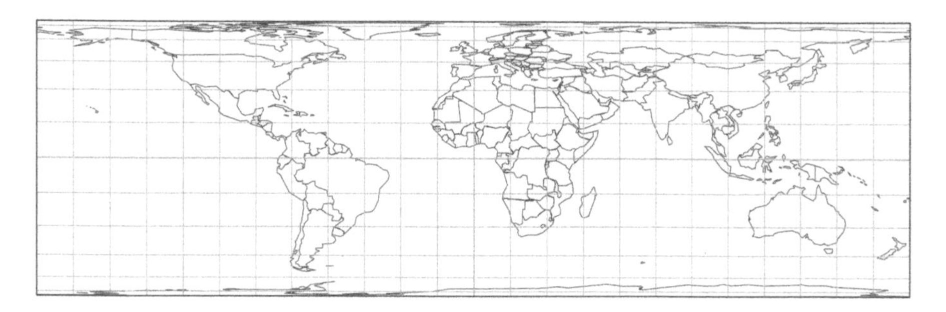

FIGURE 2.9
The Lambert equal area cylindrical projection with a 15° graticule spacing.

If we combine the cosine and sine functions together, Equations 2.10 and 2.11 result:

$$x = R * (\lambda - \lambda_o) * \text{Cos}\,\phi \quad (2.10)$$

$$y = R * \text{Sin}\,\phi \quad (2.11)$$

Using Equations 2.10 and 2.11, an unnamed polycylindrical shown in Figure 2.11 is produced. In this case, some of the graticule characteristics of the previous projections should be visible. On this projection, the curved meridians converge to the poles as shown in the sinusoidal. The parallels are parallel as with the Lambert cylindrical but are unequal lengths as shown by the sinusoidal.

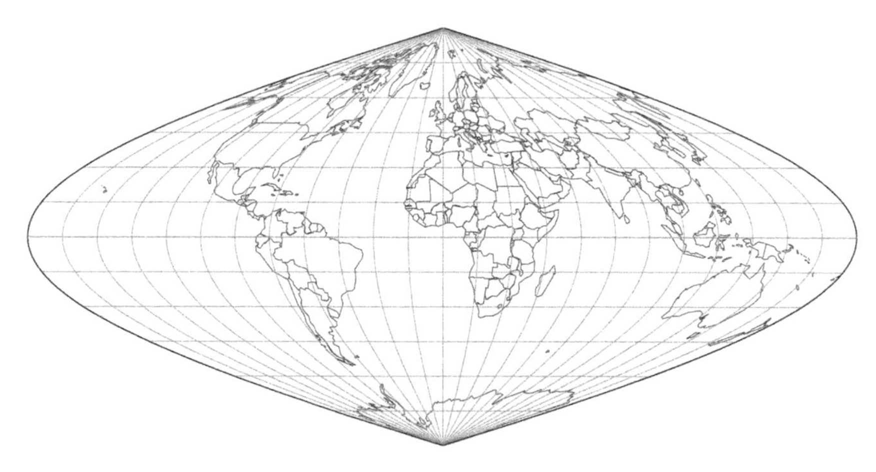

FIGURE 2.10
The sinusoidal equal area pseudocylindrical projection with a 15° graticule spacing.

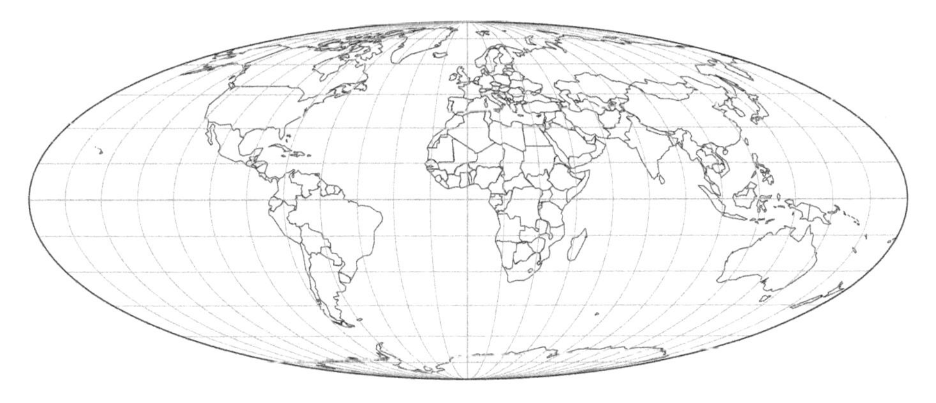

FIGURE 2.11
An unnamed polycylindrical projection with a 15° graticule.

Spherical vs. Ellipsoidal Projection Equations

All projection equations presented in this section are based on treating Earth as a sphere. A spherical Earth assumption generally simplifies the equations used in calculating the x and y Cartesian coordinates (Equations 1.1 and 1.2). Projection equations are also available that assume Earth as an ellipsoid. The ellipsoidal form of the equations used to solve for x and y are more complex and require additional calculations such as determining the scale factor along a chosen parallel and inserting the value of the semi-major axis of the chosen reference ellipsoid (Equations 1.3–1.6). Generally speaking, the difference between the use of the spherical and ellipsoidal form of projection equations results in an accuracy difference in the resulting x and y coordinates; the ellipsoidal form of projection equations will result in more accurate x and y values. The reason for this increased accuracy is that the reference ellipsoid is a closer approximation to Earth's true shape than a sphere. Maps requiring highly accurate positions (such as navigational charts) are derived using ellipsoidal form of projection equations. Thematic and general reference maps, however, generally do not have expectations of highly accurate coordinate values and can use the spherical form of projection equations.

The Map Projection Properties

Maps are generally designed to meet a specific purpose. That purpose may involve a ship navigator measuring a distance along a ship's course, a surveyor measuring the angle between two features, an epidemiologist examining the distribution of malaria across the globe or showing a river network in a country. Projections must be selected carefully so that if maps are used to measure, for example areas, those areal measurements will be the same as if those measurements were carried out on Earth's surface. Angles, areas, directions, and distances are referred to as Earth's spatial relationships and can be preserved on a map by choosing the appropriate projection property. If a projection preserves angles, areas, distances, and directions then the projection property is called conformal, equal area, equidistant, and azimuthal, respectively.

Projections also exist that possess no specific property (e.g., they neither preserve areal or angular measurements). For instance, general reference maps that show base information such as hydrologic networks or administrative boundaries are often based on a projection without a specific property. Since general reference maps do not usually involve measurement activities, there is no need to preserve any particular spatial relationship. Projections that preserve no specific property are referred to as compromise projections.

Preserving Angles (Conformal Property)

An angle can be defined as a numerical value (usually expressed in degrees) that results when two straight lines meet at a common point. Figure 2.12A shows lines a-b and a-c meeting at a vertex and creating an angle. Neither line a-b nor a-c in Figure 2.12A is aligned to the North Pole. Figure 2.12B shows that line a′-b′ is aligned to the North Pole while line a′-c′ is not. Here, the resulting angle in Figure 2.12B reports a direction measured clockwise from line a′-b′ (pointing to the North Pole) to line a′-c′. In this example, the direction is also referred to as an azimuth.

If an angle is measured on Earth's surface and needs to be preserved on a map (e.g., a surveyor laying out a property boundary or tracking a hurricane path), then a conformal projection is appropriate. Conformal projections preserve angles everywhere on the map. To demonstrate the utility of a conformal projection, examine Figure 2.13A that shows point A on Earth's

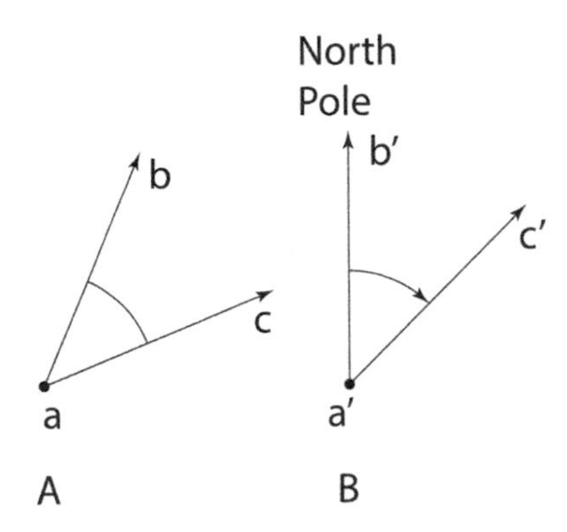

FIGURE 2.12
An angle (A) and a direction (B).

surface located over southern Florida. A line extends from point A to point B in Liberia. Another line extends from point A and terminates at point C in Sweden. The angle at point A, as measured on Earth's surface, is 32°. Figure 2.13B shows lines extending from point A and terminating at points B and C on the Mercator conformal cylindrical projection. Since conformal projections preserve angular relations, the angle at point A in Figure 2.13B is 32°. Despite the utility of preserving an angle, a consequence of preserving this projection property (e.g., angular relationships) is that shapes and sizes of landmasses are distorted (especially near the poles).

Over the years, the definition of conformal has been somewhat misconstrued in its interpretation. As presented above, the idea of conformality centers on preserving angles across a projection's surface. However, in the literature, you will find instances where the conformal property has been associated with preserving shapes of landmasses. There is no mathematical means of any projection to ensure that a landmass's shape is preserved from the globe to the map. Figure 2.13B shows the world cast on a Mercator *conformal* projection. Yet, if you were to compare a landmass's shape on this projection (e.g., Canada) to its corresponding appearance on the globe, you could see stark differences. On this and other conformal projections, many landmasses are stretched or compressed in order to preserve angles.

Preserving Areas (Equal Area Property)

One can measure an area directly on Earth's surface (e.g., the area of a country); however, to do this accurately on a map requires the selection of an equal area projection. Figure 2.14A highlights Algeria, Democratic Republic of the

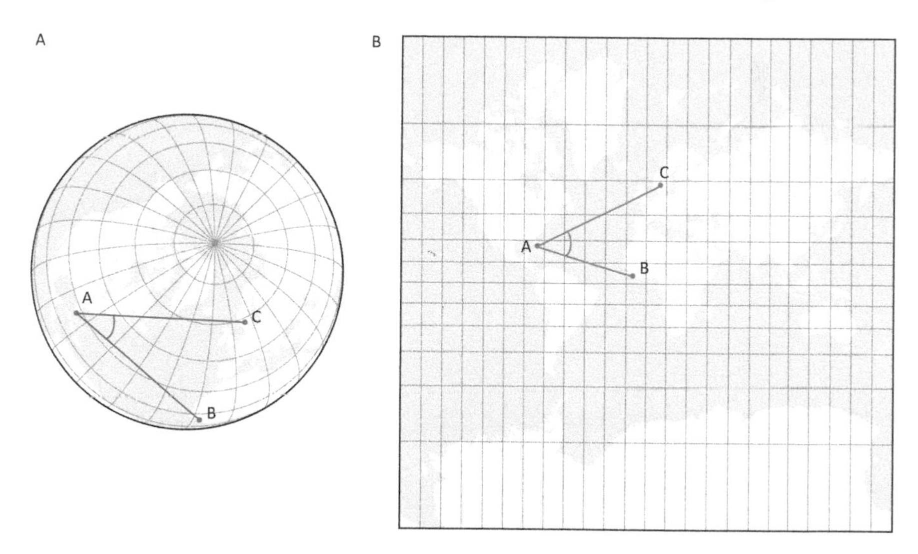

FIGURE 2.13
An angle shown on Earth's surface (A) and on the Mercator conformal cylindrical projection (B).

Congo, Greenland, and Russia as they appear on Earth's surface. Figure 2.14B and Figure 2.14C highlight those same countries on the Mollweide equal area pseudocylindrical and Albers equal area conic projections, respectively. Since these projections are equal area, the highlighted landmasses' areas are preserved as they exist on Earth's surface. In fact, the areas of all landmasses

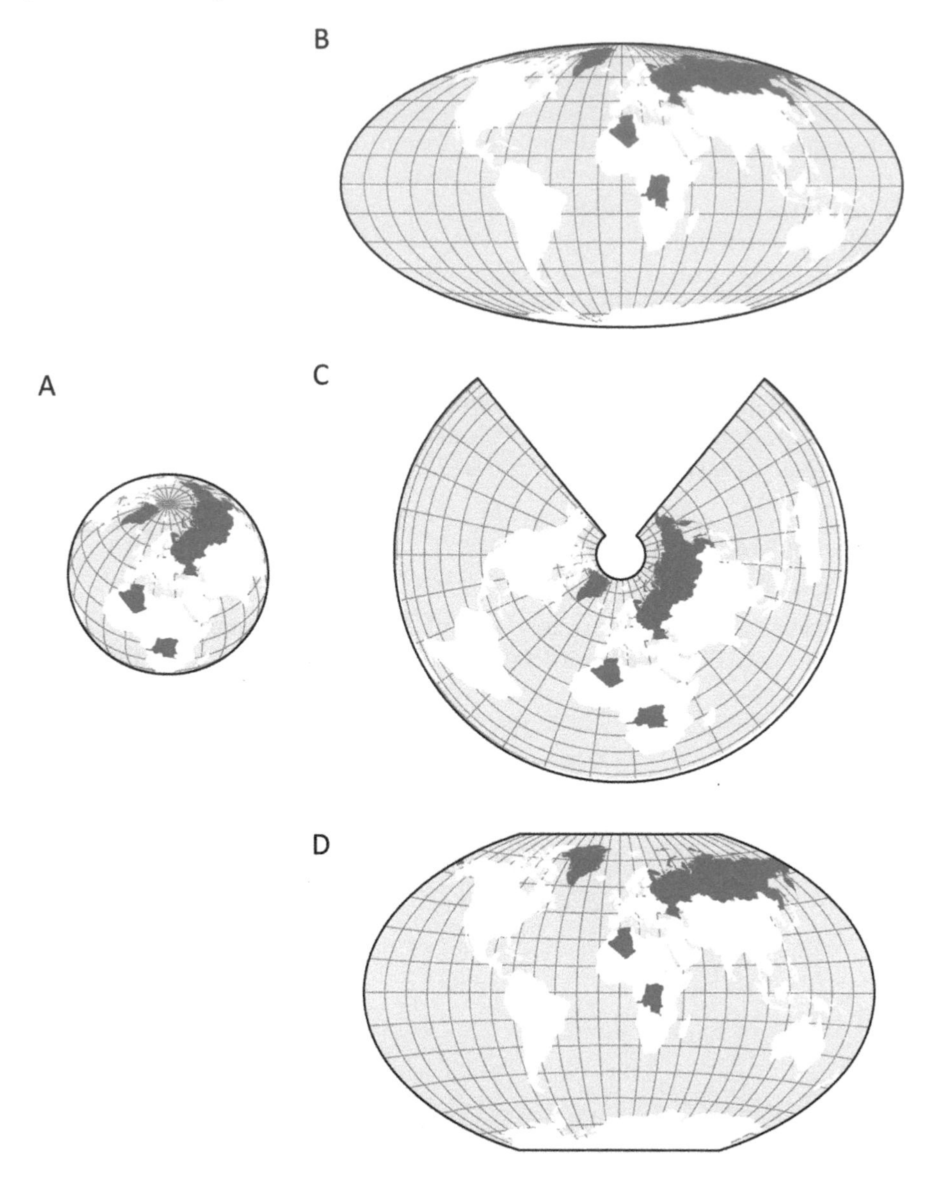

FIGURE 2.14
Algeria, Democratic Republic of the Congo, Greenland, and Russia shown on Earth's surface (A), Mollweide equal area pseudocylindrical (B), Albers equal area conic (C), and Winkel tripel compromise modified azimuthal (D) projections. All maps are shown at the same scale.

shown in Figure 2.14B and Figure 2.14C are the same as they exist on Earth's surface. In Figure 2.14D, however, the Winkel tripel compromise modified azimuthal projection is not equal area and thus the highlighted landmasses are not shown with the correct area. Table 2.3 compares the area of Algeria, Democratic Republic of the Congo, Greenland, and Russia as measured on Earth's surface, the area of each country as projected on the Winkel triple projection, and the percent change between the areas measured on Earth and the projected areas. This area difference can be seen by noting that, for example, Greenland's area on the Winkel tripel projection is exaggerated 85% (Table 2.3) compared to its appearance in Figure 2.14A. This large percent change in area is due in part to the fact that the Winkel tripel projection does not preserve the specific property of area and the high latitude of Greenland's location places this landmass where considerable distortion is present on this projection.

Equal area projections are especially important for thematic maps that show the spatial distribution of a dataset. When using symbolization methods, such as dots, to show the distributions of a phenomenon across space, an equal area projection is preferred to maintain the apparent dot density on the map. Preserving area on a map enables the map reader to correctly compare a data's distribution (through a symbolization method) across the geographic area of interest.

Preserving Distances (Equidistant Property)

If the map purpose calls for preserving distances (such as for ship or aircraft navigation), then an equidistant projection should be considered. Equidistant projections allow distances to be preserved, but only in specific ways. It is important to realize that, unlike conformal and equal area projections where angles and areas are preserved throughout the projection, no projection can accurately maintain all distances on a map. Although the name *equidistant projection* may imply differently, distances on equidistant projections are only preserved along a meridian or from a single point to any other point following great circles (the shortest distance between two points on a spherical surface). Figure 2.15 shows five lines (A, B, C, D, and E) illustrating their

TABLE 2.3

Area Comparisons Using Projections Shown in Figure 2.14 (All Areas Reported in Square Kilometers)

Country	Area Measured on Earth	Area Measured on the Winkel Tripel	Percent Difference
Algeria	2,381,741	2,055,790	−13.6%
Democratic Republic of the Congo	2,267,048	1,926,340	−17.6%
Greenland	2,166,086	4,009,296	+85.1%
Russia	16,377,742	20,885,800	+27.5%

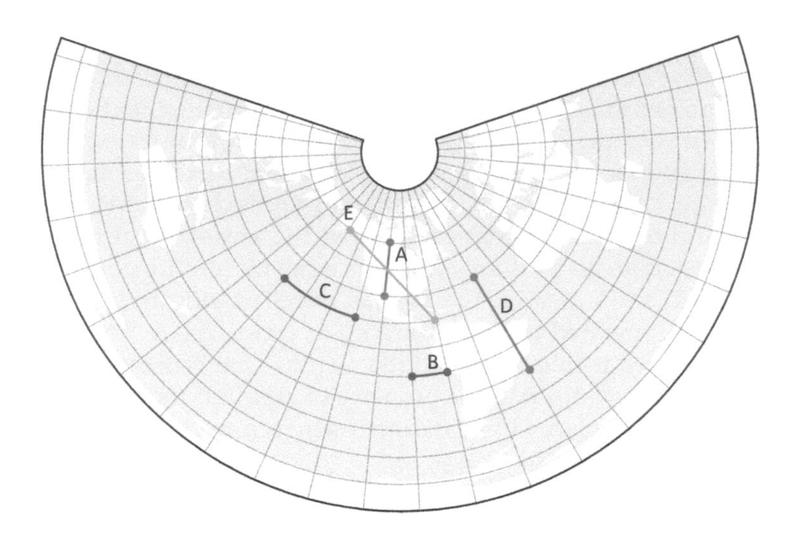

FIGURE 2.15
Five different measured lines on the Euler equidistant conic projection.

paths on the map based on the Euler equidistant conic projection. Table 2.4 shows the true distances on Earth's surface, the path distance measured on the Euler, and the percent difference between the two distance measurements. In Figure 2.15, lines A and D follow meridians. Since the Euler is equidistant, measured distances along lines A and D represent the true Earth distance (see Table 2.3). Lines C and B follow parallels. Even though the Euler is equidistant, the measured distances along parallels are not true and, in this case, represent larger distances compared to the true Earth distances. Line E on the Euler neither follows a parallel nor meridian and similar to lines C and B does not accurately represent the true Earth distance. On the Euler, line E results in a distance that is shorter compared to its true Earth distance. If you were to carry out this same exercise using non-equidistant projections, you would find that distances would generally measure larger or smaller than the true Earth distance. In other words, measuring distances

TABLE 2.4

Distance Comparisons Using Different Line Segments on the Euler Projection Shown in Figure 2.15 (All Distances Reported in Kilometers)

Distance Measured	Distance Measured on Earth	Distance Measured on the Equidistant Conic Projection	Percent Difference
A	3,335.8	3,335.8	0%
B	1,612.9	2,235.3	38.5%
C	4,830.9	5,118.0	5.9%
D	6,640.2	6,640.2	0%
E	7,970.5	7,874.5	–1.2%

on an equidistant or non-equidistant projection must be made with a great deal of caution.

Equidistant projections offer other ways to accurately measure distances. For example, the azimuthal equidistant planar projection allows any straight-line distance to be measured from one specific point (the projection's center) to any other point on the map. That measured distance will be the same as the true Earth distance. Similarly, the two-point azimuthal equidistant projection allows any straight-line distance to be measured from one or two specific points to any other point on the map. Figure 2.16 shows the two-point azimuthal equidistant projection. Continuous color tones, illustrating the overall distortion pattern on this projection, appear in Figure 2.16A. Locations exhibiting lower distortion (more faithful representations of Earth's spatial relationships) are cast in lighter shades while darker shades suggest higher distortion (less faithful representations of Earth's spatial relationships). Tones of green represent differing amounts of areal distortion, magenta represents angular distortion, and gray indicates a mixture of areal and angular distortion. This same continuous color tones convention to illustrate distortion is used throughout this book.

Figure 2.16A illustrates two white areas on the map. These areas are defined by two anchor points A and B located at 40° S, 60° W and 40° N, 80° W, respectively (Figure 2.16B). These anchor points function to allow distances to be accurately measured from these two points in all directions to any other point on the map. Figure 2.16B shows four points located on the map which will be used to compare distance measurements on

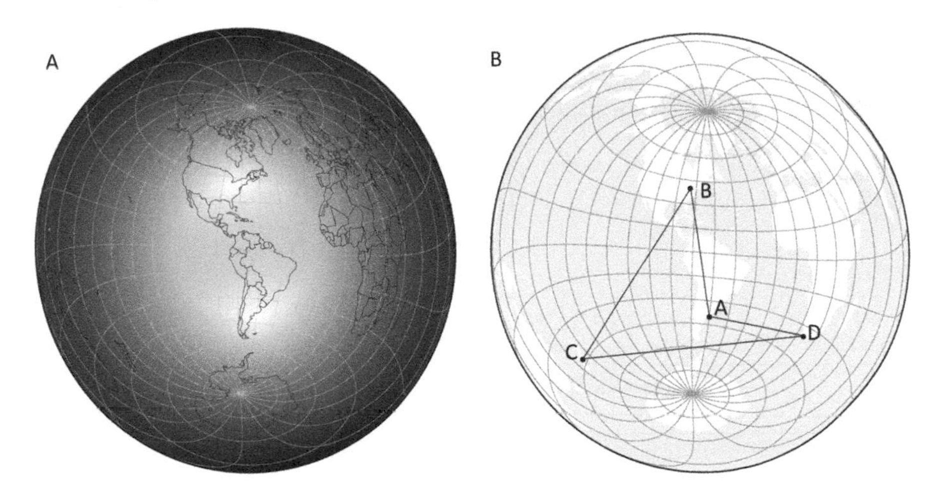

FIGURE 2.16
The two-point azimuthal equidistant planar projection. Distortion visualization showing the areas of low distortion around two anchor points that are highlighted by two white areas at 40° S, 60° W and 40° N, 80°W (A). Four points on the two-point azimuthal equidistant projection (B).

this projection. Points A and B are the anchor points. Point C is located in New Zealand and point D is located in South Africa. Table 2.5 reports the measured distances from A–B, B–C, A–D, and C–D. As shown in Table 2.5, the measured distances from an anchor point to another point are not distorted compared to the true distance as measured on Earth's surface. However, the measured distance between non-anchor points D and C is 44% shorter than the true Earth distance.

Preserving Directions (Azimuthal Property)

Projections also offer limited ways to accurately measure directions or azimuths. As we discussed earlier, a direction can be considered a special kind of angle where the angle is measured clockwise from the North Pole.[6] Projections that preserve directions are called azimuthal. On azimuthal projections, directions measured at the projection's center are correct to all other points. In Figure 2.17A, we are interested in measuring the direction from point A to point C on Earth's surface. To illustrate this direction, a line is extended from point A to the North Pole (point B). Another line is extended from point A to point C. The direction from point A to point C is 98°. Figure 2.17B shows the stereographic azimuthal conformal planar projection with the lines extending from point A to B and from A to C. On this projection, the measured direction between points A and C is 98° which is the same as found on Earth's surface (Figure 2.17A). In order to measure the direction on the stereographic or any other azimuthal projection accurately, the projection's center must be at point A—the point of interest and one of the lines connected from the center to one of the poles.

Preserving No Specific Property (Compromise)

Some projections do not preserve any specific property and, thus, are not generally recommended for maps that require specific measurements.

TABLE 2.5

Distances Measured on the Two-Point Azimuthal Equidistant Projection Shown in Figure 2.16 (All Distances Reported in Kilometers)

Points	Distance Measured on Earth[a]	Distance Measured on the Two-Point Azimuthal	Percent Difference
A to B	9,134.08	9,134.08	0%
B to C	14,178.99	14,178.99	0%
A to D	4,302.57	4,302.57	0%
C to D	17,492.84	9,814.44	–44%

[a] Earth distance measurements were obtained using the National Geodetic Agency's Geodetic Toolkit Inverse (www.ngs.noaa.gov/cgi-bin/Inv_Fwd/inverse2.prl).

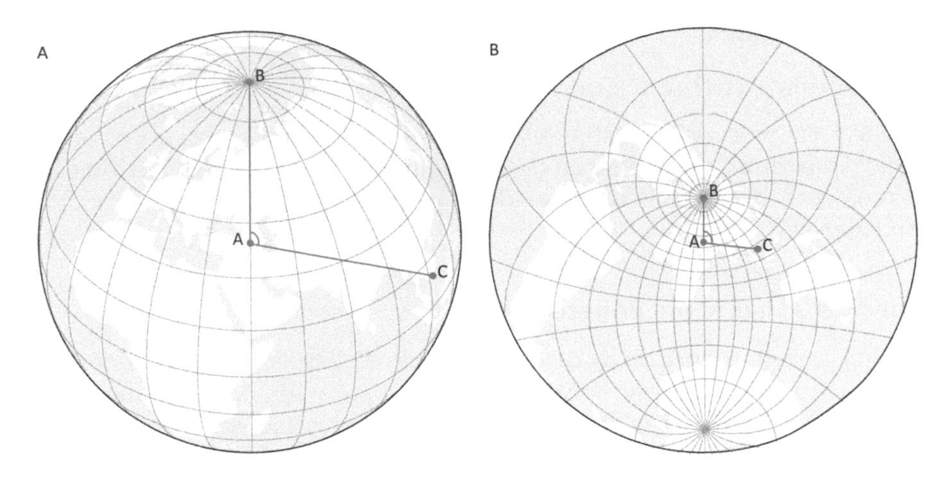

FIGURE 2.17
The direction or azimuth measured at point A on Earth's surface (A) and on the stereographic azimuthal conformal planar projection (B).

However, they are suitable for many reference maps. These projections are referred to as "compromise." Their advantage comes from balancing distortion across the projected area. In some cases, compromise projections are created to preserve the visual appearance of Earth's landmasses. Due to this visual goal, compromise projections are suitable for general reference maps. Figure 2.18 shows the Winkel tripel modified azimuthal and Eckert III compromise pseudocylindrical projections which have seen considerable usage as the basis for global-scale maps in world thematic atlases. The lack of preserving any specific projection property helps to present the world's landmasses without the excessive visual distortion that is usually associated with a projection that preserves a specific property.

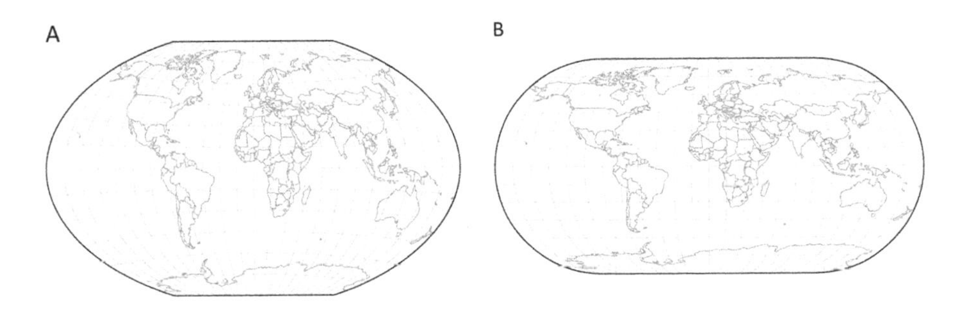

FIGURE 2.18
The Winkel tripel compromise modified azimuthal (A) and Eckert III compromise pseudocylindrical (B) projections.

Projection's Influence on Map Design

Aside from preserving spatial relationships, projections can also be viewed as an important variable when designing maps. There are two attributes of a projection that can be considered a variable when designing maps: class and aspect.

The Map's Shape: The Projection Class

In this section, we primarily discuss the technical aspects of projection class and its impact on shape, while in Chapter 4 we will address some of the cognitive and perceptual aspects of map readers' preference for particular map shapes. Projections can be described referencing a particular shape that conveys their overall appearance. Common shapes include cone, rectangular, oval, and circular and relate to the projection classes of conic, cylindrical, pseudocylindrical or modified azimuthal, and planar, respectively. Figure 2.19 gives an overview of these general projection classes. Conic projections (Figure 2.19A) appear as a "cone" shape with meridians of longitude represented as straight lines that diverge away from one of the poles (the South Pole in Figure 2.19A) and converge to the other pole. Parallels of latitude are shown as concentric circular arcs (rather than circles as would appear on a globe) that concave toward one of the poles (the South Pole Figure 2.19A). Note that the length of the parallels continually increases from the South Pole to the North Pole which is not a characteristic that is present on Earth's surface. Cylindrical projections (Figure 2.19B) appear rectangular in shape and represent meridians and parallels as straight lines that appear inside of a bounding rectangle. On this projection, meridians are all the same length as found on Earth's surface while parallels are also shown as the same length as the equator. While parallels appear as equal length on this projection, this characteristic does not occur on Earth's surface and suggests distortion. Like cylindrical projections, pseudocylindrical projections (Figure 2.19C and Figure 2.19D) appear as an oval and show parallels as straight lines and curved meridians. Note that the meridians in Figure 2.19C converge to the poles represented as points, or as lines in Figure 2.19D. While representing the poles as points is more reflective of their representation on a globe, the landmasses in the upper latitudes can become compressed, making identification difficult. Representing the poles as lines tends to stretch the landmasses in the upper latitudes which may be beneficial if the map is intended to display thematic symbols in these upper latitudes. Modified azimuthal projections (Figure 2.19E) also appear oval. However, unlike the pseudocylindricals, parallels on the modified azimuthal class are curved. Modified azimuthal projections also have utility for global-scale maps in world thematic atlases. Planar projections (Figure 2.19F) show the world (usually only a hemisphere or less) inside of a bounding circle. If the planar projection is

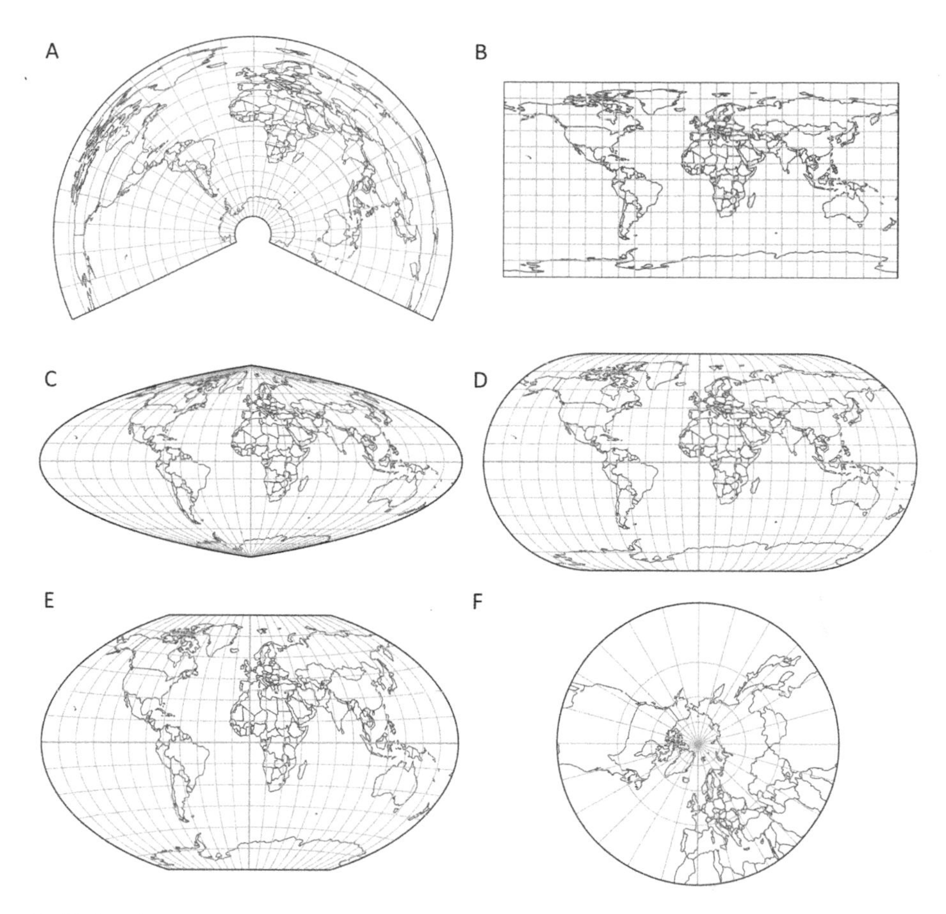

FIGURE 2.19
Illustrating the projection classes. The Euler equidistant conic (A), plate carrée equidistant cylindrical (B), quartic authalic equal area pseudocylindrical (C), Eckert III compromise pseudocylindrical (D), Winkel triple compromise modified azimuthal (E), and gnomonic azimuthal planar (F) projections.

centered on a pole, the meridians are straight lines radiating outward from a pole and parallels are concentric circles centered about the pole (as in the case of Figure 2.19E). If the projection is centered anywhere else, the parallels and meridians are represented by various types of curves.

Projection shapes should be considered in the map design process. For instance, conic projections are often used to map landmasses that have a greater east–west extent than north–south such as the United States, Russia, and China. Cylindrical projections are useful for maps appearing in rectangular formats such as computer screens or landscape-oriented page layouts. Cylindrical projections fill out the available "page" space in these formats better than other classes. Some scholars, however, have criticized the use of cylindrical projections citing their tremendous distortion in the appearance of landmasses, especially near the poles. However, when showing the entire Earth on a map, this same criticism can be raised against most other projections. Pseudocylindrical (with their curved meridians) and modified azimuthal (with their curved meridians and parallels) projections mitigate some of the criticisms leveled at cylindrical projections and more accurately portray lines of latitude and longitude lines as they appear on the globe than do other shapes. Pseudocylindrical and modified azimuthal projections are often selected for world maps where the overall appearance of the landmasses is an important design consideration as these projection shapes tend to distort landmasses less than other projection shapes. Planar projections, especially the orthographic azimuthal compromise planar, present Earth as if you are looking upon it from outer space, which is also desirable for many inset or locator maps.

Focusing the Map's Center: The Projection Aspect

The aspect of a projection controls the projection's center. On some projections, like the planar class, the center is usually defined by a point where the central meridian and central latitude intersect. The central meridian divides the mapped area east and west while the central latitude divides the mapped area north–south. One way to think about a projection's aspect is to look at the examples shown in Figure 2.20. Figure 2.20A, Figure 2.20B, and Figure 2.20C show three aspects of the azimuthal equidistant projection: polar, oblique, and equatorial. Figure 2.20A shows a polar aspect of a map centered over the North Pole. On this map, the location of the central latitude is 90° N while the location of the central meridian is 95° W which places the North Pole at the map's center and focuses the map on the Eastern Hemisphere. Figure 2.20B shows an oblique aspect of a map centered over the Indian Ocean. On this map the location of the central latitude is 40° N while the location of the central meridian is 95° E. This map's aspect is considered oblique as the center of the map is neither aligned to a pole nor centered along the equator.

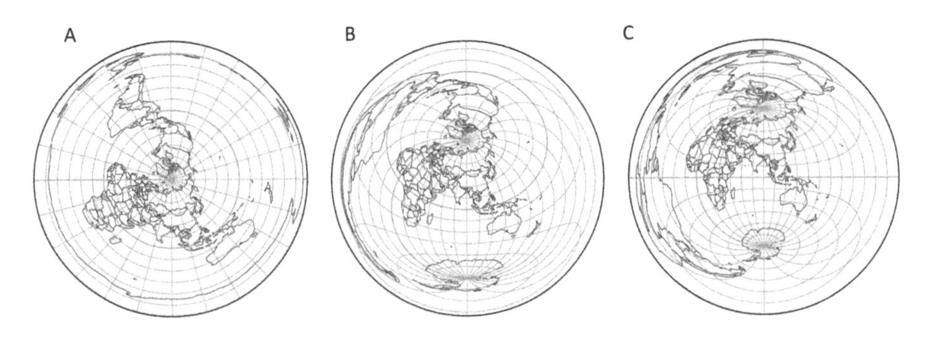

FIGURE 2.20
Three general aspects of a projection polar (A), oblique (B), and equatorial (C) shown on the azimuthal equidistant projection.

Figure 2.20C shows an equatorial aspect of a map centered over the Indian Ocean. On this map the location of the central latitude is 0° while the location of the central meridian is 95° E. This map's aspect is considered equatorial as the center of the map is along the equator.

The projection's aspect plays an important role when you want to align the map's center with the geographic area of interest. In mapping software, for instance, the map center often defaults to an equatorial aspect of (0°, 0°), which is a location positioned slightly west of Africa's Ivory Coast in the Atlantic Ocean along the Prime Meridian. This center is okay for maps whose data focus on this African Atlantic coastal location but for maps of other locations the center could be changed to provide better focus on the dataset or location of interest. For instance, consider making a map of Russia. You would specify the projection center, for example, so that the central meridian divides the east–west extent of Russia into half and then choose the central latitude so that the latitudinal extent of this landmass is equally divided.

Map Projection Distortion

So far, we have explained that a consequence of the projection process is the introduction of distortion to the map. In this section we discuss distortion, how it is distributed across a projection, and what controls that distribution. Distortion is an inevitable consequence of the projection process. Distortion can be controlled by carefully selecting the appropriate projection class, property being preserved, and location of the standard point or line(s). These three topics frame the discussion in this next section.

Distortion Patterns and Projection Class

Each projection class exhibits a characteristic distortion pattern. Figure 2.21 shows the pseudocylindrical, planar, conic, and cylindrical projection classes and their overall distortion patterns. Figure 2.21A shows the distortion pattern on the quartic authalic. On this projection, note that lighter color values of magenta, suggesting low distortion, are found along zones that surround the equator and central meridian. Darker color values of magenta, suggesting higher distortion, appear near the map's periphery (Figure 2.21A). The Lambert azimuthal equal area planar projection is shown by Figure 2.21B. On this projection, an area of low distortion is found at the projection's center. From the center, distortion increases outward reaching a maximum at the projection's periphery. The Albers projection is shown in Figure 2.21C. A zone of low distortion is located mid-latitude in the Northern Hemisphere on this projection. This zone follows parallels of latitude which coincide with the location of standard lines on this projection. Similar to the other

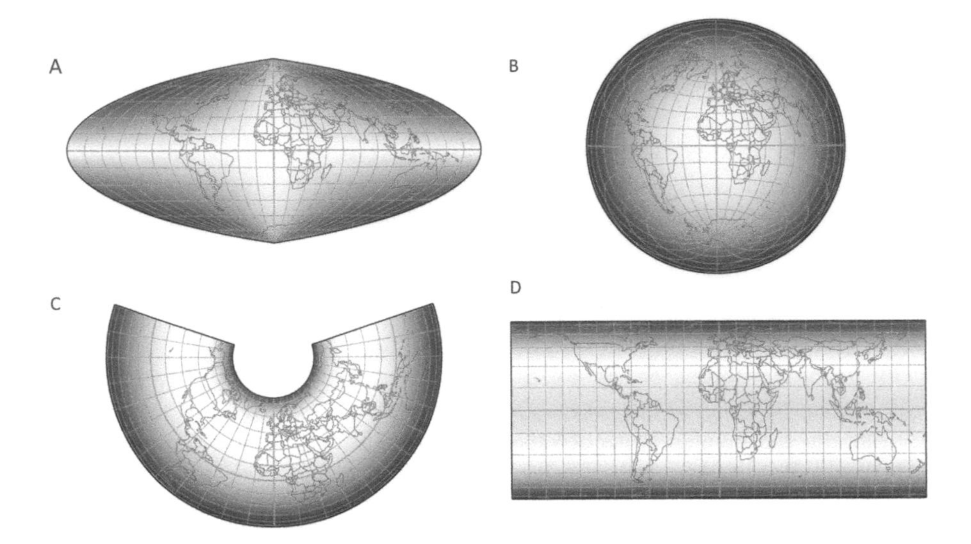

FIGURE 2.21
Overall distortion patterns on the quartic authalic equal area pseudocylindrical (A), Lambert azimuthal equal area planar (B), Albers equal area conic (C), and Behrmann equal area cylindrical (D) projections.

projections in Figure 2.21, maximum distortion is found near the poles. The Behrmann equal area cylindrical is shown in Figure 2.21D. On this projection, there are two zones of low distortion that are located approximately 30° north and south of the equator. The darkest magenta color values are located at the poles where distortion is the greatest.

Some distortion patterns are fixed, meaning that the map maker has no control over where low and high distortion on a projection is found. The quartic authalic (Figure 2.21A) and Lambert azimuthal equal area (Figure 2.21B) projections have fixed distortion patterns but the map maker can change the projection's aspect and move different geographic areas into and out of zones of low distortion. Choosing an appropriate aspect is advantageous if the map maker wants to bring a specific geographic area into the zone of low distortion or push other landmasses to areas of greater distortion. On other projections, the distortion patterns are dependent upon specific parameters that are defined by the map maker. For example, the Albers (Figure 2.21C) and Behrmann (Figure 2.21D) projections include parameters that allow the map maker to specify which parallels are assigned no distortion. By adjusting the projection parameters, the distortion pattern can be redistributed so that the geographic area of interest or data extents can be mapped with the lowest distortion.

Figure 2.22 uses the quartic authalic to illustrate the utility of changing the projection's aspect and bring different geographic areas of interest to the map's center. Figure 2.22A, Figure 2.22B, and Figure 2.22C shows each map centered over central Australia, the southern tip of the Saudi Arabian

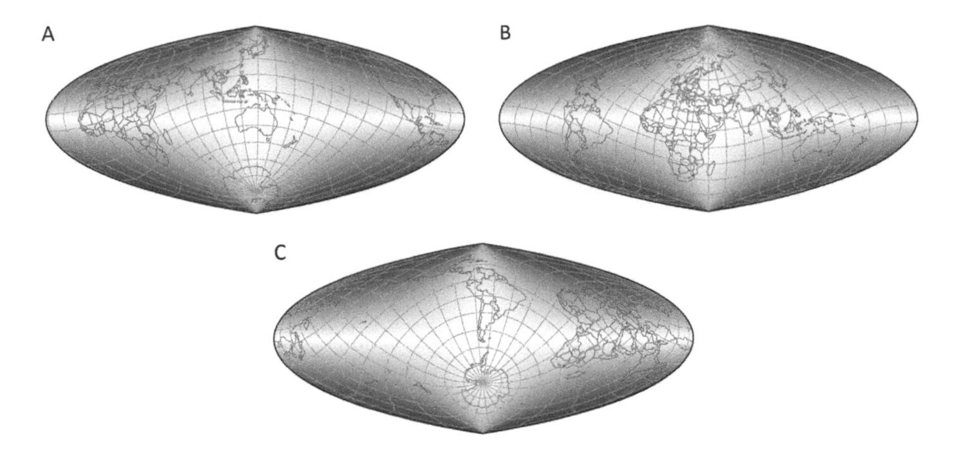

FIGURE 2.22
Visualizing distortion patterns on the quartic authalic equal area pseudocylindrical projection with three different centers: central Australia (A), Saudi Arabian Peninsula (B), and Terra del Fuego (C).

Peninsula, and Terra del Fuego, respectively. As you change this projection's aspect, different landmasses are brought into the map's center where the lower amounts of distortion are found. It is important to note that on this projection, changing the aspect does not change the distortion pattern, the distortion pattern is constant. On this projection, the areas of higher distortion are always found near the map's periphery while areas of lower distortion always coincide along lines of latitude and longitude that define the projection's center.

Distortion Patterns and Projection Property

For a projection to preserve a property, scale must be manipulated in very specific ways. For example, to preserve angular relationships, the scale about any given point must be equal in all directions (as is the case on conformal projections). On the other hand, equidistant projections preserve scale, for example, only along meridians and exaggerate or compress scale along all other lines. With compromise projections, scale is often balanced in certain ways but is not preserved in a particular way throughout the projection, avoiding the excessive distortion found in projections that preserve a single property. When a particular property is preserved, scale needs to be manipulated throughout the rest of the projection and this scale manipulation directly links to the type and severity of distortion.

Figure 2.23 illustrates the distortion patterns on a conformal and equal area projection. Figure 2.23A shows a green continuous tone color on the Eisenlohr conformal projection illustrating the areal distortion pattern that results from preserving angular relations. On this projection, the center exhibits low distortion. As you progress outward from the center, darker

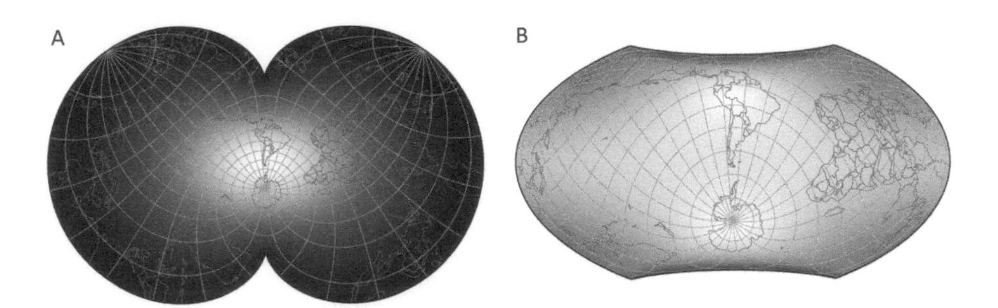

FIGURE 2.23
The Eisenlohr conformal (A) and Wagner VII equal area modified azimuthal (B) projections.

tones of green are present suggesting that greater amounts of areal distortion are present. Near the periphery of the projection, the color becomes nearly black, indicating that there is considerable area and scale distortion present at these locations. Since the projection is conformal, there is no magenta coloration in the distortion visualization, since there is no angular distortion. Figure 2.23B shows the Wagner VII equal area modified azimuthal. Its distortion pattern is different from the Eisenlohr in two ways: first, the Wagner VII is equal area, so the magenta continuous tone color illustrated in Figure 2.23B shows angular and scale distortion. Areal distortion is absent on this projection, so there is no green coloration like we see in the Eisenlohr example. Second, instead of one centralized area where distortion is low there are two areas roughly at 42° north and south latitude where distortion is low. Starting at these two locations, distortion increases outward. Near the periphery of this projection, the darker color tones suggest greater amounts of areal and scale distortion. At the poles the distortion is maximized since the poles are represented as lines on this projection. Whether or not the poles are represented as points or lines has an impact on the distortion pattern that is present in the upper latitudes, a topic we will discuss in more detail in Chapter 5.

Figure 2.24 illustrates two interesting distortion patterns where a mixture of areal and angular distortion on the Aitoff compromise modified azimuthal (Figure 2.24A) and the Robinson (Figure 2.24B) projections are displayed.

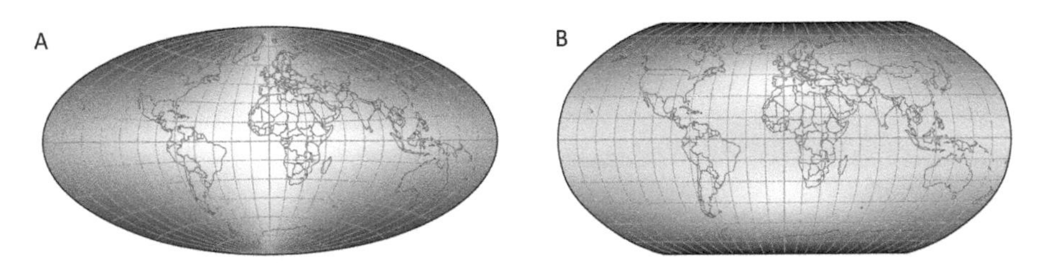

FIGURE 2.24
The Aitoff modified azimuthal (A) and Robinson pseudocylindrical compromise (B) projections.

Mixtures of both the green and magenta color tones suggest that neither projection preserves any specific projection property (doesn't preserve either areal or angular relations). In Figure 2.24A, light green tones suggest that low amounts of areal distortion are present along the equator and Prime Meridian. Beyond this areal distortion, increasing amounts of angular distortion appear toward the periphery of the projection. In Figure 2.24B, light green tones in the tropic regions suggest a band of areal distortion, but further toward the periphery, angular distortion increases.

Recall that Figure 2.19 introduced the projection classes. Figure 2.25 shows the same projection classes but displays the corresponding distortion pattern. In Figure 2.25, continuous color tones are used to represent the changes in distortion across the projection's surface. By examining each projection

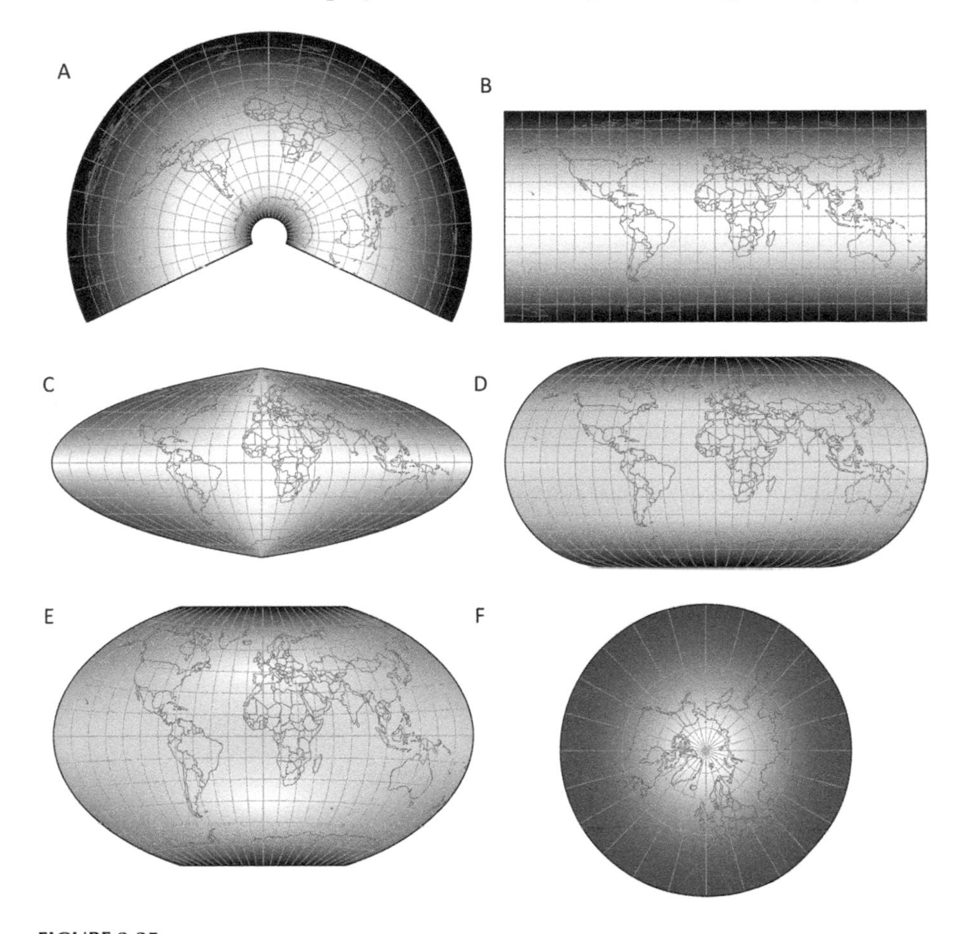

FIGURE 2.25
Comparing overall distortion patterns on the projection classes. The Euler equidistant conic (A), plate carrée equidistant cylindrical (B), quartic authalic equal area pseudocylindrical (C), Eckert III compromise pseudocylindrical (D), Winkel triple compromise modified azimuthal (E), and gnomonic azimuthal planar (F) projections.

and the color gradations shown, you can see specific distortion patterns typical for each projection class. Figure 2.25A shows the Euler and Figure 2.25B shows the plate carrée projections. Both projections are equidistant, meaning that distances are preserved along all meridians. In order to achieve equidistance, however, this projection distorts areas and angles. On both projections, severe distortion in areas and angles occurs in the upper latitudes as evidenced by the dark gray shades in these regions. Locations in the northern mid-latitudes on the Euler and the equator on the plate carrée have low area and angular distortion. This is apparent by the very light gray shading that is visible in these areas. Note that on each projection, the distortion appears in bands that are parallel to the lines of latitude, which is a characteristic feature on these classes of projections. Figure 2.25C shows the distortion pattern on the quartic authalic. Since the projection is equal area, angles are distorted, which is shown by the darker magenta shades located in the four extreme quadrants. On this projection, regions along the equator and central meridian have low levels of angular distortion. Projections in Figure 2.25D (Eckert III) and Figure 2.25E (Winkel tripel) are compromise projections that do not preserve any property. In some locations on these projections area distortion dominates (darker green shades), while in other locations angular distortion is more prevalent (darker magenta shades). For instance, in the Eckert III projection in Figure 2.25D, the areal distortion is more prevalent around the equatorial region while angular distortion is apparent from the mid-latitudes toward the poles. Note that at approximately 35° N and 35° S there is a noticeable change in the type of distortion present as indicated by the contrasting colors. Again, on this projection, distortion appears as bands that are generally parallel to the lines of latitude. The Winkel tripel projection in Figure 2.25E shows angular distortion located toward the peripheries of the projection while areal distortion is more pronounced in three regions along the central meridian. As evidenced by the white regions at roughly 45° N/S of the equator, we see two zones of low distortion. The gnomonic projection in Figure 2.25F is azimuthal and preserves directions. Note that the region of low distortion is located at the projection's center. As you move toward the periphery, area distortion increases dramatically. With this projection, distortion appears as concentric rings centered about the projection's center, which is a characteristic of the planar class of projections.

Distortion Patterns and Standard Point or Line(s)

Some projections include a point, line, or lines where no distortion is present. These locations of no distortion are referred to as "standard." Whether a projection includes a point, line, or lines is partially dependent upon the class. For example, the planar class of projections includes a point that is free of distortion (see Figure 2.25F). This point is at the projection's center. If the projection's center is altered (changing the central meridian and central latitude values) the distortion pattern does not change. The change to the projection's center makes it easy to move the geographic area of interest into the area of lower distortion (see Figure 2.21). Unlike the planar class, the cylindrical class of projections includes one or two lines that are free of distortion. The location of these standard lines can be adjusted to redistribute distortion in the projection. For example, Figure 2.26 shows three examples of the equal area cylindrical projection illustrating different standard line placements. On cylindrical projections with an equatorial aspect, the standard line coincides with parallels of latitude, so it is sometimes referred to as a "standard parallel." Figure 2.26A, Figure 2.26B, and Figure 2.26C locate the standard line at 0°, 30° north and south, and 45° north and south, respectively. The white areas in these figures represent locations of low distortion. Figure 2.26A (the Lambert cylindrical equal area) has one standard line coinciding with the equator. The tropical region has the lowest distortion while the polar areas have the highest amounts of angular distortion as shown by the magenta color gradation. This projection would be appropriate for mapping a distribution whose geographic

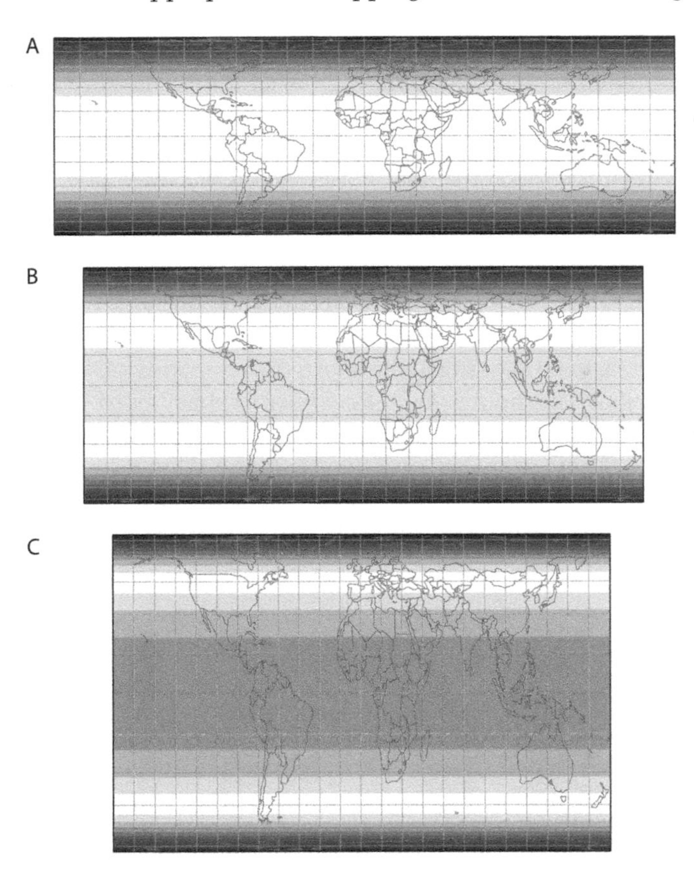

FIGURE 2.26
The Lambert (A), Behrmann (B), and Gall orthographic (C) equal area cylindrical projections with standard lines located at 0°, 30° N and S, and 45° N and S, respectively.

focus is the tropical areas. Figure 2.26B (the Behrmann cylindric) includes two standard lines that are equally spaced on either side of the equator. Angular distortion increases away from these standard lines, reaching the maximum at the polar areas. Between the standard lines, angular distortion is also present but is not as severe compared to the polar areas. On this projection, the mid-latitudes have low distortion and would be suitable for mapping distributions at these latitudes. Figure 2.26C (the Gall orthographic cylindric) places two standard lines in the high latitudes where low distortion is present. On this projection, distortion is present between and extends poleward from the two standard lines. Note however that on this projection the area between the two standard lines appears in a darker tone of magenta and that darker tone encompasses a greater latitude extent than shown in Figure 2.26B. The projection would be suitable for displaying distributions in the upper latitude areas.

The conic class of projections also offers the ability to choose the location of standard lines. Figure 2.27 shows distortion on the Albers projection. Figure 2.27A places the standard lines at 30° and 60° north. This placement creates a zone of low angular distortion between the two standard lines in the Northern Hemisphere. Areas of higher angular distortion occur in the North Polar area and in the Southern Hemisphere. Figure 2.27B places the standard lines at 30° and 60° south. In this case, a zone of low distortion is created between the standard lines in the Southern Hemisphere. Higher levels of angular distortion occur in the Northern Hemisphere and the South Polar area. Figure 2.27C uses only one standard line at 45° north. Instead of two zones with low distortion, only one zone of low distortion is created. This single standard line places greater angular distortion in the Southern Hemisphere and the extreme North Polar area.

Standard lines can be judiciously selected to minimize the distortion of a geographic region of importance. The ability to specify the location of standard lines is a great asset when trying to control distortion, for example, over a smaller geographic area. Figure 2.28 shows the Albers projection focused on the lower United States. There are several ways in which standard lines can be selected. Three of those methods will be discussed here.

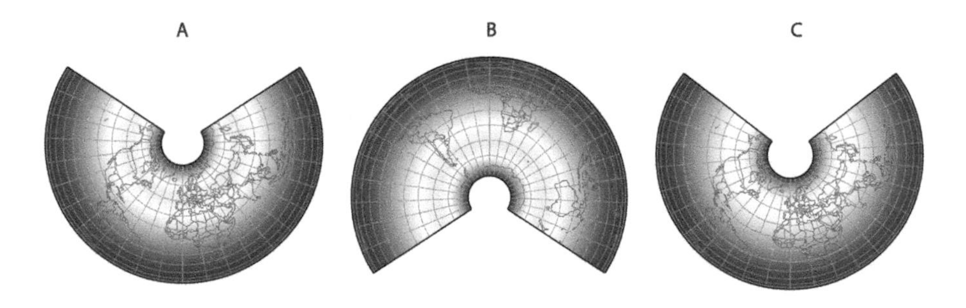

FIGURE 2.27
The Albers equal area conic projection with standard lines at 30° and 60° N (A), 30° and 60° S (B), and one standard line at 45° N (C).

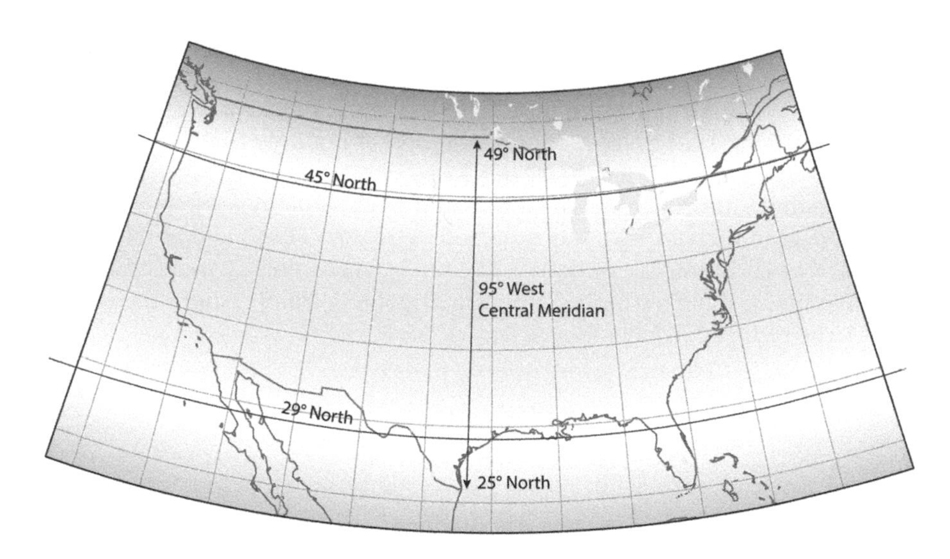

FIGURE 2.28
The Albers equal area conic with standard lines at 29° N and 45° N.

The first method divides the latitude range into thirds. For any given map there are two limiting parallels that define the extreme northern and southern geographic boundaries. In other words, these north–south boundaries define the top-to-bottom latitude limits of the map. Thus, placing standard lines at one-third the distance from the top and bottom limiting parallels allows the mapped area to be divided into thirds—one-third of the mapped area falls between the lower geographic limit and the lower standard line, one-third of the mapped area falls between the two standard lines, and one-third of the mapped area falls between the upper geographic limit and the upper standard line. While this method is simple in its execution, the results do not guarantee that the distribution of distortion is necessarily optimized throughout the mapped area.

Another method for placing standard lines based on dividing up the geographic area to be mapped was suggested by Deetz and Adams (1944, p. 80–82). Their approach divides the geographic area to be mapped with the aim of obtaining a maximum distribution of scale distortion across the mapped area. According to their suggestion, they divide the latitude range of the mapped area into sixths. For example, assume the north-to-south extent of the mapped area along the central meridian is 25° N and 49° N. Figure 2.28 shows that this latitude extent encompasses the lower 48 United States. To achieve this equal distribution of scale distortion, Deetz and Adams suggest placing standard lines one-sixth the distance from the lower and upper latitude extremes. The difference between 25 °N and 49° N is 24°. Dividing this value into sixths results in four degrees. Subtracting four degrees from 49° N produces 45° N and adding four degrees to 25° N is 29°N. Thus, the standard lines are located at 29° N and 45° N. Four-sixths of the geographic area mapped falls between the standard lines (the focus of the map) and two-sixths fall outside of the standard lines. Taking this approach, Deetz and Adams reports that the maximum scale error for the lower 48 United States when using the Lambert conformal conic is no more than 1.125 percent (1944, p. 82).

Mathematical approaches to establishing standard lines have also been developed. One such approach is reported by Maling (1992) and incorporates a constant as shown by Equations 2.12 and 2.13. The constant describes the overall shape of the geographic area under consideration where the distortion is to be reduced.

$$\varphi_2 = \varphi_N - (\varphi_N - \varphi_S) / K \tag{2.12}$$

$$\varphi_1 = \varphi_N + (\varphi_N - \varphi_S) / K \tag{2.13}$$

Where K is a constant and is varied according to the shape of the geographic area to be mapped using the following recommendations. For a country that has a…

- Small extent in latitude but large extent in longitude (e.g., Indonesia) the value of $K=7$
- Rectangular outline with longer axis north–south (e.g., Scandinavia) the value of $K=5$
- Circular or elliptical outline (e.g., Argentina) the value of $K=4$
- Square outline (e.g., the state of Wyoming) the value of $K=3$

φ_1 and φ_2 are the two standard lines under consideration, and φ_N and φ_S are the northern and southern geographical extents of the mapped area, respectively.

Distortion Impacts on Shapes of Landmasses

We have already discussed the idea that the projection process introduces distortion on a map and that distortion limits the way in which the spatial relationships are or are not preserved on a map. In addition, distortion can alter the appearance of landmasses on a map compared to how it looks on a globe. Some of the factors that control the distortion pattern across a projection's surface and thus alter the appearance of landmasses on a map include the chosen projection class, geographic extent of the landmass mapped, and the projection parameters. We have already discussed the patterns of distortion that are present on the projection classes (see Figure 2.19 and Figure 2.25), and that changing the projection's aspect does not change the overall distortion pattern (see Figure 2.22). By understanding the distortion patterns on a given projection class, you should be able to modify the projection aspect so that the geographic area of interest is aligned to the region on the projection with the lower levels of distortion. The geographic extent of the map plays a significant role in controlling distortion which influences how a landmass appears on a map. Global-scale maps generally exhibit considerable distortion in the appearance of landmasses, especially near the map's periphery. Maps of smaller geographic extents (e.g., a country or local area) generally do not suffer from these extreme amounts of distortion and thus look more similar to their appearance on a globe. In this chapter, we also explained that some projections include parameters like standard lines whose values can be specified. By specifying the lines of latitude assigned as standard lines, for example, the map maker can redistribute distortion across the projection's surface (Figure 2.28). This redistribution can then present the geographic area mapped in a zone of lower distortion. In general, global-scale maps will require more thoughtful consideration on how to manipulate the distortion pattern than will maps of local areas. But it is important to remember that no projection will preserve the true shape of every landmass. By understanding distortion patterns and carefully specifying the available projection parameters, the amount of distortion that landmasses exhibit on a map can be minimized.

Throughout the rest of this book, we will describe how distortion can affect the appearance of landmasses on a map using two terms: stretch and compress. We use *stretch* to refer to the appearance of a landmass whose overall shape has been elongated in some fashion due to the projection process. In Figure 2.29, the Miller compromise cylindrical projection stretches Canada in an east-to-west and north-to-south fashion. This stretching directly results from the lines of latitude being stretched to the same length as the equator and the spacing of the lines of latitude steadily increasing from the equator poleward. A consequence of this stretching is that landmasses appear larger compared to how they look on the globe, which can lead to erroneous mental assumptions about their size and shape. We use *compress* to refer to the appearance of a landmass whose overall shape has been shortened in some fashion due to the projection. Figure 2.30 shows the Lambert equal area conic. Note that South America is compressed north-to-south. This compression is especially noticeable in the southern portion of the continent and results from the spacing of the lines of latitude being closer than they appear on a globe, which is necessary to preserve the equal area property. With some projections, a landmass can experience stretching in one direction and compression in another direction. Figure 2.31 illustrates how Australia is stretched in an east–west direction and compressed in a north–south direction on the Albers projection. This stretching is due to the meridians diverging away from the North Pole and intersecting a circular arc representing the South Pole. Compression in this case results from the decreased spacing of the lines of latitude near the South Pole compared to what appears on the globe and is necessary in order to achieve the equal area property.

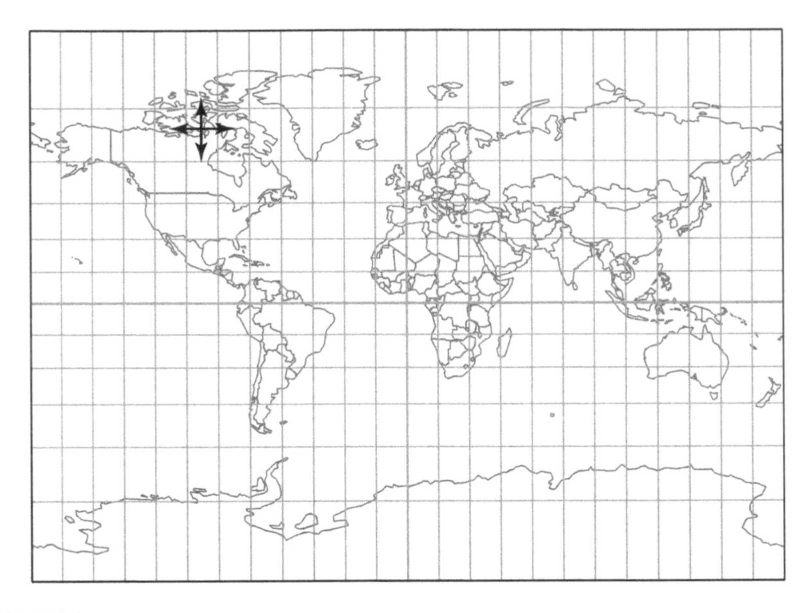

FIGURE 2.29
Illustrating how Canada experiences east–west and north–south stretching on the Miller compromise cylindrical projection.

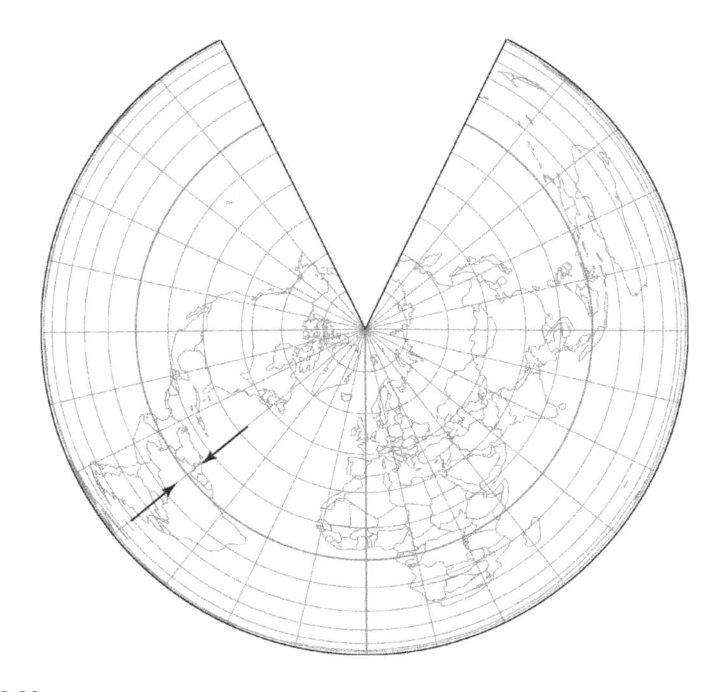

FIGURE 2.30
Illustrating how South America experiences a north–south compression on the Lambert equal area conic projection.

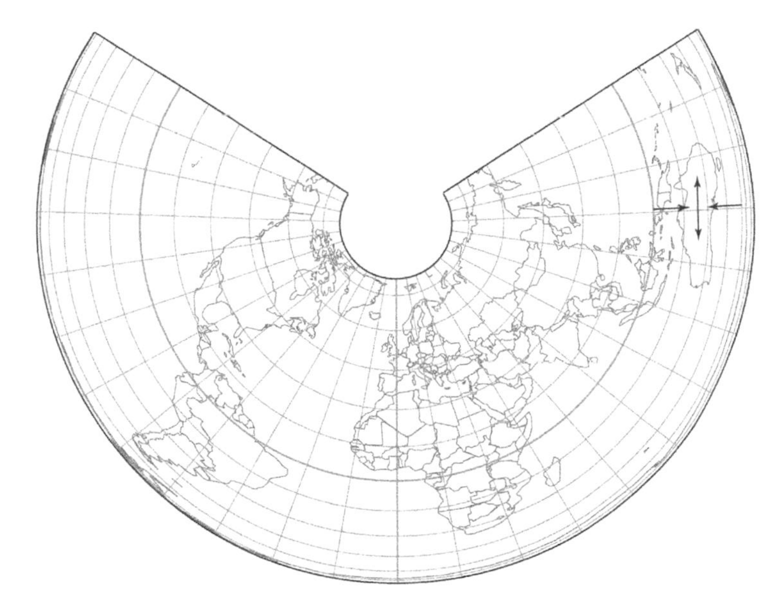

FIGURE 2.31
Illustrating how Australia experiences east–west stretching and north–south compression on the Albers equal area conic projection.

Conclusion

In this chapter, we presented what we hope is a gentle and non-technical introduction to projections, the focus of this book. We began with a discussion of latitude and longitude and how each are defined on Earth's surface. Latitude and longitude are important in mapping as they are points that define locations on Earth's surface. As we discussed, Earth's shape is very complex but can be simplified to a model based on an ellipsoid or sphere. The model choice partly depends on the desired accuracy of the final map. A more accurate map, such as a topographic map, would call for an ellipsoid Earth model while a less accurate map, such as a reference map, would call for a sphere Earth model.

The projection process is fundamentally based on mathematical equations where latitude and longitude values are "projected" to x and y plotting coordinates. Thus, the projection process projects latitude and longitude values from Earth's curved surface to a plane (a map). Depending on the chosen Earth model and intended map accuracy, ellipsoidal and spherical forms of a projection equation are available.

Projections possess properties that help the map maker create maps that fulfill specific purposes. Projections can preserve areas, angles, distances, and directions so that map readers can use maps for specific purposes.

For example, a map reader may need to measure the area of a boundary, calculate the angle between two features, measure a distance, or determine a direction from point A to point B. Projections that preserve areas, angles, distances, and directions are referred to as equal area, conformal, equidistant, and azimuthal, respectively. Equal area and conformal projections preserve areas and angles, respectively, across the entire map's surface. Equidistant and azimuthal projections preserve distances and directions, respectively, but in limited ways.

Projections can be considered a variable in map design. Two design characteristics are the overall projection shape and the aspect. Projections come in different shapes such as rectangles, circles, cones, and ovals, although not all projections fit neatly into one of these shapes. These shapes are best recognized when mapping at a global scale. Each shape can be used to the benefit of the overall map design (rectangular projections fit on a computer screen better than projections with a circular shape). The overall shape of a projection can also be used to classify projections. Rectangular, circular, oval, and cone shaped projections are organized into cylindrical, planar, pseudocylindrical/modified azimuthal, and conic classes respectively. The projection aspect controls the map's center. It is often advantageous to have the geographic area to be mapped aligned to the map's center where distortion is usually lower.

Distortion is a by-product of the projection process. Projecting Earth's curved surface to a plane surface introduces distortion that impacts a map in two ways. First, distortion limits the map reader's ability to carry out accurate measurement tasks such as distance measurements on a map. For instance, if you were interested in measuring distances across a map, you would want to choose a projection that had low-scale distortion across that area of interest. Second, distortion impacts the overall appearance of a landmass by stretching and/or compressing it. In some cases, excessive distortion renders a landmass unrecognizable. Generally speaking, distortion is most severe on global-scale maps, specifically at the periphery of the map. As the map scale becomes larger, extreme distortion decreases. Distortion can be controlled on some projections by specifying the location of the standard line (a line along which there is not distortion). For example, locating a standard line through the geographic area to be mapped will redistribute distortion so that the mapped area is located in the area of low distortion.

Notes

1. The field of geodesy is responsible for studying and describing Earth's exact shape.
2. Check out this interesting website that describes how the Prime Meridian 0° has moved over time (www.rmg.co.uk/discover/explore/prime-meridian-greenwich).

3. Sir Isaac Newton is given credit for describing this oblateness (https://history.nasa.gov/SP-4211/ch11–4.htm).
4. The literature is rather inconsistent with the use of this term and often interchanges reference ellipsoid and reference spheroid.
5. Values of f from Equation 2.1 are very small. To make values of f more comparable, the inverse of f $(1/f)$ is reported.
6. While the North Pole can be used to compute an azimuth, there is nothing special about the use of the North Pole. The South Pole is also used in reporting an azimuth. Azimuths can also be determined counterclockwise.

3

Representing Spatial Data through Cartographic Symbolization

Cartography is the process of making a map; the cartographic process blends art, science, and technology. The artistic element refers to the design and aesthetics of the map. The map needs to be designed to achieve its purpose and be legible in its medium and viewing conditions (e.g., a small digital image on a web page, a large printed poster, or something else). Likewise, the map should be aesthetically pleasing to its intended audience. The science of cartography primarily revolves around the psychology of reading and interpreting signs and symbols. Research cartographers have studied the different ways that symbols (in varying sizes, shapes, colors, and styles) are interpreted by the map reader. These cartographers have also helped to improve communication between map and reader by creating new symbolization methods.

The technology of map making has changed drastically over thousands of years, from relying on hand-carved or painted clay tablets to digital mapping with geographic information systems (GIS). Computing in the 21st century has fundamentally altered the way that spatial data is collected, symbolized, and distributed (e.g., compare Ptolemy's T-O map from circa 150 C.E. to Google Maps today). Collectively speaking, the artistic aspects, scientific research, and technological developments of cartography have changed the profession from an exclusive, specialized group of practitioners, who laboriously drew maps by hand, into a field where just about anyone has the ability to collect data on almost any topic, create maps online at no cost, and share their results to the entire world at a single click.

Cartography can be described as the abstraction and symbolization of Earth. A map cannot represent Earth in all of its detail. There is always a trade-off between what information is mapped and what is not, and for a map to be legible and communicate something to the reader, some features will be emphasized while others are diminished or eliminated entirely. This chapter will cover the basic types of maps: thematic, reference, and cartometric. The type of map refers broadly to its purpose, which can be enhanced by the projection. Understanding how features or phenomena are distributed over Earth's surface is an important initial step in the map making process. This step identifies the underlying relevant questions about how the map will be used and helps choose the projection characteristics that are most important for the map's purpose. This helps you determine an appropriate way to collect data, classify it, and symbolize it.

Types of Maps

Maps can be classified into one of three types: thematic, reference, and cartometric. Thematic maps focus on a "theme," which means that the map presents data on a specific topic, de-emphasizing nonessential information. Figure 3.1 shows a thematic map of 2014 median household income by county in the United States. Here, the "theme" is median household income. The data has been collected at the county level, so each county (or enumeration unit) has a single data value. A color sequence from light blue to dark purple helps the map reader interpret the range of income values, as light often means "less" and dark often means "more." There are more light-colored counties in the United States than dark. This pattern suggests that, generally speaking, the United States has more counties whose median household income is lower than the national median, than it has counties whose median household income is higher than the national median. It is helpful to see this type of pattern on a map, rather than looking at a table, because the spatial variation of income by county is much more apparent in a visual

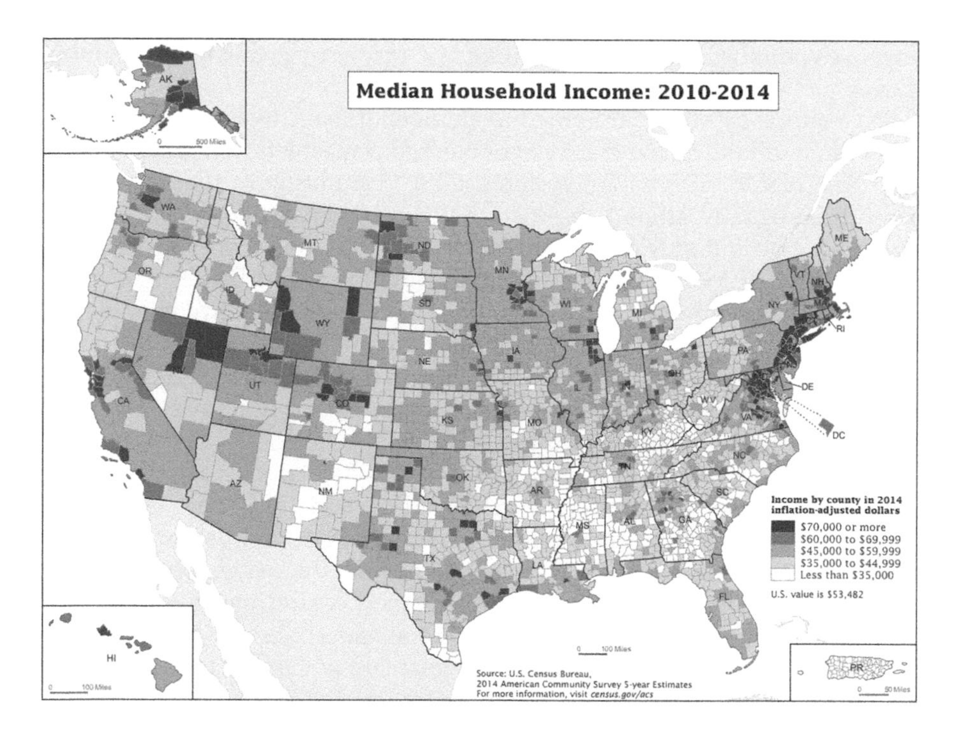

FIGURE 3.1
A thematic map produced by the U.S. Census Bureau, showing the median household income by county in 2014. (Reprinted from U.S. Census Bureau Infographics & Visualizations Library, 2015. Retrieved December 3, 2018, from www.census.gov/library/visualizations/2015/acs/2010–2014-acs-hh-income.html.)

format, and it helps the reader link median income levels to geographic setting. Thematic cartography lets us see patterns in phenomena over space and time, and helps researchers attempt to identify their causes and correlations.

Reference maps are often called basemaps, because the data they show is base-level information intended to give a general and neutral overview of the physical and/or cultural characteristics of a region. Figure 3.2 shows a reference map in The National Map online map viewer, hosted by the U.S. Geological Survey. This map shows fairly basic geographic data: roads, rivers, towns, administrative boundaries, and shaded-relief in the southwestern United States. With reference maps, the various base data are symbolized to show broad categorical differences. However, within these categories, other levels of symbolization indicate levels of hierarchy or importance. Population, or relative importance, of different cities may be hinted at with different label sizes. For instance, in Figure 3.2, "Denver" is larger than "Boulder," which directs the map reader to infer that Denver has a larger population than Boulder. One can consider the role of reference maps as a source for looking up geographic reference information on demand (e.g., where is Fort Collins? Is Aurora larger than Pueblo? And so on).

Cartometric maps are designed to provide accurate measurements. Figure 3.3 shows a portion of a nautical chart produced by the National Oceanic and Atmospheric Administration (NOAA). The term *chart* is used here because these types of maps are used to chart a course (*chart* meaning the physical process of drawing a navigation course directly on the map). Cartometric maps are often filled with detailed information and accurately mapped features that help the ship or plane navigator safely travel from one location to another. Charts will usually have a graticule (lines of latitude and longitude) and/or a grid (such as State Plane Coordinate System) drawn across the map. The graticule and/or grid is used to pinpoint a location or determine distance. Charts will also have information such as magnetic declination symbols (see the magenta circular symbol shown in Figure 3.3) that assist in wayfinding and orientation.

The Map Abstraction Process

The map abstraction process (Figure 3.4) turns observable phenomena on Earth's dynamic and 3-dimensional (3-D) surface into a 2-dimensional (2-D) map. It begins with the map maker conceptualizing the physical phenomena to be mapped, and involves the map maker understanding its spatial distribution, frequency of occurrence, and magnitude. In this step, the map maker also must determine what data should be collected from the observations of the phenomenon of interest. Once data is collected, attention turns toward describing the data as either qualitative or quantitative, which helps

FIGURE 3.2
A reference map of the southwestern portion of the United States available from The National Map Viewer, showing basic geographic information. (Reprinted from U.S. Geological Survey, The National Map, n.d. Retrieved December 28, 2018, from https://viewer.nationalmap.gov/advanced-viewer/.)

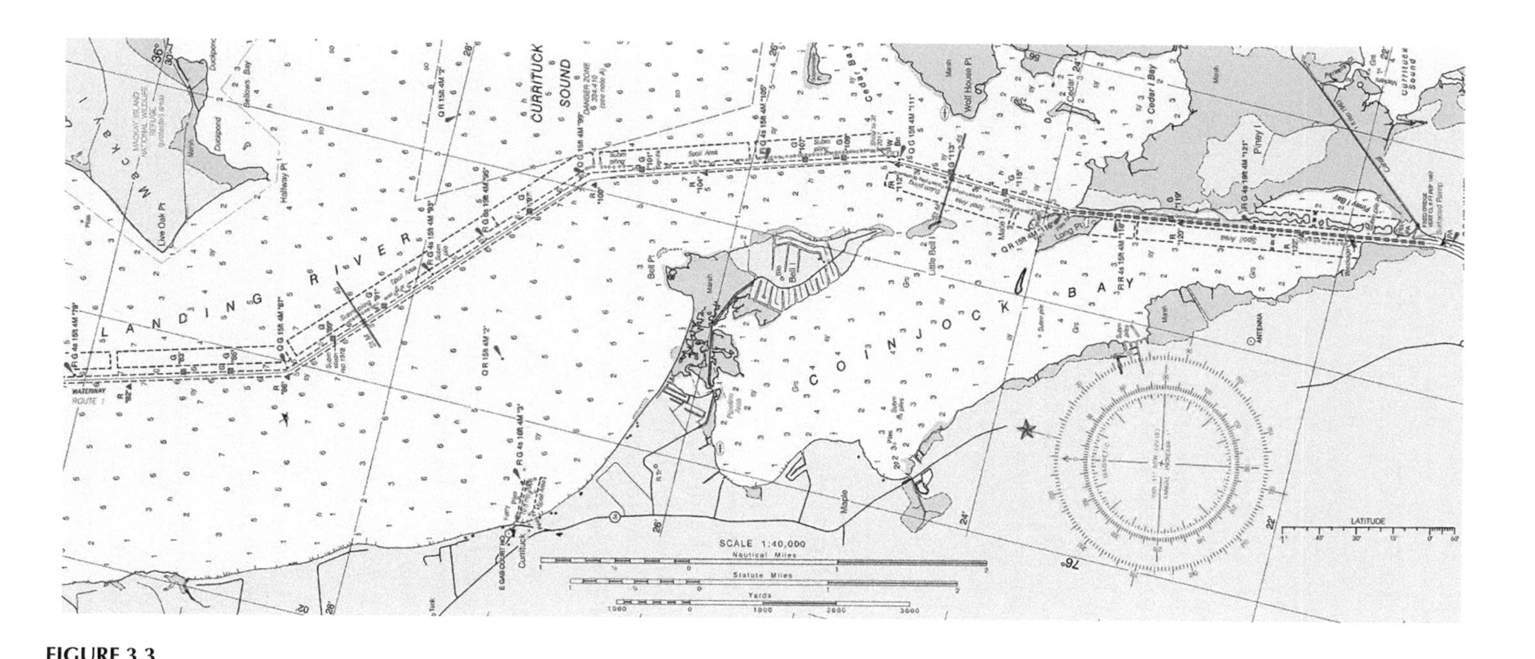

FIGURE 3.3
An example of a nautical chart produced by the National Oceanic and Atmospheric Administration. (Reprinted from Office of Coast Survey National Oceanic and Atmospheric Administration's Electronic Chart Locator, n.d. Retrieved December 28, 2018, from www.charts.noaa.gov/InteractiveCatalog/nrnc.shtml.)

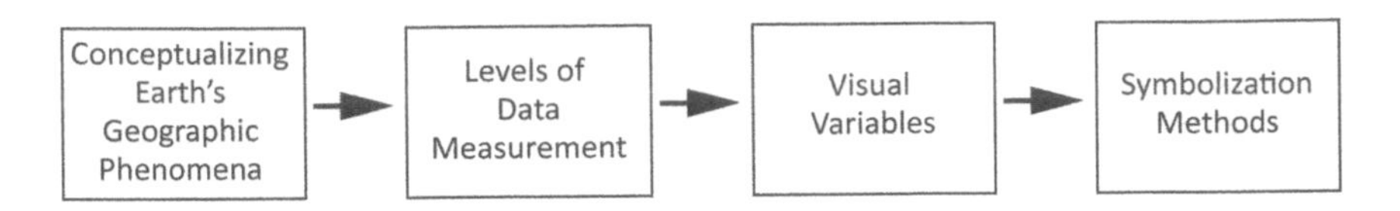

FIGURE 3.4
The map abstraction process.

to identify the data measurement level. It is necessary to understand the data measurement level, as different levels require different visual variables to symbolize them. Visual variables are the graphical building blocks of symbolization methods used to visually represent the data values. The final step in the map abstraction process is to select a symbolization method and, thus, a visual variable(s). When data is symbolized, and put into a map with necessary base information, map readers can understand where and sometimes when this data was observed, and can further infer patterns within the data. This section will discuss the map abstraction process in detail.

Conceptualizing Earth's Geographic Phenomena

Assume for a moment you are interested in making a map that shows the distribution of the world's population. Initially, you may begin to think about how different places on Earth are populated. This is important as it helps you understand in some measure that population values will be abstracted across the final map. There are two aspects to phenomena that are important at this stage of the mapping process: spatial dimension and typology.

Spatial Dimensions

Natural features or phenomena occupy one of five spatial dimensions: point, line, area, 2.5-dimension (2.5-D), or 3-D. Point phenomena exist at defined locations within a coordinate system, most often latitude and longitude, where a coordinate pair describes an infinitely small point. Some examples of these points are water wells, weather-recording stations, earthquake epicenters—just about any single monitoring station or source can be a point. Although they may have a tangible physical width (e.g., the structure that houses the weather-monitoring apparatus), representing these features on a map requires them to be abstracted to a zero spatial dimension (0-D). Linear phenomena have multiple coordinate point locations linked together—a more complex way of saying *lines*. Lines can be administrative boundaries, rail networks, or shipping routes. Lines are considered 1-dimensional (1-D) because the line's existence on Earth has a defined length, but no width. Abstracting the real line to depict on a map necessitates the assignment of a width for the line, but that width has no tangible association to that line's real characteristics. Areal features are defined areas enclosed by lines; altogether, these are called polygons. Examples are administrative areas (including their linear boundaries) such as national forests or city limits. Areal phenomena are 2-D because they have a measurable length and width.

Surfaces and volumes have 2.5-D and 3-D, respectively. Spatial phenomena that are 2.5-D have defined coordinates and an associated value that is above or below some datum creating a surface. One simple example of 2.5-D data is elevation. At any given location across Earth's surface, elevation can be expressed as a height above or below mean sea level. A topographic map is a good visualization of a 2.5-D surface, where any point on the map has a *single* elevation value. While 3-D phenomena also have defined coordinates, they can have an *infinite number* of values associated at a single location. For example, consider the concentration of carbon monoxide (CO) in the atmosphere, which extends thousands of miles above Earth's surface. For any point on the surface, going above that point out through the atmosphere yields multiple different levels of CO as one gets higher and higher above the surface.

Typology of Geographic Phenomena

The next step is to consider how a phenomenon varies across space: either continuously or discretely. Most climatological phenomena, such as temperature, are continuous. In other words, temperature exists everywhere in some quantity. There is no place on Earth's surface where temperature does not exist. In contrast, phenomena such as people or trees are discrete. There are places on Earth where people and trees do not exist.

Phenomena can also be described as varying abruptly or smoothly. Abrupt data has values that change markedly along a boundary. For example, population density can change abruptly across a country boundary or a city limit. Smooth phenomena show gradual changes from one location to another. Barometric pressure typically changes very smoothly from one location to another.

MacEachren and DiBiase (1991) developed a typology of geographic phenomenon along a continuum that helps us classify natural phenomena. The first continuum looks at where a phenomenon is found, that is, if it is continuous or discrete. The second continuum describes the variation between locations as abrupt or smooth. Figure 3.5 shows a map of average global sea surface temperature for 2009. Sea surface temperature (SST) exists everywhere across Earth's oceans, which means this is a continuous phenomenon (it is found everywhere in the sea). At the same time, SST usually changes slowly between any two given locations, which means it varies smoothly. Thus, SST is a continuous and smooth phenomenon.

Figure 3.6 shows a map of population density for the United States. Population density, the number of people per unit area, is calculated for a specified enumeration unit. The entire unit is assigned the same population

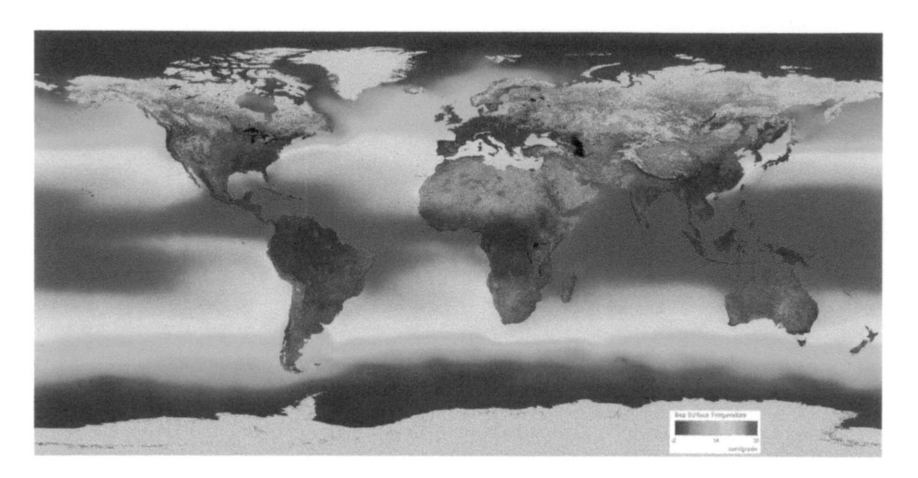

FIGURE 3.5
A map showing global average sea surface temperatures, a continuous and smooth phenomenon. (Retrieved from National Aeronautics and Space Administration's Scientific Visualization Studio, in ***Sea Surface Temperature, Salinity and Density***, by Helen-Nicole Kostis. Retrieved December 28, 2018, from https://svs.gsfc.nasa.gov/vis/a000000/a003600/a003652/.)

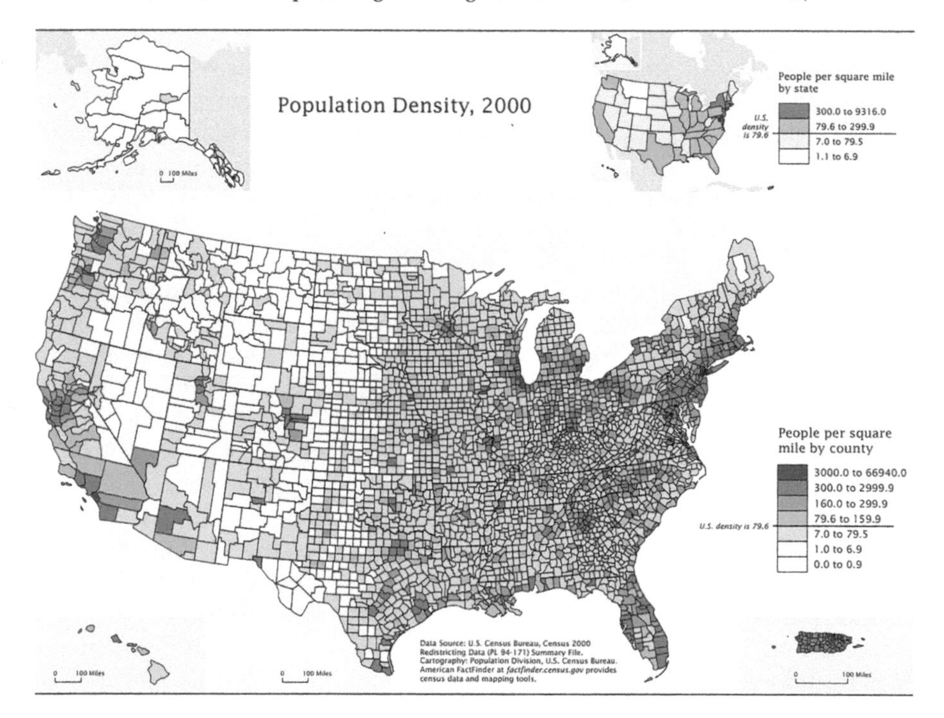

FIGURE 3.6
A map showing county-level population density for the United States, a continuous and abrupt phenomenon. (Reprinted from U.S. Census Bureau's Mapping Census 2000: The Geography of U. S. Diversity, 2000 (www.census.gov/population/www/cen2000/atlas/index.html). Retrieved December 28, 2018, from www.census.gov/population/www/censusdata/maps/density1.jpg.)

density value, but the value can change abruptly across the unit boundary (see Figure 3.6). Population density is a continuous, abrupt phenomenon.

Figure 3.7 shows a map of the total consumption of oil, gas, and coal between 1960–1964 by state. Consumption of these organic commodities is considered discrete since the consumption occurs at individual locations (e.g., a residence, an industry, or a business). For purposes of mapping, the consumption of these organic commodities by the different locations in a state is aggregated. As no two states share the same population, consumption amounts also vary abruptly between states. Thus, consumption of these organic commodities is considered discrete and abrupt.

Figure 3.8 shows the distribution of cows across the United States. Since a cow is a discrete entity, the data are considered discrete . The symbolization method uses dots to represent cows. In this case, one dot represents 2,000 cows. The resulting dot distribution gives the appearance of smooth changes between locations where a concentration of cows appear to an area where no cows are present.

Levels of Data Measurement

Data can be further classified as qualitative or quantitative. Qualitative data are non-numerical. For instance, categories of religions or types of languages

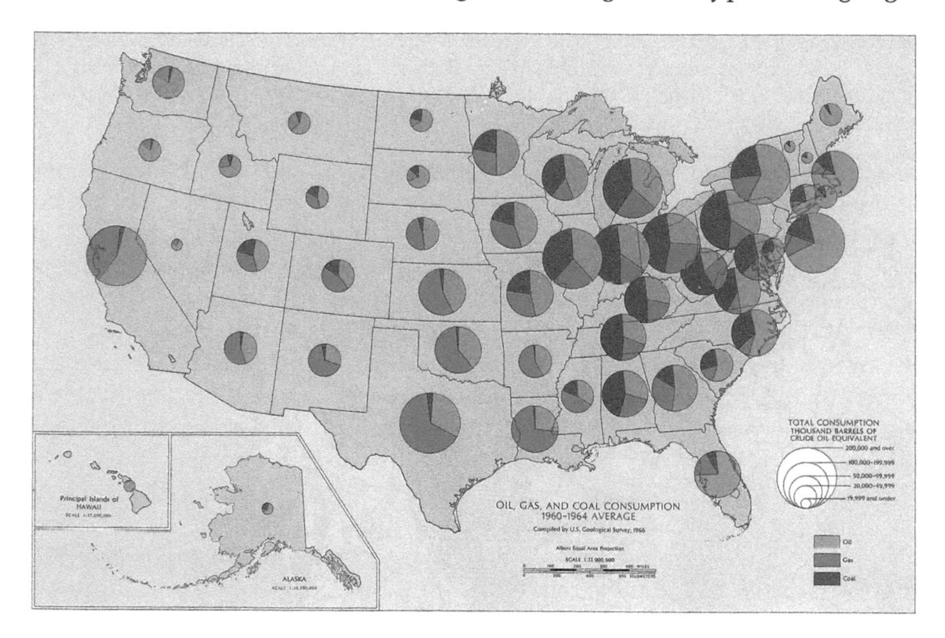

FIGURE 3.7
A map showing the total consumption of oil, gas, and coal (1960–1964) by state in the United States, a discrete and abrupt phenomenon. (Reprinted from the U.S. National Atlas, 1970, in Library of Congress, n.d. Retrieved December 28, 2018, from www.loc.gov/resource/g3701gm.gct00013/?sp=136).

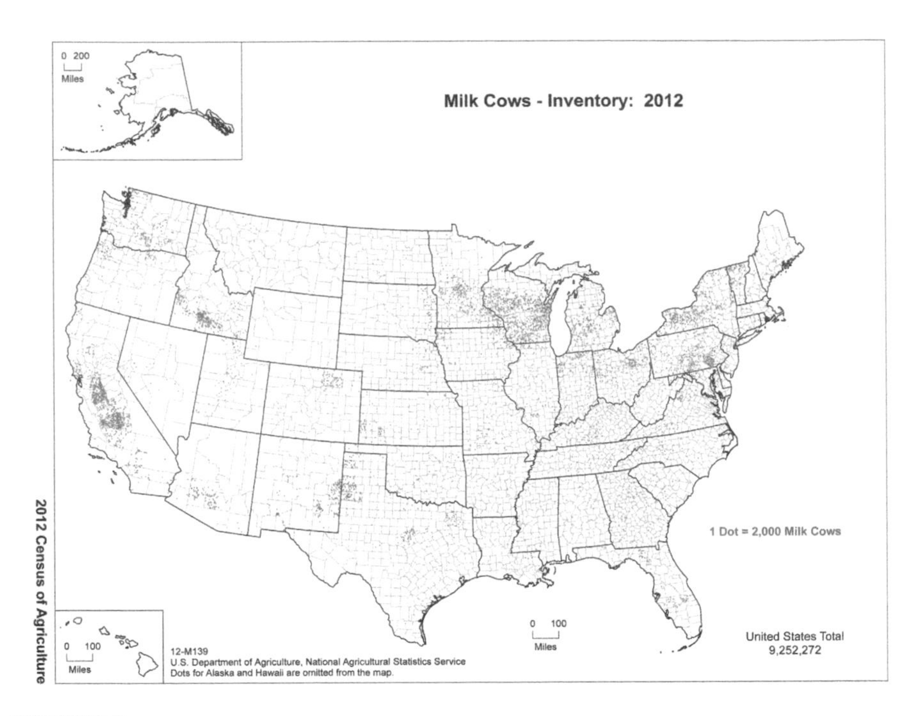

FIGURE 3.8
A map showing the distribution of cows across the United States, a discrete and smooth phenomenon. (Reprinted from Map # 12-M139 from U.S. Department of Agricultre's National Agricultural Statistics Service, 2012.. Retrieved March 2, 2019, from https://www.nass.usda.gov/Publications/AgCensus/2012/Online_Resources/Ag_Atlas_Maps/Livestock_and_Animals/.)

cannot be measured or described with numbers. On the other hand, quantitative data are numerical. For example, the amount of precipitation recorded at a weather station is numerical.

There are additional terms that can be used to more precisely classify qualitative and quantitative data: nominal, ordinal, interval, and ratio. These terms define the four levels of data measurement. Each subsequent measurement level assumes the characteristics of the previous measurement level. Qualitative data are nominal. Originating from the Latin word *nomine*, nominal data involve a categorization based on one or more (non-numeric) attributes. Returning to our religion example, Catholics have some beliefs that are different from those held by Buddhists, but, as with any nominal level data, there is no ordering present in a classification of religious belief nor is any implied.

The other three terms—ordinal, interval, and ratio—define types of quantitative data. Ordinal data is ranked in some way. For example, assume we examined world happiness by country as a measure of how happy a country's population is. Let's further assume we organized responses into a ranking of very happy, happy, neutral, unhappy, and very unhappy and mapped this classification. This classification would constitute ranked or ordinal data. Interval data have meaningful numbers associated with them but do not have a true zero point, or their zero point is arbitrary. The classic example of interval data is temperature. Temperature is interval because higher temperature has a different numeric value than a lower temperature. However, although temperature is expressed as a number, it cannot be compared arithmetically. Rather, a temperature of 60°C cannot be compared to 30°C in a meaningful arithmetic way (stating that 60°C is two times as hot as 30°C). Furthermore, 0°C does not mean that there is no temperature. We can say, however, that 60°C is warmer than 30°C and that 0°C is a useful though an arbitrary zero point. In contrast, ratio data is numeric and does have a true zero point: values can be calculated and compared to each other, and numbers are arithmetically meaningful. A zero value in a ratio dataset means the absence of something. Precipitation measurements are a good example of ratio data. Location A records 5 cm of rainfall while location B records 10 cm. Because this is ratio data, we can truly say location B has twice the amount of rainfall of location A.

Visual Variables

A map uses symbols to represent features. Every map symbol contains elements or graphical building blocks that can be changed. These elements are called visual variables. Bertin (1983) is credited with creating the idea and classification of visual variables, which have in turn been expanded upon by other cartographers and artists. Bertin's visual variables were related to data measurement levels—some variables (and thus types of symbols) are best suited for different levels. Visual variables that are recommended for qualitative data include shape, hue, orientation, and arrangement, while size, value, and saturation are often recommended for quantitative data. MacEachren (1995) explores the visual variables in a cartographic context.

Shape

Shape refers to the perceptual difference in a regular geometric form. Circles, squares, and triangles have different forms. There is no quantitative ranking implied between different shapes. For example, Figure 3.9 shows a map where different shapes (e.g., a triangle, circle, and square) are assigned to represent different types of industrial and chemical minerals in the United States; namely, sulfur, phosphate, feldspar, and barite minerals.

Hue

Hue is the dominant wavelength reflected by an object that is perceived by the viewer (e.g., red, green, and blue). Hue is often called color, but as we point out, there are other dimensions of color that must be distinguished. Figure 3.10 shows a map that uses hue to represent land cover category; brown is used for bare soil, cyan for mangroves, and light yellow for croplands.

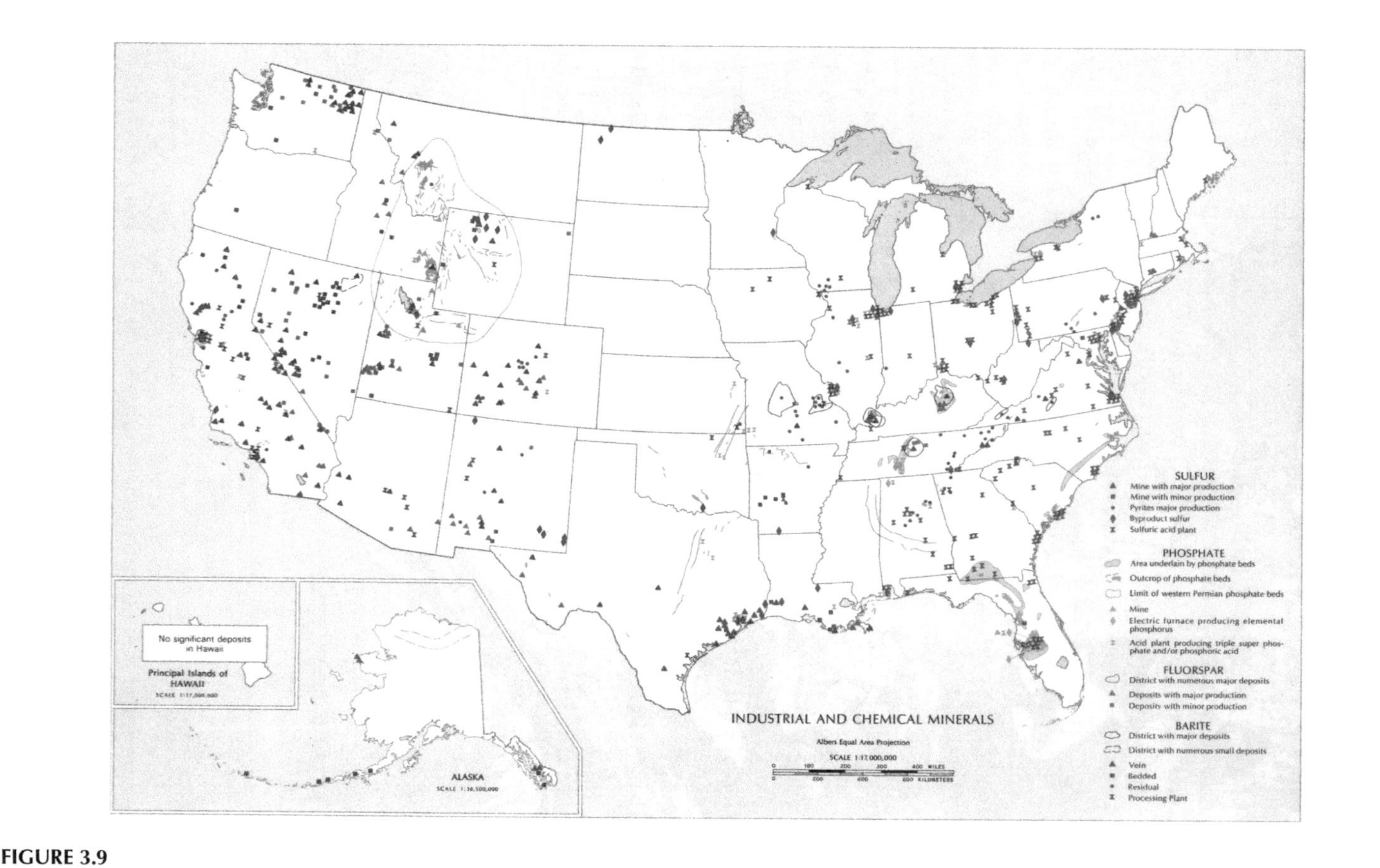

FIGURE 3.9
Geometric symbols used to represent types of industrial and chemical minerals found in the United States. (Reprinted from the U.S. National Atlas, 1970, in Library of Congress, n.d. Retrieved December 28, 2018, from www.loc.gov/resource/g3701gm.gct00013/?sp=133.)

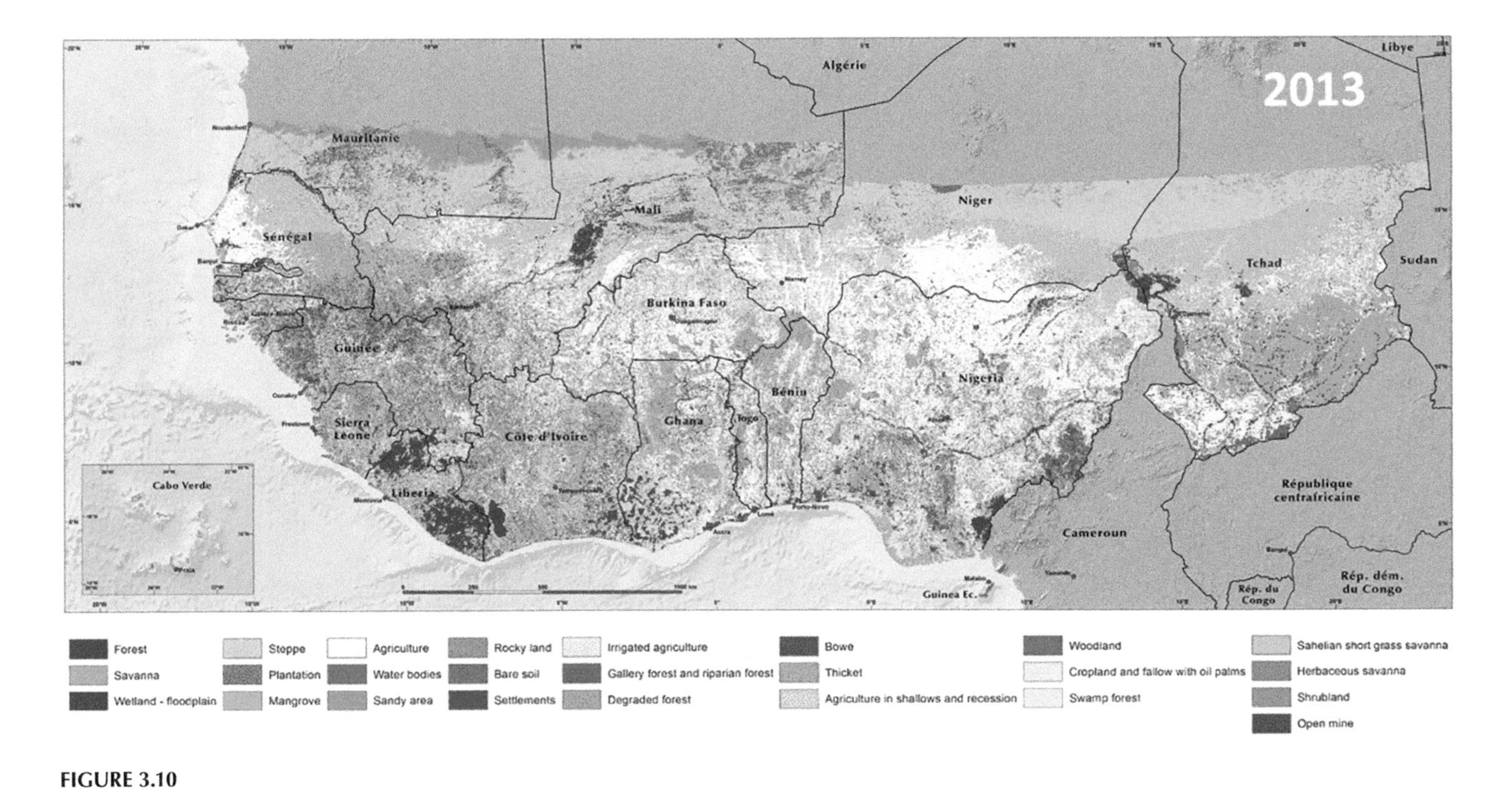

FIGURE 3.10
Hue being used to indicate different land cover categories, an example of nominal data. (Reprinted from Land Use and Land Cover Trends in West Africa, in West Africa: Land Use and Land Cover Dynamics, U.S. Geological Survey, 2014. Retrieved December 28, 2018, from https://eros.usgs.gov/westafrica/land-cover/land-use-and-land-cover-trends-west-africa.)

In some cases, due to traditional use or other factors, hue is intuitively associated with a specific land cover type (e.g., green is nearly always associated with some form of vegetation).

Orientation

Orientation is the way in which marks or shapes are aligned according to some attribute. Figure 3.11 shows how the location of geologic faults is indicated by the orientation of small, individual line segments.

Pattern

Pattern usually refers to how areas are filled and lines are symbolized on maps. Patterns are used to help the map reader identify or distinguish between qualitative or quantitative phenomena on a map. Interestingly, pattern can be used to represent both qualitative and quantitative phenomena. Figure 3.12 shows a series of area fills that are used to symbolize volcanic and igneous rock types on maps—qualitative data. Commonly used symbols to represent these rock types include hachures, "v," and "+." Figure 3.13 shows a series of lines that are used to represent differences between the quantity of dissolved solids found in areas in southwest Texas. The lines fill the areas that contain similar levels of dissolved

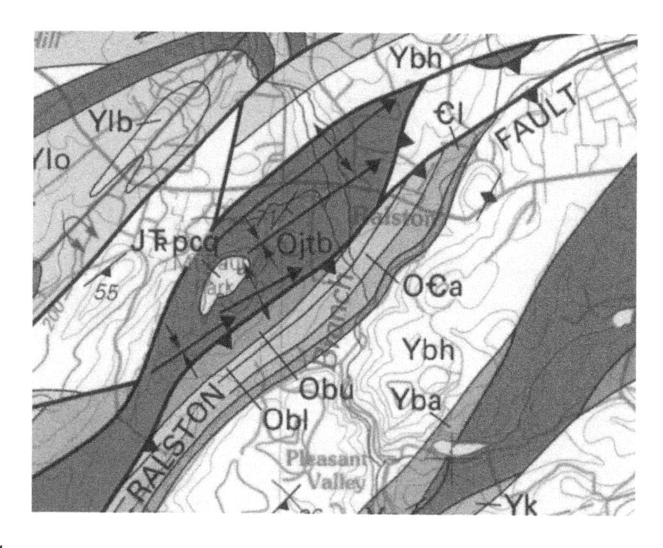

FIGURE 3.11
Line symbols indicating both the location and orientation of geologic faults. The small number indicates the dip angle of the fault. (Reprinted from Bedrock Geologic Map of Northern New Jersey, Miscellaneous Geologic Investigations Map I-2540-A, in U.S. Geological Survey's National Geologic Map Database, 1996. Retrieved December 28, 2018, from https://ngmdb.usgs.gov/Prodesc/proddesc_13025.htm.)

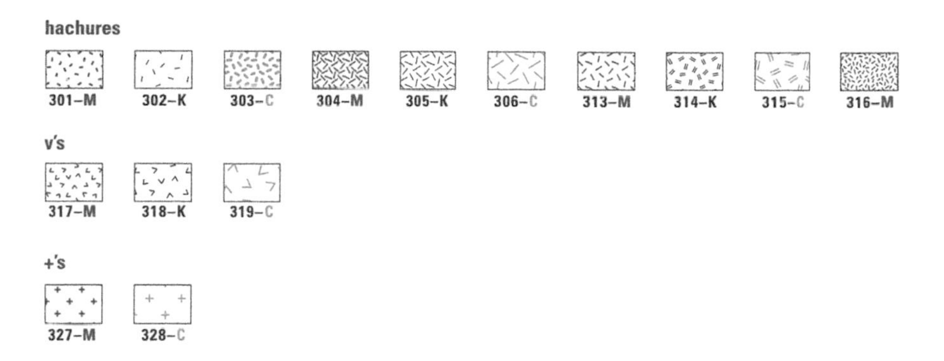

FIGURE 3.12
A series of area fills that are used by the U.S. Geological Survey to represent different volcanic and igneous rock types on maps. (Reprinted from Selection of Colors and Patterns for Geological Maps of the U.S. Geological Survey, Techniques and Method 11-B1, 2005. Retrieved December 28, 2018, from https://pubs.usgs.gov/tm/2005/11B01/pdf/TM11-B1.pdf.)

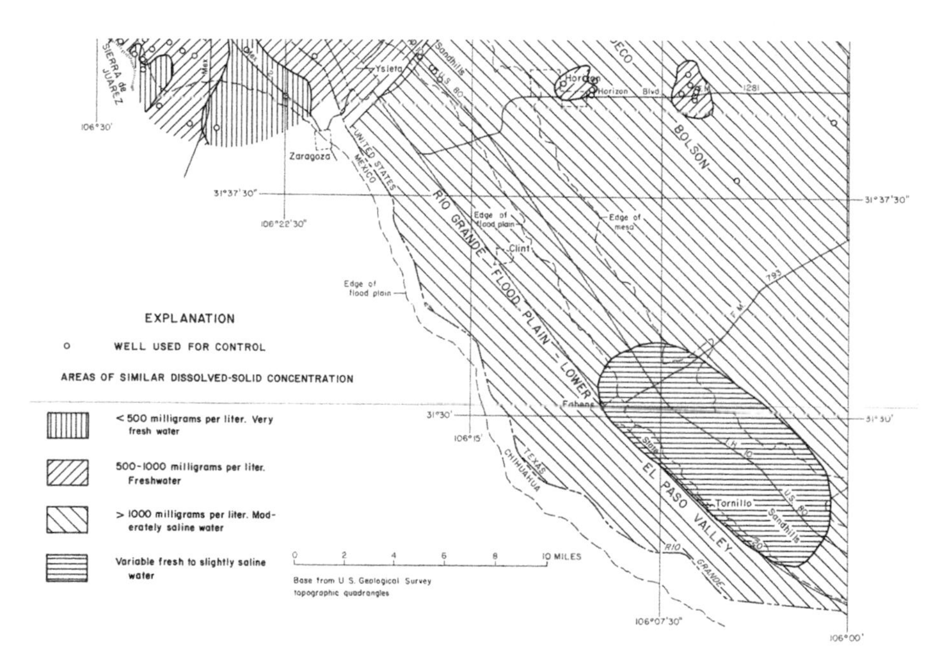

FIGURE 3.13
Line symbols used as area fills to represent similar concentrations of dissolved solids in southwest Texas. The result of the lines used to fill the areas are patterns. (Reprinted from Report 300—Summary of Hydrologic Information in the El Paso, Texas, Area, With Emphasis on Ground-Water Studies, 1903–80, in Texas Water Development Board's Numbered Reports, n.d. Retrieved December 28, 2108, from www.twdb.texas.gov/publications/reports/numbered_reports/doc/R300/Fig18.jpg.)

solids, which helps to form patterns that distinguish one concentration from another. Using lines as area fills results in a map in which it is hard to distinguish between the data classes and is not very aesthetically appealing.

Hue

Hue is appropriate for qualitative data but can also be employed for quantitative data. Figure 3.5 shows average sea surface temperature symbolized in hues that have a logical connotation. In this figure, red represents warmer temperatures, an intuitive association expected of the reader, and blue shows cooler temperatures. Orange and yellow, between the red and blue extremes, represent intermediate temperatures as they shift from warmer to cooler. In another context, hue has often been associated with terrain, in a symbolization method called hypsometric tinting. Figure 3.14 shows a map of the terrain in Oregon's Willamette Valley. Low elevations (approximately 5 meters) are shown in green hues, gradually transitioning to tans, yellows, oranges, and browns (1,300 meters). Aside from its aesthetic appeal, this hue assignment is designed to also hint at vegetation cover that would be present at these elevations. Lower elevations would likely have vegetation present while higher elevations would be tree barren.

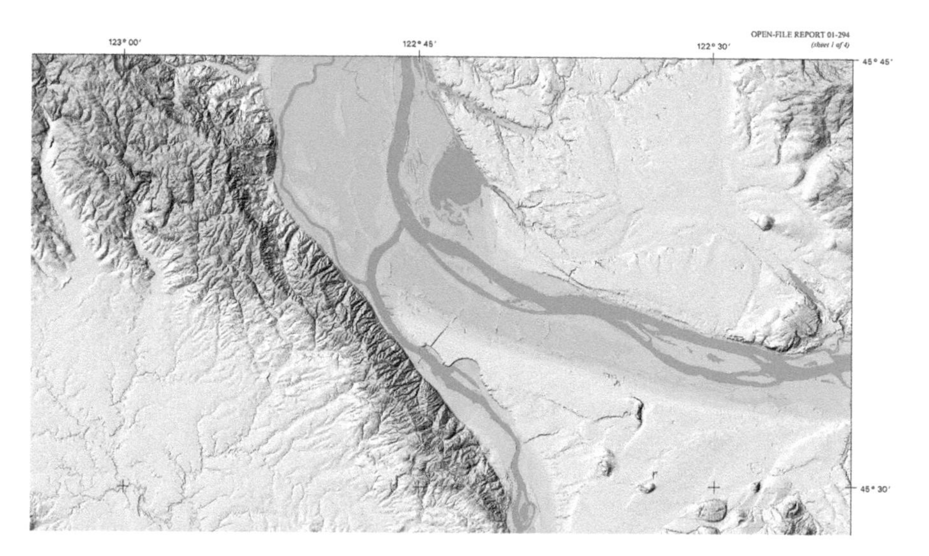

FIGURE 3.14
Hues assigned to represent change in elevations through hypsometric tinting in Oregon's Willamette Valley. (Reprinted from Shaded-Relief and Color Shaded-Relief of the Willamette Valley, Oregon, in U.S. Geological Survey Open File Report, 2001-294. Retrieved December 28, 2018, from https://pubs.usgs.gov/of/2001/0294/pdf/wvc125.pdf.)

Value

Value is the lightness or darkness of a hue. Gray is also a hue, so a gray scale (or other single-hued) map changes in value rather than in hue. Value is often used to show quantity, where lighter is less and darker is more. The map in Figure 3.15 uses a single purple hue, whose values show mortality rates due to stroke deaths for each county in the continental United States. Counties with low death rates are light purple, while higher rates are shown in dark purple.

Saturation

Saturation is the purity of a hue; it is sometimes incorrectly called brightness. A pure hue is fully saturated, while desaturated hues tend toward gray (the value—lightness—does not change with changing saturation). The more desaturated the hue, the grayer it will be. Terms such as "brilliant red" or "vivid red" might describe a saturated red, while "muted red" or "dull red" might describe a desaturated red. By itself, variation in saturation is not a heavily used symbolization technique. Usually, saturation is used in combination with hue and value to emphasize or de-emphasize colors of symbols.

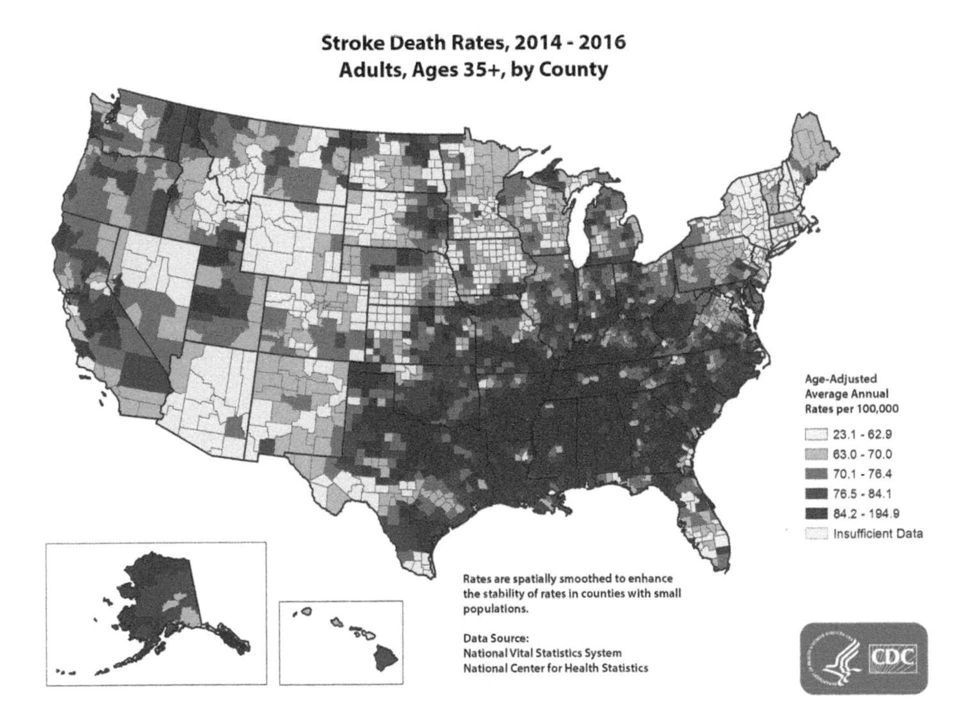

FIGURE 3.15
The use of value to indicate different stroke mortality rates by county. (Reprinted from Stroke Maps and Data Sources, in Centers for Disease Control and Prevention Maps and Data Sources, 2017. Retrieved December 28, 2018, from www.cdc.gov/stroke/maps_data.htm.)

Size

Size is the physical dimension of a shape. Larger symbols suggest greater amounts or higher observed values, while smaller symbols suggest the opposite. Size is a commonly used visual variable for quantitative data, in the form of proportional symbols. The map in Figure 3.7 uses proportional circles to represent consumption of oil, gas, and coal by state in the United States; as the legend shows, smaller circles represent lower consumption, and larger ones show higher consumption.

Cartographic Symbolization Methods

In this section we discuss six basic symbolization methods as described by Slocum et al. (2009). These symbolization methods are choropleth, dot density, proportional symbol, isarithmic, dasymetric, and cartogram.

Choropleth

This symbolization method is commonly used when data have been collected from an enumeration unit. Each enumeration unit is assigned a single data value and is then symbolized with a single color. This value can be either qualitative or quantitative. One challenge with choropleth mapping is that within a given enumeration unit, the data value is continuous throughout but changes abruptly at the enumeration unit boundary, which may not reflect real-world data. Figure 3.6 is an example of a map that uses a choropleth method to represent quantitative data, population density. While the real-world patterns of change for population density are not well reflected by the abrupt changes in a choropleth map, as population can be regularly dispersed across boundaries, other attributes with single values across the entire enumeration unit, such as sales tax by county, show changes that are well reflected by a choropleth map. At every county border the sales tax value changes abruptly based on the sales tax rate in the adjacent county.

Ideally, choropleth maps work best for visual analysis when the enumeration units are similar in shape and size. This makes it easier for the reader to compare values across enumeration units, ensures that the visual impact of any given area is strictly based on the visual variable (e.g., hue) used to encode its value, and minimizes problems of different "container" sizes when discrete phenomena are aggregated for mapping (e.g., larger countries are more likely to have larger populations than smaller countries). The challenge of differing enumeration sizes is particularly important when working with simple counts that have been aggregated.

While it would be ideal to have enumeration units of similar shape and size, this is rarely the case in real-world data. To help minimize the problem of uneven unit size in a choropleth map, it is common to *normalize* or *standardize* count values. Standardizing data mitigates the problem of area differences and makes it possible to directly compare values across enumeration units. There are many ways to standardize data. Figure 3.6 shows population density in people per square mile, which is calculated by dividing the population of an enumeration unit (here, a county) by its area. Other standardization methods include calculating a percentage of a total (e.g., percentage of African-American population is calculated from the count of African-American population divided by total population count).

With respect to the type of data mapped, choropleth maps are quite robust. They can be used to map nominal, ordinal, interval, or ratio level data. With choropleth maps, we rely on visual variables, usually the color-related ones (hue, value, and saturation), to determine the appropriate way to symbolize their polygon fills and emphasize differences between enumeration units.

Dot Density

This symbolization method is used for a dataset of discrete objects whose density or concentration changes smoothly across a geographic space. Generally, one dot stands in for an aggregated number of objects, such as 10,000 sheep, 5 tons of soybeans, or any aggregate number. These dots are placed randomly or semi-randomly within a region where the phenomenon occurs, and are not indicative of the true location of the actual objects. Figure 3.16 is a dot density map of the number of acres where wheat was harvested for grain across the continental United States. Here, one dot equals 10,000 acres of harvested wheat. A higher density of dots means more harvested acres and thus higher wheat yield (or more acres were given to wheat cultivation than to other crops).

Proportional Symbol

This symbolization method scales symbols in proportion to the magnitude of data values. Proportional symbol maps rely on discrete and abrupt data that are usually collected from point locations. With this method, symbols associated with larger data values appear larger on the map than do symbols associated with smaller data values. Proportional symbols can be geometric shapes or pictorial icons. The circle is a commonly used shape, although squares and bars have seen some use. Pictographic icons may also be used to represent the entity being mapped. For example, a proportionally scaled beer mug could be used to represent the number of brewpubs in a state.

Proportional symbols also can be used to represent flow or movement between locations. Flow lines are scaled so that wider flow lines indicate larger values, and thinner flow lines represent smaller values. If the range of values is very great, and each line is proportionally scaled, the numerous flow line sizes can become overwhelming and difficult to discern.

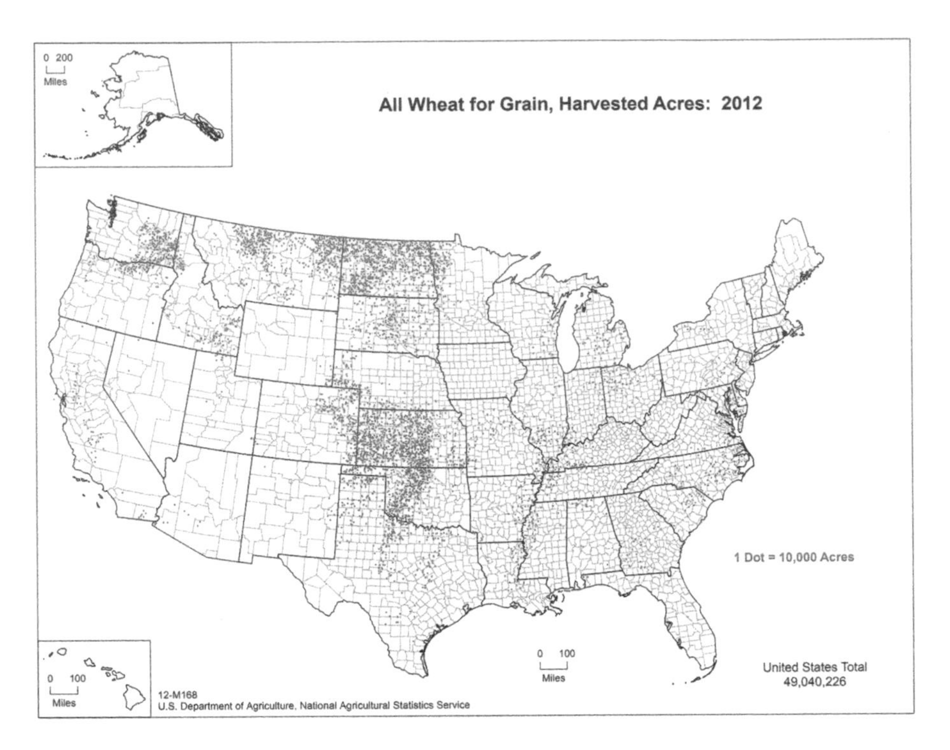

FIGURE 3.16
A dot map showing the distribution of wheat harvest per acre. (Reprinted from Map # 12-M168 from U.S. Department of Agriculture's National Agricultural Statistics Service, 2012. Retrieved December 28, 2018, from www.nass.usda.gov/Publications/AgCensus/2012/Online_Resources/Ag_Atlas_Maps/Crops_and_Plants/.)

For this reason, some flow maps use data that has been aggregated into classes, which reduces the total number of line widths to display. Using classes to group similar data values together reduces visual clutter and allows the reader to more easily reference the line sizes in the map legend. Flow lines can also illustrate flow direction and route. This is often accomplished by using directional arrows attached to the end of each flow line.

The map in Figure 3.17 uses flow lines to represent the volume of intermodal rail freight throughout the United States in 2017. The freight volume data has been aggregated into five classes. This aggregation reduces the visual complexity on the map, whereas if each flow line was scaled in direct proportion to a data value, it would be likely impossible to understand what values were represented by each line.

Isarithmic

Isarithmic maps require data collected from a continuous and smooth phenomenon. Data collected for an isarithmic map can be from true locations (a weather-recording station) or conceptual locations (from polygon centroids). Technically

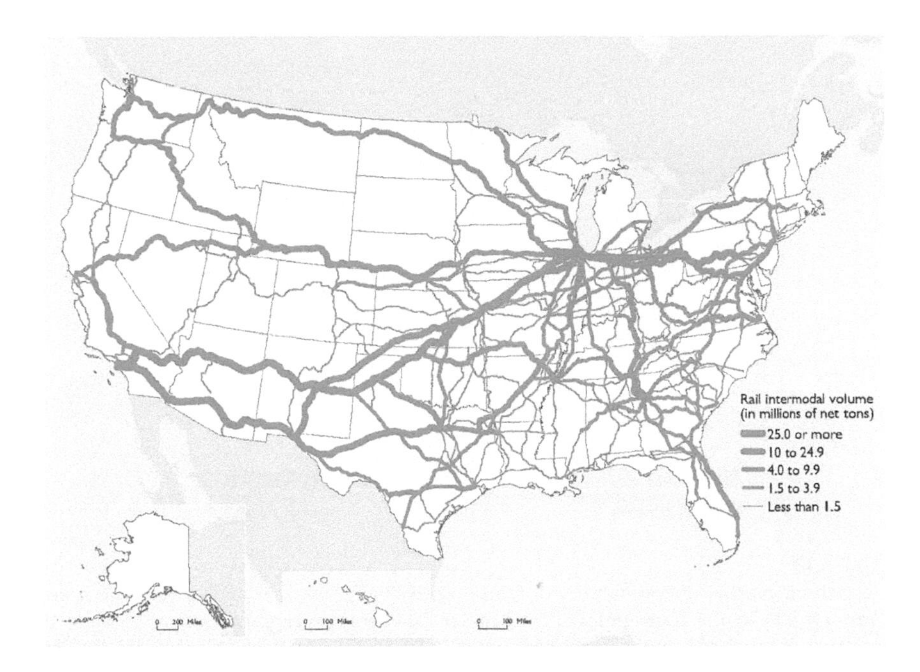

FIGURE 3.17
A proportional symbol map using flow lines to show the volume of intermodal rail freight in 2017 throughout the United States. (Reprinted from Freight Facts and Figures 2017, Chapter 3 The Freight Transportation System, in Bureau of Transportation Statistics, n.d. Retrieved December 28, 2018, from www.bts.gov/bts-publications/freight-facts-and-figures/freight-facts-figures-2017-chapter-3-freight.)

speaking, data collected from true point locations results in an iso*metric* map, while data collected from conceptual point locations produces an iso*pleth* map.

Isarithmic maps are symbolized using isolines. An isoline is a line whose data value is equal along its entire length. One of the more common datasets to map using isolines is surface elevation, which results in the familiar topographic map. Some isolines that show specific types of data have their own names. For example, isotherms represent temperature, isohyets represent rainfall amounts, and isotachs represent wind speed. Figure 3.18 shows an isometric map of the probability of temperatures deviating from normal, both warmer and cooler. The isolines are the borders where the blue and red colors with changing value and saturation meet. This added coloring is designed to emphasize the temperature trends. Note that on this map, the colors do not represent actual temperature but probability of temperature departures from normal.

Dasymetric

Previously, we discussed how choropleth symbolization relies upon continuous and abrupt data collected from or aggregated to enumeration units. Those

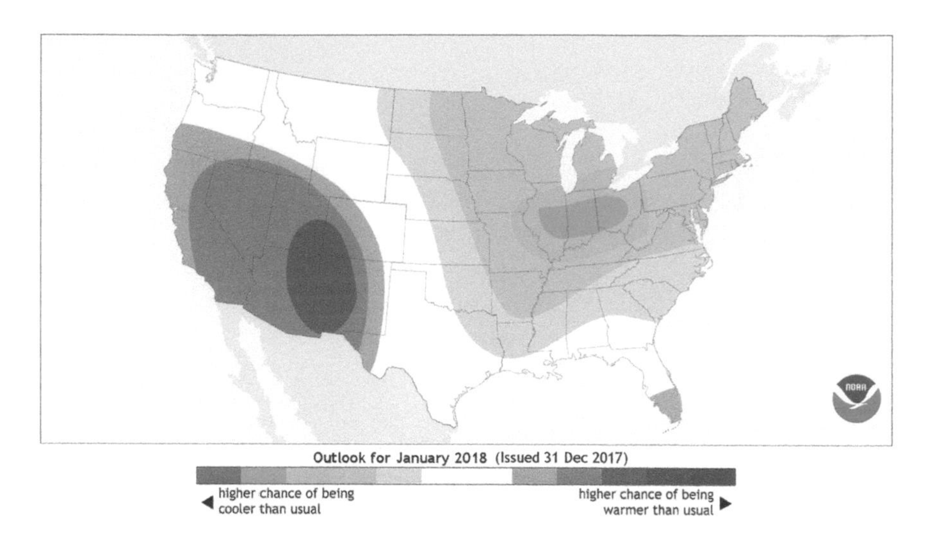

FIGURE 3.18
An isarithmic map of the probability of temperatures being warmer and cooler than normal in January 2018 in the continental United States. The isolines represent equal chances that a cooler or warmer occurrence will happen. The probability pattern is enhanced by using blue and red colors. (Reprinted from Data Snapshots, in National Oceanic and Atmospheric Administration Climate.gov, 2014. Retrieved December 28, 2018, from www.climate.gov/maps-data/data-snapshots/tempoutlook-monthly-cpc-2017-12-31?theme=Temperature.)

units serve as data "containers." The difficulty with using the choropleth method is that in most cases, the underlying spatial variation of the dataset does not adhere to unit boundaries. Data naturally or forcibly spreads from one enumeration unit into an adjacent unit, meaning that unit boundaries are somewhat arbitrary, and choropleth maps can outright conceal the true pattern of the data. The dasymetric symbolization method also uses continuous and abrupt data but does not restrict symbol boundaries to fixed enumeration units. Rather, this method relies upon the data's geographic extent and distribution to create the boundaries. In this regard, the dasymetric symbolization approach can be said to more accurately reflect the distribution of the data.

The map maker must use ancillary information to help delimit these boundaries. A popular dataset to map dasymetrically is population, and one of the more helpful ancillary information sources is land cover data that characterizes where people live. Thompson and Hubbard (2014) used the dasymetric method to map population densities across Afghanistan. In their approach, they used land cover, human settlement locations, and province and district boundaries as their ancillary data. There are several ways that other datasets can be merged with population data to create new boundaries. Thompson and Hubbard used the binary method, where they created a binary raster layer from the combination of land cover and human settlements locations, that was coded as either inhabited or uninhabited. This binary raster layer was then overlaid on the population data. They then assigned a population value to each inhabited raster cell. Figure 3.19 shows a comparison between the patterns shown by choropleth mapping (Figure 3.19A) and by dasymetric mapping (Figure 3.19B). The choropleth method suggests that the country's population fills the provinces, while the dasymetric reveals a much more disparate pattern. Thompson and Hubbard reported that the dasymetric method reveals that the inhabited areas of Afghanistan are mostly in flat terrain near water, rather than in the high, arid mountains.

Cartogram

Cartograms share similar assumptions with choropleth maps, such as that the underlying data are continuous and smooth, and are collected from enumeration units. Cartograms scale the entire enumeration unit according to its data value. For this reason, the cartogram symbolization method is often referred to as value by area. Enumeration units that appear large on a cartogram reflect greater data values, while small enumeration units have lower values. Figure 3.20 shows a cartogram of the population of the United States, by state, in three different years.

While choropleth symbols and proportional symbols represent data with a visual variable (e.g., choropleth maps use value and proportional symbol maps use size), cartograms use the enumeration unit as the symbol itself. This method offers some advantages. Unlike choropleth maps that require standardization, cartograms can map raw counts or ratios. Since cartograms scale each enumeration unit in proportion to the data, there is no need for data classification.

Cartograms can be divided into two categories, contiguous and noncontiguous. This distinction is based in large part on how the enumeration units are spatially connected. For example, if one unit physically touches another, then a contiguous cartogram will preserve that connectivity, as shown in Figure 3.20. However, the shapes of the enumeration units are usually distorted. Noncontiguous cartograms do not preserve any connectivity and will show size differences but with gaps between units.

Cartograms have often been used to map election returns in the United States by state and by county. There are concerns about the reading and interpretation of these cartograms. A given unit's shape may have a distorted appearance. Thus, map readers who are unfamiliar with the mapped geographic area may not be able to recognize some of the enumeration units.

Map Reading Tasks

Maps function in many ways, from providing location information to allowing the accurate measurement and calculation of distance and area. We end this chapter by looking at map reading tasks. It is important to understand

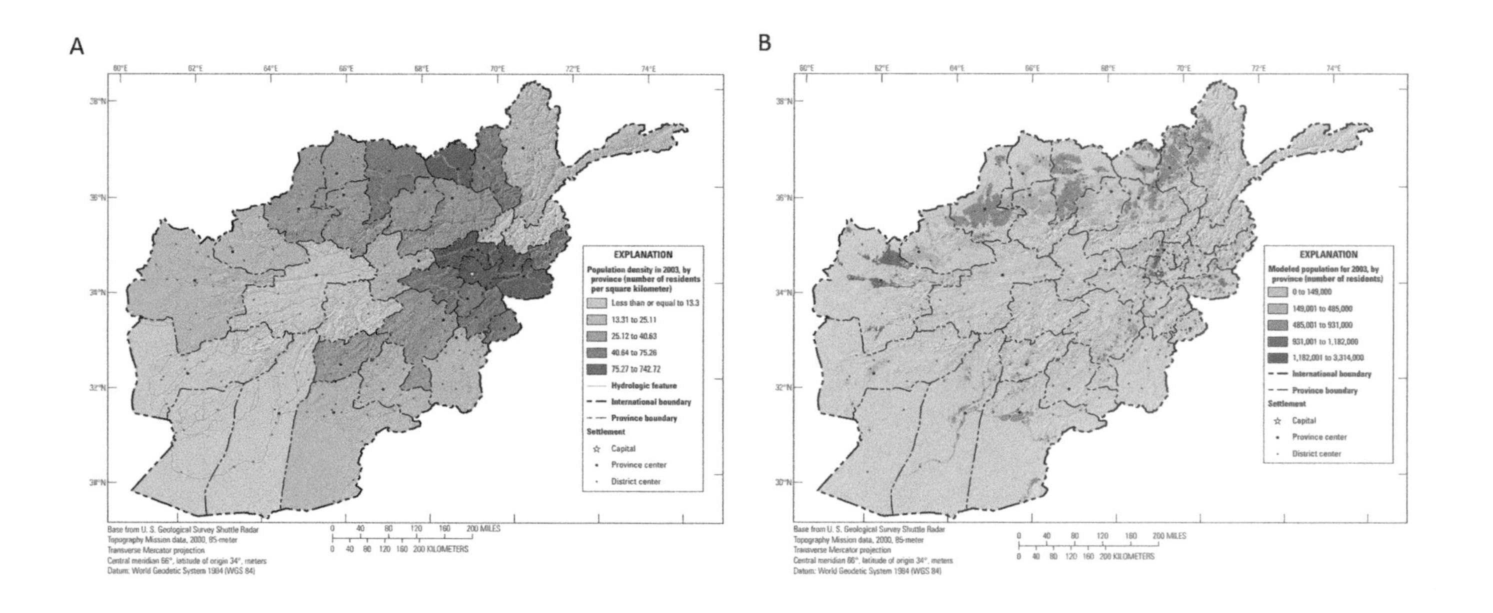

FIGURE 3.19
A choropleth map of Afghanistan's 2013 population distribution (A), and a map of the same data created with the binary dasymetric method (B). (Reprinted from A Comprehensive Population Dataset for Afghanistan Constructed Using GIS-Based Dasymetric Mapping Methods, by Allyson Thompson and Bernard Hubbard, in U.S. Geological Survey's Scientific Investigations Report 2012-5238. Retrieved December 28, 2018, from https://pubs.usgs.gov/sir/2013/5238/pdf/sir2013-5238.pdf.)

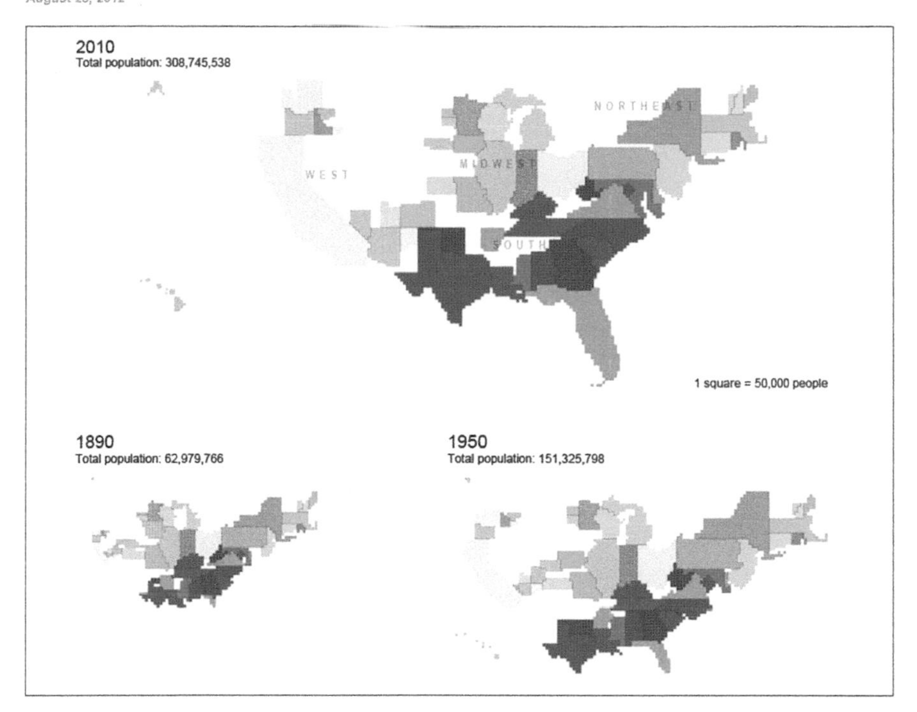

FIGURE 3.20
A contiguous cartogram of the population of the United States in 1890, 1950, and 2010. (Reprinted from Cartograms of State Populations in 1890, 1950, and 2010, in U.S. Census Bureau Infographics & Visualizations Library, August 2012. Retrieved December 28, 2018, from www.census.gov/dataviz/visualizations/021/.)

these tasks, because choosing an appropriate symbolization method (and as we will discuss later, an appropriate projection) can impact the function of the map and thus the map reader's ability to perform map reading tasks. Morrison (1978) presented a framework of basic map reading tasks, categorizing them as pre-map reading, detection and recognition, and estimation.

Pre-Map Reading

Pre-map reading is essentially the reader orienting the map. Although it is common for north to be at the top of the map, this convention is not always followed. Orienting the map is important for the map reader to align their mental map with the map they are viewing. Most symbolization methods do not necessarily impact the reader's ability to orient the map. However, cartograms often distort enumeration units to the point of unfamiliarity. Orientation is something the map maker should be aware of and it is not always recommended to include a north arrow. Consider Figure 3.21, which

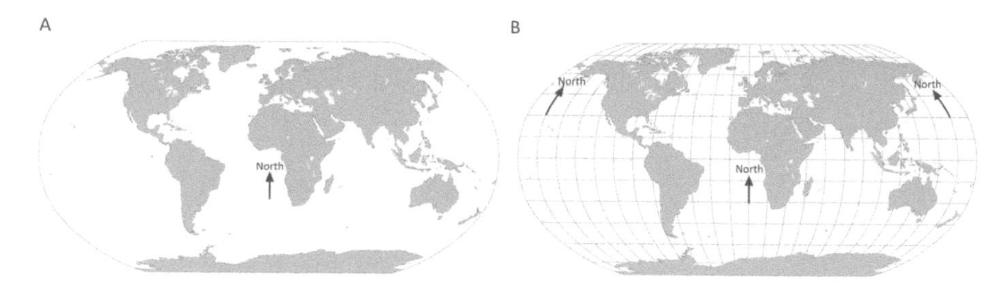

FIGURE 3.21
The Robinson projection with a single north arrow and no graticule (A), and with three north arrows and a graticule (B).

shows the Robinson pseudocylindrical projection. Figure 3.21A does not include the graticule but there is a north arrow. Here, the north arrow is only valid along the meridian to which it is aligned (the central meridian). If that north arrow were moved to another location, it would no longer actually point north due to the characteristics of the projection. Figure 3.21B includes two additional north arrows near the left and right edges. These additional north arrows are aligned to meridians, leading the reader to infer that north is found by following each meridian to the top of the map. However, it is not common to include multiple north arrows, and it adds unneeded complexity to the map. Projections with curved meridians should dispense with the north arrow, and include the graticule only, as the meridians will serve as directional indicators.

Detection and Recognition

Detection and recognition tasks include searching, locating, and identifying features. According to Morrison (1978), the map reading task begins with the map reader searching the map for a target. Assuming the map is well designed, the intended target can be located. Once located, that target is discriminated against other targets to determine if the target in question meets one or more predefined desired characteristics. The ability of the map reader to complete these tasks may be confounded by the selection of an improper symbolization method and/or projection. Figure 3.22 shows the global distribution, by country, of the measles virus genotype. Proportional circles represent the number of specimens possessing a genotype, while the pie slices within the circles represent the different genotypes. The Robinson projection used here has curved meridians and straight parallels (note that no graticule is visible). As a result of this projection choice, there is considerable visual congestion across Europe on the map. It would be difficult for a map reader to search, locate, and identify a specific country's data value. There are three reasons for this congestion. First, Europe has several small compact countries in a rather limited geographic extent. Second, the data values for the European countries vary considerably, resulting in circles in a wide range of sizes. Third, the curved meridians tend to compress landmasses in the upper latitudes. To rectify this, the map maker could choose a projection with straight meridians instead. The inset map in Figure 3.22 appears to use a different projection with straight meridians to help show the symbols in a less congested manner.

Estimation

Estimation tasks include counting, comparing, and measuring. A counting task might involve counting the number of dots on a dot density map to estimate the number of people in a given county. A comparison task could be examining the pattern of population distribution for a given year in one region on a proportional symbol map, doing the same on another proportional symbol map for a different time period, and drawing conclusions about any observed changes that occurred over time and space. A measuring task could be using the information contained in a legend to determine the associated data value of a flow line by measuring its width.

To demonstrate how to tailor a map to a counting task, let's assume we are interested in making a dot density map representing the number of females in Germany, Belgium, and the Netherlands. The purpose of this map is to permit the reader to look at a given country and estimate its female

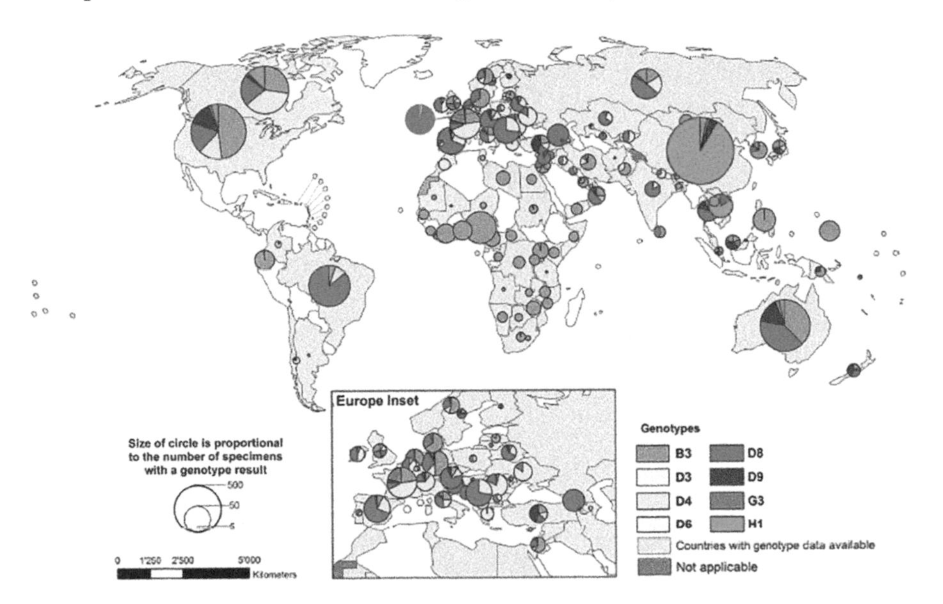

FIGURE 3.22
A proportional symbol map of the global distribution of measles virus genotypes by country from 2010–2015, using the Robinson projection. (Reprinted from Global Measles and Rubella Laboratory Network Support for Elimination Goals, 2010–2015, in Morbidity and Mortality Weekly Report, by Mulders et al., May 2016, 65(17):438–442. Retrieved December 28, 2018, from www.cdc.gov/mmwr/volumes/65/wr/mm6517a3.htm.)

population. To begin making this map, we start a session in a GIS program and visualize the data on three projections: Lambert equal area cylindrical, Albers equal area, and Mercator conformal projections. Furthermore, when visualizing the data, assume that we accept the default parameter values (e.g., standard line, aspect, etc.) for each projection and simply zoom in to the area of interest on a world map rather than customize the projection parameters for the area of interest. We do not recommend this practice, but we are describing this approach as an example of why it is important that you do not choose a projection's default parameter values.

Figure 3.23 shows the dots on the Lambert equal area cylindrical (A), Albers equal area (B), and Mercator conformal (C) projections. All maps are shown at the same map scale (1:25,000,000) and report the same dot value (1 dot = 200 females). Customarily, we would select an equal area projection for dot mapping, as it is important to ensure that map readers will correctly interpret the relationship between the population distribution represented by the dots, and the area in which those dots are contained. However, the distortion on the Lambert cylindrical projection with the standard line set at the equator is great and could affect the ability of the map reader to accurately estimate the population. Here, the landmasses are compressed along the north–south axis but are stretched out along the east–west axis. On the other hand, by adjusting the parameter values on the conic projection, the landmasses are portrayed in a way that better mirrors their appearance on a globe (the meridians converge at the poles), and without the compression and stretching on the cylindrical projection. Values for the standard lines on the conic projection were selected so that the north-to-south geographic extent was divided into thirds. As we discussed in Chapter 2, this division into thirds is one way to select standard lines. The central meridian value was

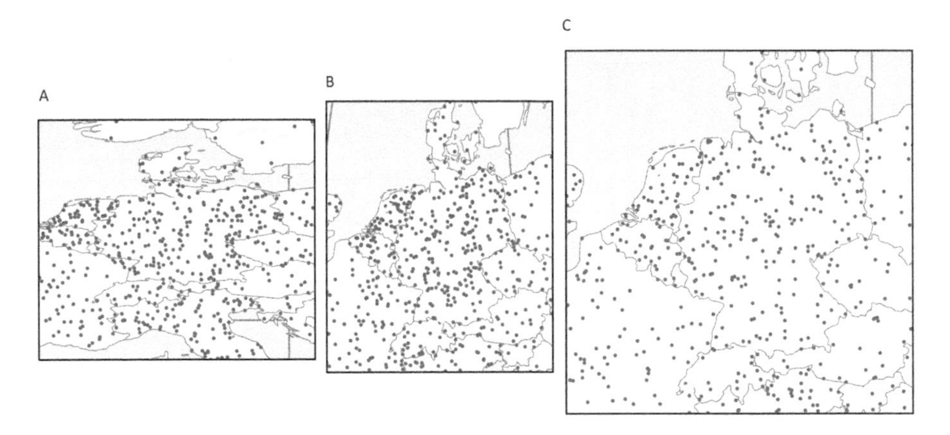

FIGURE 3.23
Dot density maps of population using the Lambert equal area cylindrical (A), Albers equal area conic with adjusted parameters (B), and the Mercator cylindrical conformal (C) projections. All maps are shown at the same scale.

selected to coincide with the meridian that splits the east–west geographic extent of the mapped countries in half. The Mercator projection, being conformal and standard line at the equator, exaggerates the landmasses' size. This exaggeration presents the dot distribution as less dense compared to the Lambert or Albers equal area projections and may not convey the correct density of females to the map reader.

Conclusion

In this chapter, we presented a basic overview of the map abstraction process. The abstraction process takes phenomena from Earth's surface and creates a map illustrating their features. Although simple in concept, the abstraction process is rather challenging from two standpoints. First, Earth is extremely detailed; no map can preserve all detail of a phenomenon (at least not yet). Second, because of this, the map maker must decide what is included and left off the map. Whatever phenomenon is included on the map, the map maker must decide how it should be symbolized.

There are three basic map types. Reference maps include base information that reports features such as hydrologic networks, administrative boundaries, or transportation routes. Reference maps are often used as "look-up" tools to locate features. Thematic maps represent specific "themes" on a map (e.g., distribution of rainfall). To accomplish this, thematic maps incorporate symbolization methods that represent the spatial distribution of a phenomenon. Cartometric maps are designed to facilitate measurement activities such as navigation or surveying. As such, cartometric maps are based on accurate ellipsoid models of Earth. It is important to understand the kind of map to be made as the map type influences the different choices made during the map abstraction process.

Spatial data can be described according to one of four dimensions. Point data (e.g., a temperature sensor) exist at a single location that is defined by a point (e.g., a latitude and longitude pairing). Since a point does not have a length or breadth, it is considered 0-D. Linear data is defined as a series of inner-connected point data. Linear data has a length but no width and is described as 1-D. Areal data has a bounding line that encompasses a given geographic area. Areal data has a width and breadth (2-D). An example of 2.5-D data is a temperature sensor set upon Earth's surface to record daily temperatures. The temperature sensor is located at a point (defined by latitude and longitude) and possesses a single data value at that surface location. On the other hand, 3-D data assumes a point location for data recording but has many values associated with that location. In most cases, 2.5-D data is used to create isarithmic maps (e.g., topographic maps showing elevation) and 3-D data is used to create maps showing volumes.

Phenomena on Earth's surface can be considered to exist along continuous–abrupt and discrete–smooth continuums. The continuous–abrupt continuum describes where objects exist on Earth's surface. Continuous phenomena are assumed to exist everywhere (e.g., temperature) while discrete phenomena exist at individual locations (e.g., trees). The abrupt–smooth continuum describes how objects change across Earth's surface. Abrupt phenomena have definite breaks (e.g., median household income changes between census tracts) while smooth phenomena change values gradually (e.g., barometric pressure) across Earth's surface.

An important aspect of collecting data from a phenomenon of interest is to describe the data's measurement level. The data measurement level describes the nature of the values assigned to a variable. There are four data measurement levels that can be classified as either qualitative or quantitative. Nominal data classifies data according to qualitative differences (e.g., types of languages are an example of a nominal data level). Quantitative data are described by ordinal, interval, or ratio. Ordinal data involves ranking. Data collected from quality-of-life, happiness, or satisfaction surveys are examples of ordinal data. Interval data possesses a relative amount of difference between data values. Temperature is often used as the example of interval data (e.g., 60°C is described as being 20° warmer than 40°C). Ratio level data includes the idea of a meaningful 0. The 0 allows for a ratio to be determined (e.g., 4 cm of snowfall is twice as much as 2 cm of snowfall).

In order to communicate the specific characteristics about the data, symbolization methods rely on the visual variables. The visual variables discussed in this chapter include shape, orientation, pattern, color hue, color value, color saturation, and size. Visual variables are so defined as they are assumed to intuitively associate the qualitative or quantitative characteristics of the data. For example, using different geometric shapes such as triangles, circles, and squares allows the map reader to intuitively understand that the symbols communicate qualitative differences in the data categories. On the other hand, color value can be applied to reflect quantitative differences in data. For instance, a lighter blue color value represents low precipitation values while a darker color value is associated with high amounts of precipitation.

Given a specific dataset, six different symbolization methods are available to represent the data on a map. Choropleth symbolization represents data that are continuous and abrupt and are collected by enumeration units (e.g., total number of forested acres by county). The enumeration units are filled with an appropriate visual variable such as color value to show differences in data quantities. Proportional symbols represent data that are considered discrete and abrupt and are collected at point locations (e.g., number of physicians in cities). Proportional symbols are scaled in proportion to the data values reflecting differences in the quantity of data. The dot density method assumes data that are discrete and smooth. Dots are scaled so that one dot represents some finite number of discrete objects that exist in reality (e.g., one dot = 5,000 cows). Isarithmic symbolization is associated with the common isoline map and represents continuous and smooth data collected from point locations. Similar data values at each point location are connected to form an isoline map (e.g., temperatures). Dasymetric mapping uses continuous and abrupt data. However, unlike the choropleth symbolization method that relies upon enumeration units to create the spatial pattern, dasymetric mapping allows the data to form the boundaries thus creating a more realistic distribution in the data's pattern. Cartograms use continuous and abrupt data like the choropleth method. However, with cartograms, the area of the enumeration unit is scaled in direct proportion to the data values rather than using a color fill to represent the data's distribution.

The chapter concluded with a discussion of various map reading tasks and the impacts that projections have on the ability of the map reader to successfully carry out these tasks. There were three map reading tasks discussed: pre-map reading, detection and recognition, and estimation. Pre-map reading involves the process of the map reader orienting the map to their own mental map. While orienting the map is usually accomplished by including a north arrow on the map, projections with curved meridians minimizes the utility of the north arrow. On such projections, "north" is not aligned to one but multiple meridians. When including a single north arrow, make sure that the meridians are straight so that one north arrow points in a consistent direction across the map's surface. Detection and recognition tasks include searching, locating, and identifying features. A projection, for example, that includes meridians that converge to points representing the poles compresses landmasses in the upper latitudes. This compression can create a congested display of the symbols and limit the map reader in, for example, recognizing the landmasses and identifying the pattern that exists in these latitudes. Estimation tasks deal with the map reader counting, comparing, and measuring. The effectiveness of symbolization methods can be enhanced by selecting projections that facilitate estimation tasks. The example given in this chapter dealt with the dot method, how the projection distortion can alter the size of the enumeration unit, and how that alteration of size impacts the visual perceptibility of the dot distribution.

4

Map Projections' Influence on Data Representation

In Chapter 2, we detailed the projection process, properties of projections, and various ways that these geographic properties may be distorted when data are projected. In this chapter we move into discussion of how *you* may interpret and apply your knowledge of projection properties and distortions and how it might influence the people who are reading your map. Specifically, we discuss issues surrounding:

- How people intuitively, though sometimes incorrectly, interpret projected data and how this may lead to incorrect assumptions about geographic distributions or spatial relationships.
- Aesthetics of projections and how the design of the projection may help or hinder map reading.

These issues are important in map design because no matter how carefully you approach the problem of selecting what *you* see as the ideal projection, someone less familiar with projections is likely to be *reading* the map. The cognitive biases or misconceptions of that individual will shape the way that they interpret the data on the map, possibly in ways that hinder evaluation of valid spatial patterns. Understanding the challenges of how people intuitively think about projections is helpful in explaining your projection choice, annotating and documenting the map making process, and identifying potential map reading challenges in advance.

As we saw in Chapter 2, projections distort spatial relationships in such a way that planar representations of spatial data do not behave in the same way as we would expect "in the real world." Because of these distortions, the projection that you choose for your map—and the resulting distortions—can significantly influence the map reader or spatial data analyst. As smaller-scale (continental- to global-scale) maps present more substantial distortions than large-scale maps (local-scale) with parameters appropriate to the map extent, much of the discussion in this chapter will be related to considerations for the smaller-scale maps. However, there are still interesting design points related to large-scale mapping that we will discuss throughout the chapter.

We will also provide general discussion of how you can optimize your projection choice based on the map type, purpose, and audience. While there are clear guidelines that will steer you to specific projection types (e.g., equal area or conformal), ultimately the final decision on what projection to select for a mapping project, after appropriate technical consideration of the properties of the projection, may be heavily influenced by aesthetics and on how successfully readers will be able to interpret the patterns on the map. From a purely aesthetic standpoint, there are questions of how the mapped area will look with respect to the space allocated (e.g., fit on the page), addressing requirements of local, regional, or organizational standards, and general visual preference. From the cognitive and perceptual standpoint, there are questions of how the map reader recognizes (or not), interprets, and understands the distorted spatial relationships on the map. This chapter will not provide an answer to the questions of what projection is most correct, or easiest to interpret, or will be the most aesthetically pleasing, but we hope that it will help you better understand some of the visual challenges and projection-related preferences that map readers have, as well as the potential influences of projections on data representation. In Part II of this book we apply this knowledge and will dive into more detail of identifying projections based on specific map types or tasks.

Interpreting Map Projections

One of the memories that we (Fritz and Sarah) hold onto from our childhood is of a large, colorful classroom map hanging on the wall in elementary school. As we learned about the world, we often wondered why so little of our learning was spent on Greenland—clearly it was one of the largest and most centrally located countries in the world. Shouldn't it have played a more important role in history?? Alas, we fell prey to one of the classic problems of mapping—most people simply interpret the map as is, without consideration of the fact that Earth is not flat and that the spatial relationships that we interpret on the map (e.g., the size and central location of Greenland) may not be reflective of the true relationships on Earth.

In this section, we discuss several more serious examples of perceptual challenges and what they mean in terms of selecting a projection. The short story is that many map readers have no background in projections, and it is challenging, if not impossible for most people—even those with some level of projection experience—to identify the projection type (e.g., equal area, conformal, etc.), its distortion pattern, and the distribution of distortion across a map. This doesn't sound like we are painting a promising picture of the projection selection process, so what does it mean for you in your process of selecting an appropriate projection? As we'll repeat several times through this book—there is rarely a single, right projection that will solve all of your needs. The projection selection process is an art of compromise.

Understanding the perceptual challenges of how people may misinterpret your map is an important part of the problem-solving process when debating different projection options. It also helps to have a handy set of guides that you, or your map readers, can use to help decode and understand distortion in projections.

While there is ample (and often very dense) reading material available on the technical aspects of projections, there are few guides to help the average map reader decode a projection on-the-fly. One set of guidelines that we particularly like is from Judy Olson (2006), in her article "Map Projections and the Visual Detective: How to Tell if a Map Is Equal-Area, Conformal, or Neither." Her work in perception of projection distortion has led to a suggestion of a set of cues that can be used as part of "visual detective" work to interpret distortion for large geographic areas. Her idea is that without having to have extensive knowledge of the mathematics behind projections, the specific details for a projection, or a separate map to quantify and visualize the distortion (discussed in Chapter 2), it should still be possible to identify ("with good probability, not with absolute certainty"; Olson, 2006, p. 15) basic characteristics of distortion to area and angles simply by looking for specific cues in the graticule (lines representing latitude and longitude). Olson's "visual detective" process is based on evaluating deviation from the following visual characteristics of latitude and longitude on Earth's surface:

- "parallels and meridians meet at right angles
- lines of latitude (parallels) are equally spaced
- lines of longitude (meridians) converge, and lines of latitude get shorter, toward the poles
- areas (latitude/longitude cells) marked off by equal increments of longitude between a given pair of parallels are equal in size
- latitude/longitude cells marked off by equal increments of latitude between a given pair of meridians get smaller and smaller toward the poles
- latitude/longitude cells marked off by comparable latitude lines on either side of the equator are equal in size
- the 60° N and S lines are ½ the length of the equator" (Olson, 2006, p. 16)

Using these characteristics, for instance, you can approximate projection type based on whether the parallels and meridians meet at right angles (suggesting that the projection is conformal), whether the meridians converge toward the poles or not (non-converging meridians suggest that there is stretching of the east–west scale), or examining how regions demarcated by equal increments of latitude and longitude vary across the map.

While Olson presents simple heuristics for evaluating distortion, do we see these in practice in everyday work with projections? In general, the research on this subject doesn't paint an encouraging picture about general knowledge of projections. This makes it all that much more important that you consider what projection is most appropriate with respect to how your readers will see the data, and what sorts of questions they will try to answer with the map—because the map reader may not be able to accommodate for the distortion when evaluating spatial patterns on your map. To put this in context, let's look at a few interesting examples of how distortion in projections may be misunderstood by readers.

As a simple example of the problems that may arise with how projections influence data representation and interpretation, consider the geometric principle that the shortest distance between two points is a line segment connecting the points. For many people, this principle is learned relatively early in life—whether in the classroom, or in cutting across the neighbor's yard to shorten the walk home from school. On the plane, this shortest path is a straight line. On the sphere, the shortest path is a great circle. The great circle path over long distances is rarely a straight line on the projected plane (Figure 4.1), though for relatively short distances the straight line on the plane may sufficiently visually approximate the great circle arc on the sphere. The path distance at which the deviation is noticeable between great circle arc and straight line on the plane depends on the scale of the map, the

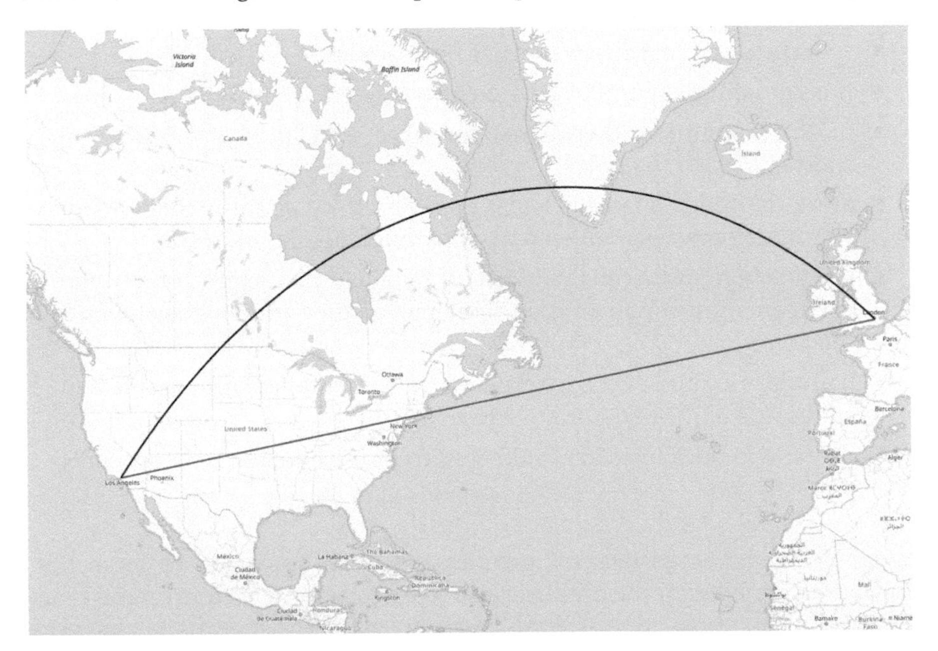

FIGURE 4.1
Great circle path (black) versus straight line (red) on the web Mercator projected plane to connect two locations.

characteristics of the projection, and the location of the path with respect to the patterns of distortion from the projection.

In a research project by Anderson and Leinhardt (2002), focusing on map interpretation challenges for both experts and novices, they looked specifically at this problem of identifying shortest paths on maps. In their work, they found that most people, including many mapping experts, had difficulty with this task. One of the interesting suggestions in Anderson and Leinhardt's work is that not only were people mostly incorrect in their assumptions, some were actively *trying* to adjust based on their knowledge of projections. However, map readers often applied incorrect rules or heuristics to solve the task. For instance, stating rules that are appropriate to planar, but not spherical, math such as "the shortest distance between two points is a straight line" (Anderson and Leinhardt, 2002, p. 308), or the line should be curved (presumably because they have seen curved great circle lines on Mercator or other projections before), but the map readers were not able to apply the rule correctly. Both of these misapplied rules are interesting, but for different reasons. One assumes that planar and spherical spatial relationships are equivalent, and the other assumes that a pattern seen on one projection is applicable to other projections. Other research indicates that these misconceptions are fairly widespread, suggesting that unless our map readers have detailed knowledge of projections, we should assume that they will take any spatial relationships shown on the map at face value. As Egenhofer and Mark (1995) suggested in their definition of *"Naïve Geography,"* perhaps maps are "more real than experience" and there is a naïve assumption that the map is a true representation of geographic space.

The interpretation of a map as a true representation of geographic space can lead to concerning results—particularly when the spatial relationships being measured are related to threats to human life. Consider the challenge of mapping distance that a missile can travel from a launch point; missile attacks are a serious issue, and most map readers would very much like to know whether they are in danger. If a map designer assumes that the missile range can be measured on a map by drawing a circular buffer around the origin—and the map does not preserve distances from this location—the results will be incorrect. Figure 4.2 and Figure 4.3 demonstrate this on the web Mercator projection with three circular buffers of different distances drawn around a fictitious launch site set in North Korea (Figure 4.2) and accurate buffers showing the true distances from the same location, as projected in web Mercator (Figure 4.3). The circular buffers themselves are not the problem; with an appropriate projection circles *are* a perfectly reasonable way to present the range of the missiles (Figure 4.4). While the azimuthal equidistant projection used in Figure 4.4 may be less familiar looking than that of the Mercator, the accuracy of the data representation makes it preferable. Of course, it would also be reasonable to use a more familiar looking projection provided the missile ranges are drawn appropriately to accommodate for the distortion of distance across the map.

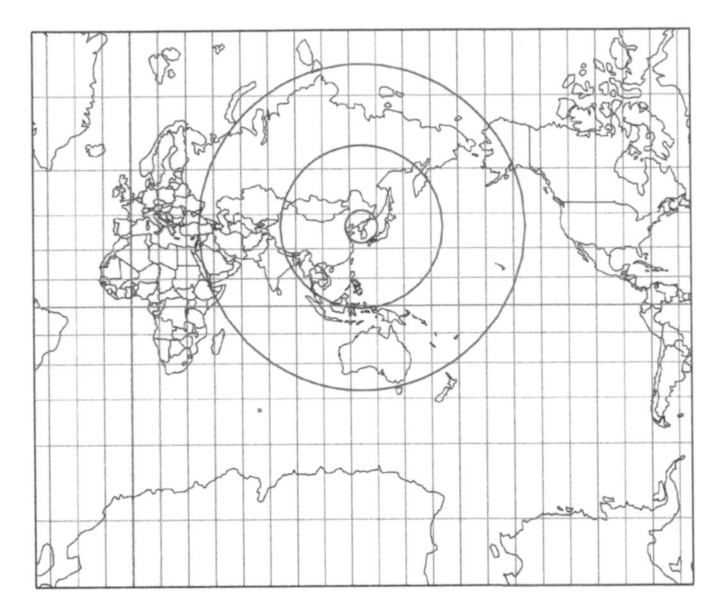

FIGURE 4.2
Circular buffers incorrectly used to depict three different missile ranges (1,000 km, 5,000 km, and 10,000 km). Using a web Mercator projection (near conformal) means that the circular regions shown are incorrect representations of the actual range for each missile.

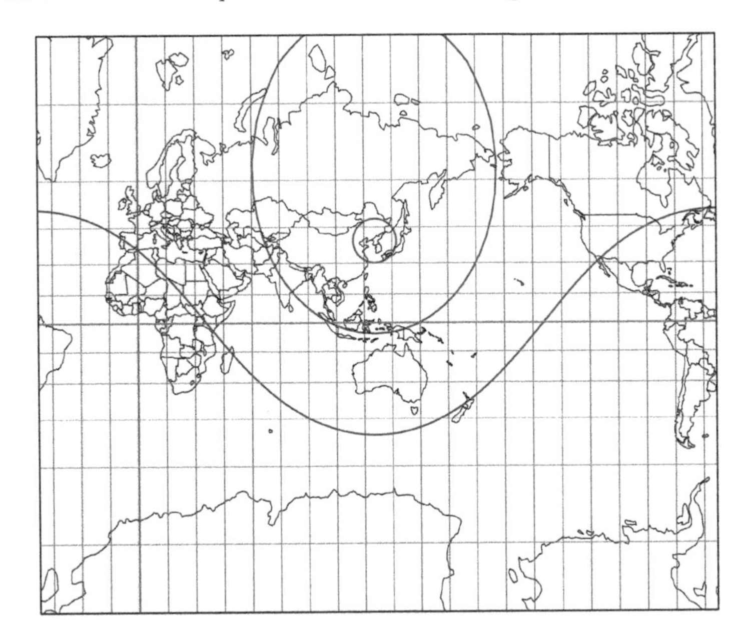

FIGURE 4.3
The same missile range distances (1,000 km, 5,000 km, and 10,000 km) as shown in Figure 4.2, but using correct calculations of distance that adjust for the distortion inherent in the web Mercator projection.

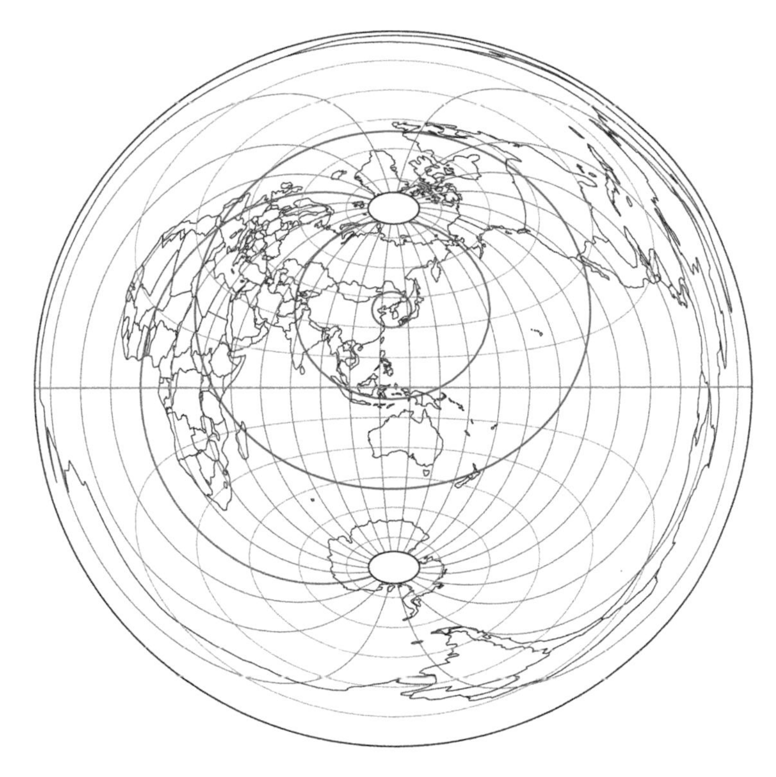

FIGURE 4.4
Missile range as drawn on an azimuthal equidistant projection. The use of an equidistant projection centered on the presumed missile launch location means that the distance represented by the circular buffers is mapped correctly in this example.

While Anderson and Leinhardt (2002) addressed understanding of angular distortion in projections, others have identified similar problems in map readers' ability to understand distortion of geographic areas. A 2009 study by Battersby (2009) showed similar patterns of map readers assuming that spatial relationships on the plane are "truth." In her study, she asked map readers to use one of three references (globe, conformal map (Mercator), or equal area map (sinusoidal)) to estimate the land area of several countries or territories around the world. Participants in the study were specifically informed that spatial properties such as area may be distorted due to projection, and were asked to provide their estimate of *actual* land area, not the projected size on the map. For all three of the conditions in the study (globe, Mercator, and sinusoidal map), the correlation between land area estimation and area *as shown on the reference* was approximately 0.91—a good indication that the map is reality, even though a second study with the same participants showed that almost 90% of participants identified that the Mercator projection showed "exaggeration" of area in the high latitudes. Even though the estimates made using the Mercator map as a reference showed a small effort by participants to try to accommodate for the distortion of the projection, it was not a sufficient scaling of area to negate the actual distortion.

Later work by Battersby and Kessler (2012) to explore how map readers identify distortion in projections provides additional evidence for the assumption that known patterns of distortion in a familiar projection are true on all projections. We believe that often this "golden reference" is the Mercator projection, since this is one of the most common projections used as an educational example when teaching about projection distortion. In their study, Battersby and Kessler (2012) found surprising reliance on cues from a single projection (Mercator; a conformal projection) when map readers were asked to identify distortion patterns on several common and uncommon global-scale projections. This suggests that even if a map reader has some basic knowledge about projection distortion, heuristics for evaluating distortion are not sufficiently developed to be able to translate to other projections.

The Edge of the Map (Periphery)

Another interesting problem that comes up is in understanding what happens at the edge of a map, or the periphery. All global-scale maps are interrupted in that they have a distinct edge somewhere along the periphery. For most maps, this interruption surrounds the entire mapped area, but in some cases, such as with web map tiles that allow continuous east–west panning, the interruption is only at the north and south extents of the map. The interruption at the periphery presents an unavoidable problem that may have substantial impact on a map reader's ability to perceive spatial patterns. While there is little empirical evidence of the true challenges, a study by Hruby, Avelino, and Ayala (2016) has suggested that map distances are estimated more accurately when learned from a map centered so that the shortest path between locations *does not* cross over the periphery. This suggests that the selection of the central meridian is more important than just centering your region of interest. For comparisons between locations, you should consider optimal centering of a projection to minimize the need to compare regions at the periphery and on opposite sides of the map. Even though the regions may be geographically close on Earth, their location on the periphery gives them the appearance of being a great distance apart. Though the edge of the map has received limited attention in projection research, it is an important consideration in map design.

On a more positive note about the periphery on the map, the shape of the projection and the spaces along the periphery that are left over in the overall map design can lead to some wonderful design decisions. Consider how many world maps have made creative use of the space outside the periphery for inset maps to show related details (Figure 4.5).

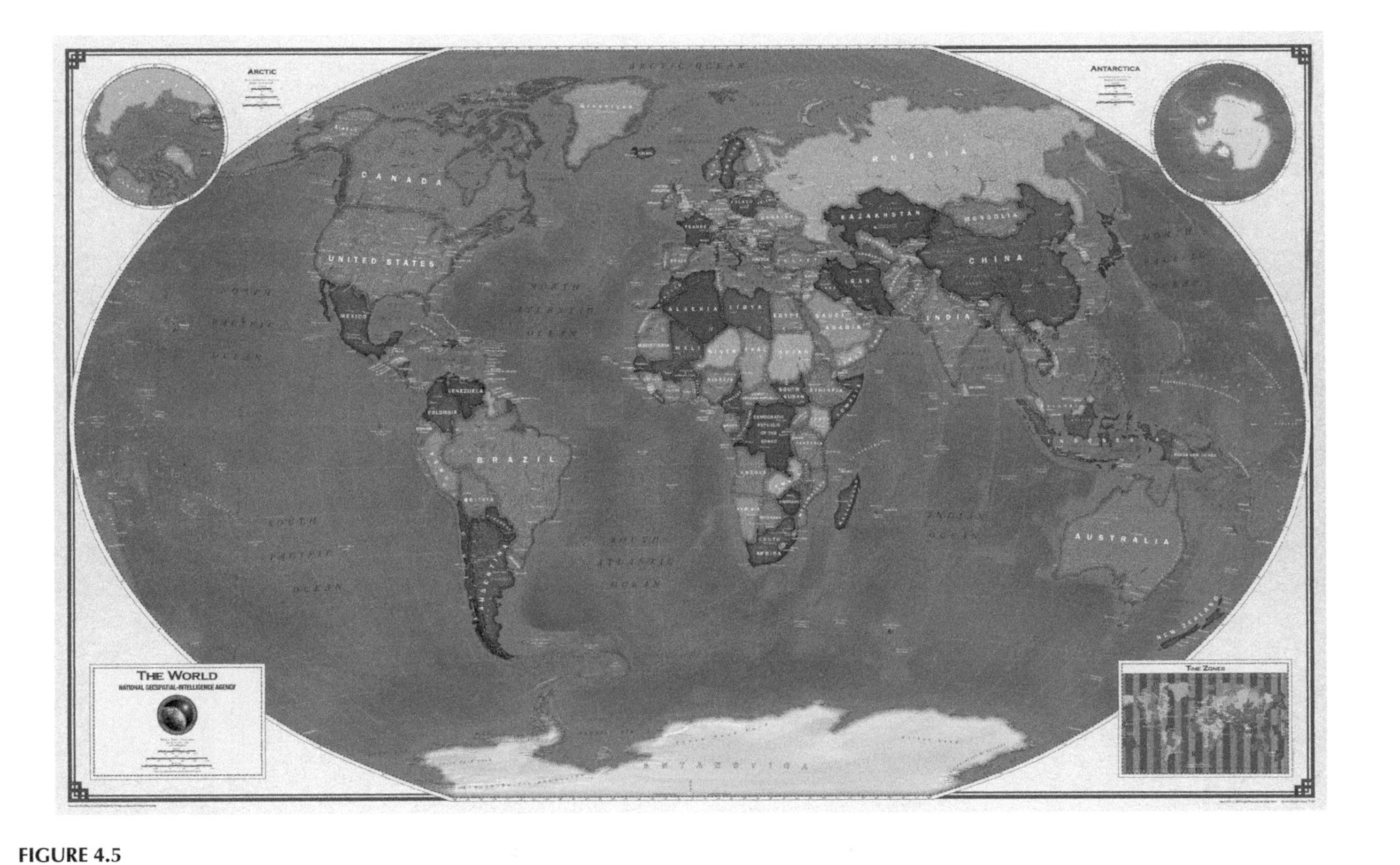

FIGURE 4.5
A world map using two hemispheric views in the upper corners and a world time zone map in the lower corners beyond the map's periphery. (Image retrieved December 28, 2018, from www.nga.mil/Partners/Academic_Opportunities/Pages/NGA-Maps.aspx.)

The fact that it is difficult for map readers to interpret projected data makes it all that much more important that *you* consider what projection is most appropriate with respect to how your readers will see the data, and what sort of questions they will try to answer with the map to try to minimize misinterpretation of spatial patterns. However, we can also offer a reassurance at this point—*there is no perfect answer to the challenges of projections.* There are only better or worse choices based on your data, the region you are mapping, and the tasks that your map readers need to accomplish. Our stories about the challenges of estimating spatial relationships on projected maps are meant to simply enlighten you about some of the interpretation problems that you may expect from your map readers.

Long-Term Impacts of Map Projection?

Outside of how people directly interpret what they are seeing on a map, there is an interesting longer-term consideration of how people remember and understand spatial relationships at a global scale. The only possible way to see the entire world in one view is using a planar map, and there is no single undistorted, or even minimally distorted, planar map that can be used. Because most of the graphical source material that we use to learn about global-scale space is distorted, it is important to consider how common projections may influence map readers' global-scale cognitive maps. A cognitive map is a mental representation of geographic space that people can use to store and recall information about relative locations and other characteristics of the spatial environment. The idea here is that if people are mostly familiar with distorted visual representations then these distortions may impact the master image underlying our cognitive maps.

Why should we care? Our cognitive map:

1. Is used when evaluating unfamiliar projections and deciding if it "looks right" or "like expected" or is a "good" representation of the world. This has notable impact on how well people like or trust the map that you create.
2. Is used whenever we are evaluating spatial information in non-map form to mentally assess spatial patterns. This has notable impact when map readers *recall* spatial information from a map held in memory.

One of the most notable projection-related concerns is the potential impact of the Mercator projection. Earlier, we humorously recounted a story of wondering why Greenland, a very large and centrally located region as seen in the classroom Mercator projection, didn't play a more significant role in world history.

While we have broken ourselves of this faulty idea over the years, it serves as an example of how the perceptual problems that we discussed in the last section can have a lasting effect on our cognitive models of the world if they become sufficiently familiar that they serve as our model for spatial patterns and we don't actively counter them with information from more accurate sources.

The challenge of familiarity of specific maps leading to incorrect assumptions about projection properties has been debated extensively in the cartographic community, particularly with respect to the Mercator and web Mercator projections for use in general purpose global-scale mapping. In the not-so-distant past, the Mercator projection has seen considerable use in wall maps and atlases. Many of these have been used as classroom references. As a cylindrical projection, it fit nicely into a wall- or page-sized image with little wasted space. However, as a conformal projection, it presents significant inflation of area (relative to other regions) in the areas closer to the poles. Because this projection has been used in high-visibility locations like the classroom, cartographers developed a theory that it would encourage students to become too familiar with the shape and the distortion patterns of the projection, and Mercator would become ingrained in our minds as the true shape of the world.

With respect to this fear, cartographers have gone so far as to suggest that "we have been 'brainwashed' by the rectangular Mercator" (Robinson, 1990, p. 103) and that it has become a "master image" for world maps (Vujakovic, 2002). By the late 1980s, the discussion around the negative influence of the Mercator projection had become sufficiently heated that seven cartographic associations banded together to write a resolution against rectangular (or cylindrical) projections, with Mercator being named specifically (American Cartographic Association et al., 1989). The debate around the Mercator projection has been quite heated at times. If you are interested in more detail in the interesting stories around the Mercator debate, we can recommend Monmonier's (2004) excellent, in-depth discussion of the controversy.

Several research projects have explored the potential for Mercator to have influenced people's cognitive maps. Saarinen and others (Saarinen et al., 1996; Chiodo, 1997; Saarinen, 1999) have suggested a "Mercator Effect" and area-distortion related artifacts of Mercator as seen in global-scale sketch maps that they have collected in several studies. Though, the comparisons are primarily qualitative in nature. Quantitative analysis by Battersby and Montello (2009) to examine memory-based land area estimates found little suggestion of an overwhelming Mercator influence on cognitive maps; however, the limited relationship found between Mercator land areas and estimated land areas may have been more related to improved education on projections and the limited familiarity that the subject pool may have had with Mercator wall maps—as many of the Mercator wall maps had been replaced in classrooms by that time. In comparison, at the time of Saarinen's work when a Mercator effect was suggested, many classrooms still relied on Mercator projection wall maps for reference. While many classrooms have moved away from the projection, there are still many modern examples of schools making the switch away from Mercator in favor of equal area projections. For instance, the Boston public schools adopted new world maps to move away from Mercator in 2017.

Another interesting Mercator impact was found when Battersby and Kessler (2012) asked map readers to identify distortion patterns on several common and uncommon global-scale projections. While landmass shape was a notable cue used (e.g., a region looked "wrong," presumably in comparison to the shape of the region on the reader's cognitive map), there was also a strong influence of the distortion pattern in the Mercator projection, where the areal distortion increases at higher absolute latitude. Across all projections that the map readers were asked to evaluate, the top regions noted as indicative of being "distorted" were near the poles (Antarctica, Greenland, and generic "polar regions").

While there have been many efforts to limit the impact of the Mercator projection on our cognitive maps, it still has a lasting impact. Part of this is likely due to its use as an exemplar when teaching about projections—open almost any geography textbook or atlas with a section on "learning about projections" and you will find a description of the distortion in the Mercator projection. This is likely due in part to the resurgence of the projection for use in web mapping (e.g., the mapping services from Google, Bing, Yahoo, Mapbox, etc.). This familiarity may challenge map readers with their abilities to assess the accuracy of non-Mercator projections.

Does the Projection Look "Right" or "Wrong"?

Without consciously thinking about it, a map reader will assess whether your map looks "right." This can be based on a perceived pattern in the data (e.g., there shouldn't be high values over there), or based on the overall shape of the projection. Whether your map is for the entire world or just a small part, people tend to have strong feelings about what shape is "right" for any given region. A large part of this is due to familiarity—for instance, is this a common projection that the reader would have seen many times before? However, it can also be a factor of whether the shape of regions of interest just seems odd in some way (e.g., squished, stretched, or compressed), if the map fits poorly in the space allotted and leaves a large amount of empty space, or if there are visual conflicts between design elements due to the projection (e.g., visual conflict between the graticule and the map data or the bounding box for the map). In this section we consider the problems of familiarity, visual preference for certain shapes of projection, and overall aesthetics and fit of projection in your map design when considering if a projection looks "right" or "wrong."

One humorous way of looking at the accuracy of the look of a projection is to go back to the debates over the use of the Mercator projection and its potential for becoming the master image for "right." In the early 1970s, a historian named Arno Peters introduced the Peters projection, a "new" equal area projection that would provide a better and fairer perspective on the world (Figure 4.6). Since the projection had already been defined earlier by James Gall in 1855, it is now more commonly referred to as the Gall–Peters projection. Peters's re-introduction and suggestion for its use as an improvement over other rectangular projections (e.g., Mercator) caused quite an outcry with some cartographers. Arthur Robinson even went so far as to describe it as "somewhat reminiscent of wet, ragged, long winter underwear hung out to dry on the Arctic Circle" (Robinson, 1985, p. 104). We believe this is a proper academic way of stating that a projection "looks wrong."

In more serious terms of assessing a projection purely based on familiarity, Battersby and Kessler (2012) examined the topic by asking map readers to identify patterns of distortion on different projections. In their study, they found a general familiarity and like for the Robinson projection, and it was frequently rated by participants in the study as having no or minimal distortion of area, as well as being more along the lines of a "normal map that

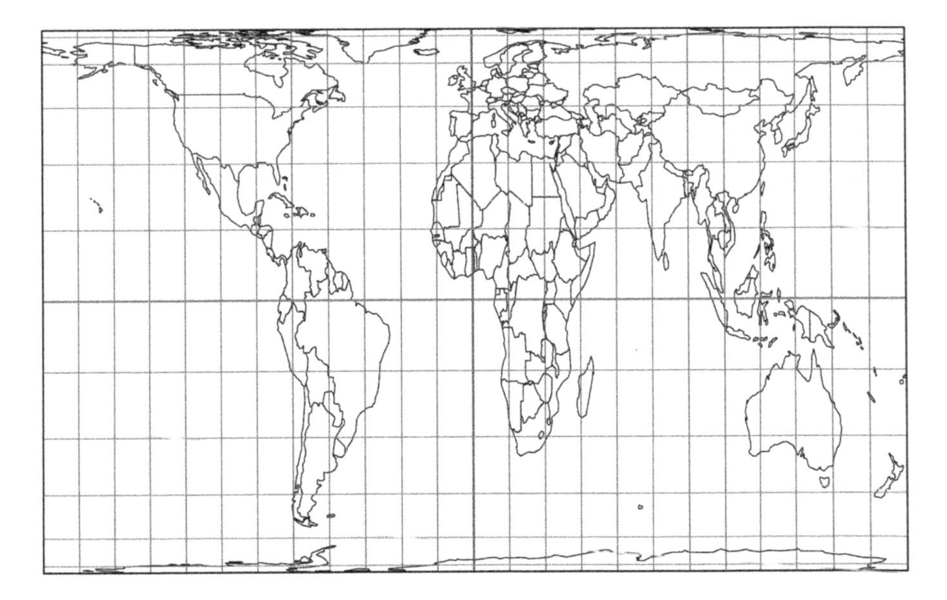

FIGURE 4.6
Gall–Peters projection (equal area). Robinson suggested that this projection looks like wet, ragged winter long underwear. Compared to other projections that you might be familiar with, what do you think?

you'd regularly see" (Battersby & Kessler, 2012, p. 95). In terms of interpreting distortion, however, the Mercator distortion pattern was seen as being the "familiar" pattern, and the pattern of area distortion (e.g., to the polar regions) was applied to almost all of the projections in the study. This presents both a benefit and challenge for mapping—familiarity is likely to lead to a map being seen as more accurate, but it also means that readers may incorrectly assume that any distortion in a new projection is just like that in a familiar one. This means that familiarity may be a bit blinding, and that readers may either assume a map is undistorted or interpret distortion incorrectly.

Of course, familiarity with projections isn't just a challenge with global-scale maps. Shape recognition is a factor in design and interpretation for larger-scale mapping projects as well. Consider the difference in how the United States appears in three different projections (Figure 4.7). The plate carrée projection (Figure 4.7A) is the typical display for "unprojected" or "Geographic Coordinate System" data—this simply means that latitude and longitude coordinates are being displayed on a plane without any additional translation of location; it is as if, oddly speaking, the spherical coordinates were planar coordinates. This leads to horizontal stretching of distances that increases in magnitude for locations farther from the equator (instead of the lines of longitude converging on the poles, they are stretched out so that one degree of longitude appears the same distance in all locations of the map). To us, casting the map in this projection appears unappealingly elongated along the east–west axis. Figure 4.7B shows the same region in the web Mercator projection (what is typically seen on a web map). While the shape of the region may look (debatably) more reasonable in a global context, when the conterminous United States is isolated in the map it gives the opposite problem to plate carrée—now it looks squashed along the east–west axis. To us, the map in Figure 4.7C using the Albers projection is the more aesthetically pleasing, so we would be likely to select it over these other two options. The curved border with Canada and the nice balance in the shape along the east–west and north–south axes feels more aesthetically pleasing. It also has the benefit of being an equal area projection, which often works well for thematic mapping.

FIGURE 4.7
Conterminous United States in plate carrée (A), web Mercator (B), and Albers equal area conic (C) projections.

Is the Overall Shape Appealing?

Outside of familiarity and aesthetic interpretation of specific regions (or all of the world) on a map, there is a broader area of research on *general* shape of maps (e.g., rectangular vs. more round, aspect ratio, etc.) that is interesting to consider. It turns out that people not only prefer more familiar looks for individual regions on the map, but that there are measurable preferences for overall shape and proportion of the entire projected area.

The proportions and shape of the graticule are a common target for evaluating user preference for the look of specific projections. For instance, Gilmartin (1983) examined user preference for proportions and shapes of graticule on global-scale maps. One of the questions that she set out to explore was whether or not there was a preference for projections that had a proportion close to the "golden proportion" of 1:1.6. In her work, she found general preferences for elliptical graticule as opposed to rectangular, for more "compact" shapes to be preferred over elongated ones, and no clear evidence for a "golden graticule." With cylindrical projections, squares were most liked (Figure 4.8) and elongated rectangles with a 1:2.5 ratio were least liked (Figure 4.9). With the elliptical shapes, the most preferred ratio was 1:1.3 (Figure 4.10), not the most compact shape (circle).

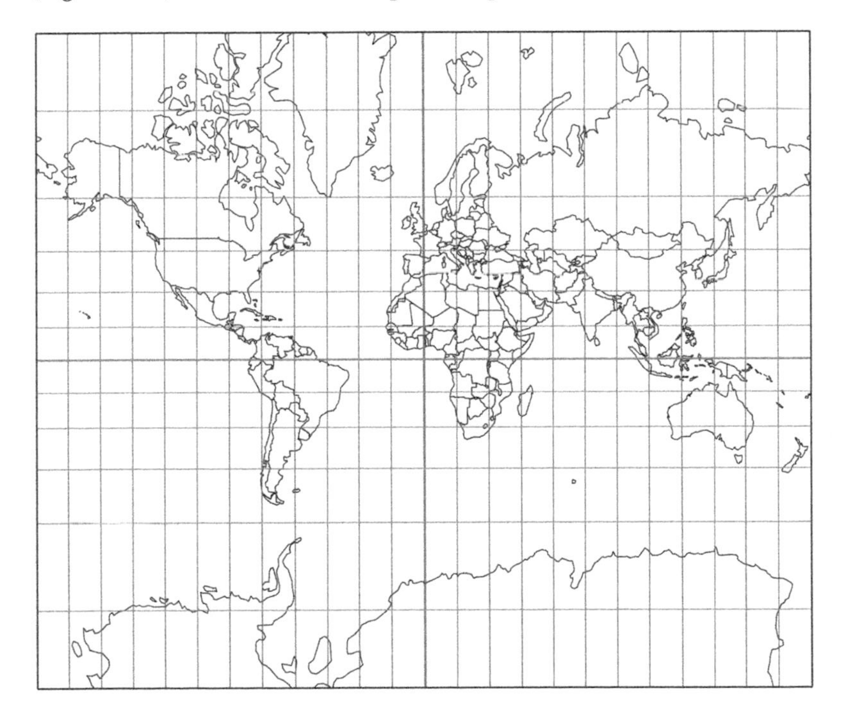

FIGURE 4.8
A web Mercator projection as example of the most preferred cylindrical projection ratio.

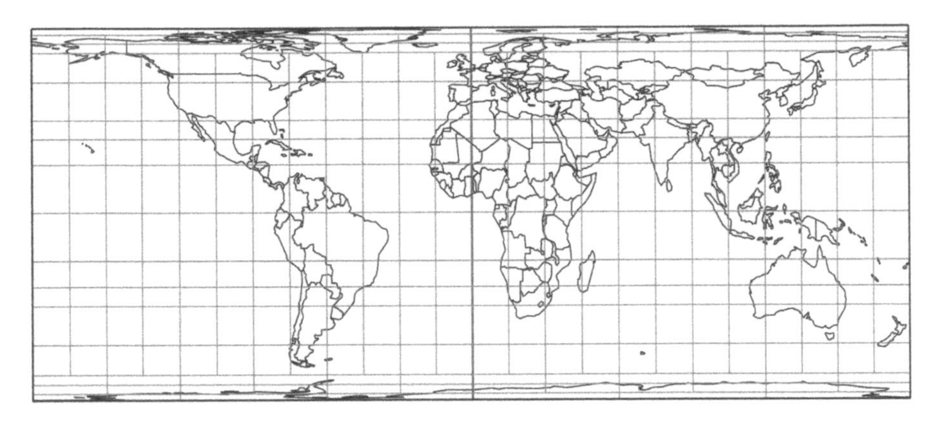

FIGURE 4.9
A Behrman equal area projection as an example of one of the least favored cylindrical projection ratios (near to 1:2.5).

FIGURE 4.10
A modified Hammer projection as an example of one of the most favored elliptical shaped projection ratios (near to 1:1.3).

More recently, Šavrič et al. (2015) conducted a survey of 496 map readers to evaluate user-preferred graticule characteristics of small-scale projections. Through their work evaluating characteristics of nine projections for global-scale maps, they found preferences for elliptical rather than sinusoidal shapes, and for meridians and straight rather than curved parallels (Figure 4.11). There were no clear user preferences for how poles were

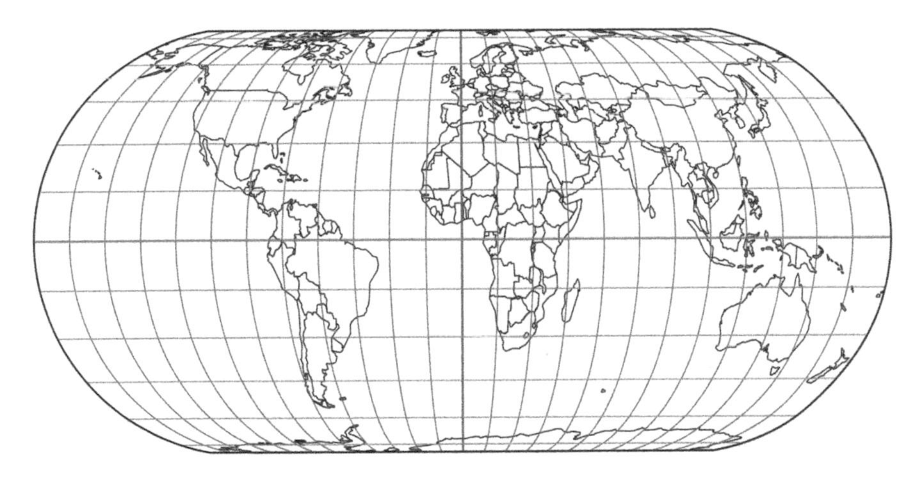

FIGURE 4.11
Eckert IV projections, showing meridians with an elliptical shape and straight parallels.

represented (as points vs. lines) or whether the edges of pole lines were rounded or not.

While these studies examined preferences for representing the graticule, a map typically shows more than just lines of latitude and longitude. Map readers also show strong preferences for projections based on the shape of land and ocean features. Using projections with graticule and continent outlines, Werner (1993) explored map reader preferences for projections on nine equator-centered global-scale maps. From a survey of 60 map readers, he identified that pseudocylindrical were preferable to cylindrical (Mercator, Miller, and Peters) projections. Of the pseudocylindrical projections used in the study, uninterrupted projections (Voxland and Robinson) were preferred over interrupted projections (Mollweide with three interruptions and Goode Homolosine). An interesting aspect of Werner's study is the comments received from map readers regarding the projections. Werner noted that many people commented specifically on "roundness" being a preferred characteristic, suggesting that it was more in line with the shape of Earth, while the cylindrical projections were less preferable because they were "flat" (Werner, 1993, p. 37). While familiarity with a projection wasn't found to be an influence on preference, there was still an element of "looking right" that seemed to underlie the preferences. Specifically, Werner notes comments on the dislike of certain projections due to them being unlike "my mental picture of the world" (Werner, 1993, p. 38; in reference to the Peters projection).

The visual preference that individuals have for shape and proportion of projections is mostly a factor to consider with global-scale mapping and should be balanced with other design considerations such as the size and shape of the space available to display the map.

Aesthetics and Fit

There are also important considerations of aesthetics and fit in the space allotted for the map. Assuming that you aren't constrained by an organizational standard for your projection selection, once you have established what properties are most critical to preserve in the map (e.g., equal area, conformal, or a compromise of distortion to both area and angles), and have a sense of what parameters will help you optimize for those properties, you can focus on the aesthetics of your different projection options. Given the properties of interest, focus on what projection preserves those properties and looks the best in the space allotted for your map. Your goal should be to allow the reader to focus their eyes on the *data* and not on potential distractions due to the projection.

Whether or not a projection looks "right" to the reader makes a notable difference on the aesthetic feel of a map; the projection may detract from the visual appeal of a map by making it look clunky or awkward (e.g., the U.S. maps in Figure 4.7), or, on the other hand, it may *add* visual interest to a design by representing geography in a novel perspective. Unfamiliar or odd-shaped projections also have an interesting impact on how well map readers can identify distortion on projections.

In addressing aesthetics of your projection choice, you should also consider how the projection impacts the background data. An example of this is seen with the relationship between the graticule and land areas on the three maps of Chile shown in Figure 4.12. The projection choice introduces two interesting challenges—ensuring that the shape of the region being mapped looks good *and* minimizing visual conflict that may slow or distract map readers in their task. The three maps of Chile show very similar shapes for the country, though with slightly different rotations in the data frame. In terms of making a final decision on which to select, it them comes down to

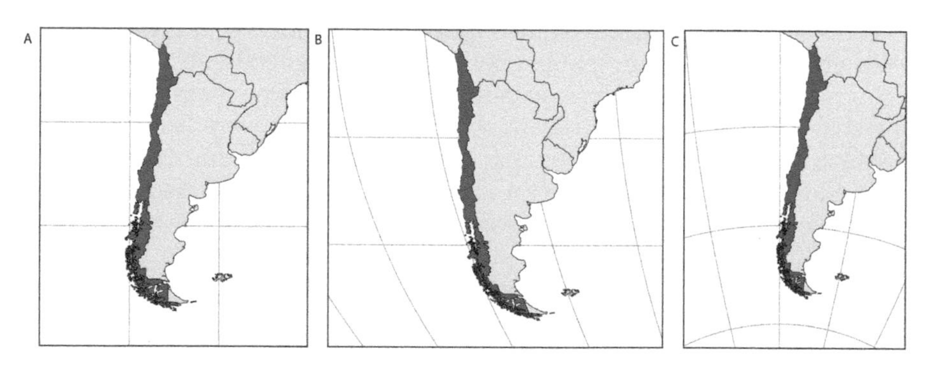

FIGURE 4.12
Chile in the Robinson (A), Miller cylindrical (B), and Chile UTM 18 South (C).

the aesthetic. While this may just be a personal preference of ours, we think the version using the Robinson projection (Figure 4.12A) has a nicer shape with the curvature of the graticule matching the curvature of the southern part of the country, as opposed to the rigid perpendicular graticule of the Miller cylindrical (Figure 4.12B) or the non-parallel graticule of the Chile UTM 18S (Figure 4.12C). But, again, this is a personal preference, so from a purely aesthetic standpoint, we would opt for the Robinson projection if we had to choose between these three because for this projection there is a better balance with the graticule in the background and the spatial data in the foreground. To us, this feels like it minimizes the visual distraction in reading the map.

Conclusion

In this chapter, we discussed the idea that projections can have an impact on the way people form their cognitive maps of the world. At the core of this chapter is the basic cognitive and perceptual issue that people struggle with: projecting Earth's spherical surface to a map's planar surface. The projection is inherently coupled with distortion. Areas, angles, distances, and directions can be adversely affected by the projection. The amount of distortion in any of these characteristics can be quantified and illustrated across a projection's surface, which you have already seen in Chapter 2 (e.g., the color gradations showing different kinds of distortion on a projection's surface). Aside from showing the quantitative aspects of distortion, the projection can alter the appearance of landmasses to the extent of being unrecognizable to the map reader.

This chapter pulled from research that examined how people understood or misunderstood projections. Olson's (2006) study described different visual cues that map readers use to characterize the distortion that was present on a map. The map reader needs to understand distortion, so they have a better grasp of how the landmasses appear on the map and the consequences that distortion has on the distribution of the data. Olson suggested that map readers could use readily identifiable cues presented by the graticule shown on the map to make educated assumptions about the nature of distortion on a map's surface. In another research study, Anderson and Leinhardt (2002) examined how successful map readers were at estimating distances on a map. They concluded that map readers have problems when taking "rules" that work on a spherical surface, such as "the shortest path between two points is a straight line," and applying those rules to the map's surface without understanding how the intermediary distortion alters those rules. To determine which path is shown on a given map accurately requires the map reader to understand how the projection portrays accurate distances. Battersby (2009) reported results of a study where she had participants use one of three references to estimate land area of continents or territories. Overall, participants tended to view what was shown on the map as the correct area—irrespective of the distortion that was present. Even when participants acknowledged the presence of distortion on one of the references, their accommodation was not enough to counter the effects of distortion. Battersby and Kessler (2012) examined whether people used specific evidence from several common projections as cues to understanding distortion patterns. They found that participants relied on cues from the Mercator projection but incorrectly applied those cues when interpreting other projection distortion patterns. Hruby et al. (2016) conducted a study that examined how map readers reconcile a continuous path's distance when that path crosses a map's edge or periphery. For example, different projections control how a path "wraps" around and re-enters the map. Distance estimation along a path is estimated more accurately when the entirety of that path is shown on a map in an uninterrupted form. Several studies have examined the curious "Mercator" effect that suggests people have encoded the appearance of landmasses and area estimation based on the Mercator projection as the framework for their cognitive maps. A recent study by Battersby and Montello (2009) offers that the Mercator effect is not as valid as earlier studies suggested that were largely based on evidence derived from sketch mapping. While the "Mercator effect" may be more imagined than real, the Mercator projection (and its direct descendent, the web Mercator) remains present in our minds through its adoption into mapping services like Google Maps, Bing Maps, and MapQuest.

The chapter concluded with three additional reflections of projections. First, "how 'correct' does the projection look?" The idea of correct is certainly relative to the map reader. This correctness could be the appearance of the data, how strongly people imagine that a given landmass should be shaped on a map, or the aesthetics by which a projection presents a map. Much of this correctness, we feel, is based on the map reader's familiarity with how landmasses really appear on a globe. Second, what are people's preferences for certain overall projection shapes. Gilmartin (1983) found that people preferred projection shapes that were elliptical compared to more rectangular shapes and compact shapes in contrast to elongated ones. In another study, Werner (1993) examined preferences for equatorial-centered world projections. Results of his study found that participants preferred pseudocylindrical compared to cylindrical projections. Uninterrupted projections were preferred to interrupted ones. More recently, Šavrič et al. (2015) studied the preferences that map readers had toward the graticule arrangement on common projections. They found that people preferred elliptical curves for meridians and straight lines for parallels. Participants did not prefer the poles being represented as lines or points. Third, how does aesthetics play into appreciation of the projection? Some ideas discussed included the extent to which the projection adds or detracts from the overall map design. The projection should not distract the map reader's eyes when looking at the patterns in the data. However, a projection could be selected to enhance the appearance of the data or create an eye-catching look to the map.

5

Assistance Using Projection Selection Guidance Tools

There are an infinite number of projections possible; however, it is estimated that only perhaps 400 have been formally described, and only about 50 of these are commonly used (Maling 1992; Kessler, 2018). This makes the problem of choosing an appropriate projection quite a bit easier, since we can constrain the search for the right projection to a relatively small list of possibilities. However, within the list of potential projections, there are still many choices to make to maximize the utility of a projection for the map area of interest *and* the map reading tasks anticipated for the map. For example, Šavrič et al.'s (2016) Projection Wizard relies on 30 different projections, where users can adjust the parameters of these projections to tailor to the area of interest.

To help navigate the common projections, several guidelines have been suggested. One of the most frequently noted guidelines for selecting appropriate projections was developed by John Snyder (1987). Snyder proposed a hierarchical system where the cartographer sequentially works through a series of questions or steps regarding the map extent (e.g., hemispherical, global, etc.) and the projection properties of importance, and ultimately is presented with a list of potential projections to consider what would be suitable for the combination of map extent and property of interest.

Snyder's general guiding steps are presented in Table 5.1 and the complete hierarchical projection selection guideline is presented in Table 5.2.

While guidelines such as Snyder's are helpful in narrowing down the options, they tend to be limited in the list of projections suggested, and many important choices related to the fine details of cartography (e.g., aesthetics, finding compromise between multiple spatial properties of importance, and optimizing the settings of parameters on projections) are left out of the discussion. Their application benefits from a strong working knowledge of projections to navigate through them, which many users don't necessarily possess.

More recently, Šavrič et al. (2016) have developed an online Projection Wizard to aid in selecting optimized projections for small- or large-scale mapping projects. Their Projection Wizard uses a combination of spatial extent and user-selected distortion property of importance to suggest specific projections and parameters (e.g., latitude of origin, standard lines, and/or central meridian) to optimize to a specified location. Results are provided

TABLE 5.1
Snyder's Basic Overview of His Hierarchical Projection Selection Guideline

Guiding Steps to Selecting a Projection
Select the geographic scale of the region to be mapped.
Select single projection property of primary interest.
Select any special characteristics for the property of interest—depends on property; for instance, for a conformal map, the options are where correct scale is desired.
Select from a list of suggested projections. Note that the projections provided in Snyder's guidelines are not intended to be an exhaustive set of the possibilities.

as Proj.4 code that can be copied for use in a GIS or with mapping programming libraries (Proj.4 is discussed in Chapter 11). Table 5.3 presents a listing of projections contained in Šavrič et al.'s (2016) Projection Wizard.

Through this chapter, we will follow a similar process outlined by Snyder's basic guidelines—considering the geographic extent of the landmass to be mapped and the projection properties—but we will focus on simplifying the concepts and explaining the broader decision-making process so that you can more easily, and confidently, select projections and explain the decision-making process. In the remaining chapters of the book, you will see these selection decisions applied, with deeper explanation, in many different mapping scenarios.

References and resources for projection selection decision-making guidelines are provided in Chapter 11.

Projection Properties of Importance

In earlier chapters, we covered the basic properties of projections, and the fact that no flat map can preserve all the spatial relationships that are true on a globe (e.g., measurements of angles and areas). Because all projections will introduce some sort of distortion, and distortion to the projection can significantly impact spatial and visual analyses, the first consideration when selecting a projection is to identify the map properties of importance in your mapping project, and to narrow your selection of projections to the ones with properties that will best suit the needs of the map maker and map reader. For many thematic maps, the choice is often an equal area projection, because the visual pattern interpretation is often driven by comparisons of size across different regions (e.g., how many acres of land are planted with a crop in different locations). However, equal area projections aren't always the right solution for all map purposes. For instance, if the map reader will be evaluating flight distances for missiles from a launching location, an equidistant projection centered on the launch zone would be more appropriate (see the discussion

TABLE 5.2

Snyder's Projection Selection Guidelines

Region Mapped	Property	Characteristic	Named Projection
World	Conformal	Constant scale along equator	Mercator
		Constant scale along meridian	Transverse Mercator
		Constant scale along oblique great circle	Oblique Mercator
		Entire Earth shown	Lagrange
			August
			Eisenlohr
	Equal area	Non-interrupted	Mollweide
			Eckert IV & VI
			McBryde or McBryde-Thomas variations
			Boggs eumorphic
			Sinusoidal
			Misc. pseudocylindricals
			Hammer
		Interrupted	Any of above except Hammer
			Goode homolosine
		Oblique aspect	Briesemeister
			Oblique Mollweide
	Equidistant	Centered on pole	Polar azimuthal equidistant
		Centered on a city	Oblique azimuthal equidistant
	Straight rhumb lines		Mercator
	Compromise distortion		Miller cylindrical
			Robinson
Hemisphere	Conformal		Stereographic (any aspect)
	Equal area		Lambert azimuthal equal area (any aspect)
	Equidistant		Azimuthal equidistant (any aspect)
	Global look		Orthographic (any aspect)
Continent, ocean, or smaller region	Predominant east–west extent	Along equator	Conformal: Mercator
			Equal area: Cylindrical equal area

(*Continued*)

TABLE 5.2 (CONTINUED)

Snyder's Projection Selection Guidelines

Region Mapped	Property	Characteristic	Named Projection
		Away from equator	Conformal: Lambert conformal conic
			Equal area: Albers equal area conic
	Predominant north–south extent		Conformal: Transverse Mercator
			Equal area: Transverse cylindrical equal area
	Predominant oblique extent		Conformal: Oblique Mercator
			Equal area: Transverse cylindrical equal area
	Equal extent in all directions	Center at pole	Conformal: Polar stereographic
			Equal area: Polar Lambert azimuthal equal area
		Center along equator	Conformal: Equatorial stereographic
			Equal area: Equatorial Lambert azimuthal equal area
		Center away from pole or equator	Conformal: Oblique stereographic
			Equal area: Oblique Lambert azimuthal equal area
	Straight rhumb lines		Mercator (principally for oceans)
	Straight great circle routes		Gnomonic (for less than hemisphere)
	Correct scale along meridians	Center at pole	Polar azimuthal equidistant
		Center along equator	Plate carrée (equidistant cylindrical)
		Center away from pole or equator	Equidistant conic

Source: Adapted from Snyder 1987.

in Chapter 4, and Figures 4.2–4.4). Or, if there is no specific spatial relationship that is a priority, identifying the projection that is a good compromise between different types of distortion, highlights the regions of interest, fits the space allocated for the map (e.g., portrait or landscape printed page, space available in a web page, etc.), and is most aesthetically pleasing.

TABLE 5.3

Available Projections in Projection Wizard

Extent	Distortion Property	Projection
World	Equal area	Mollweide
		Hammer (or Hammer-Aitoff)
		Boggs eumorphic
		Sinusoidal
		Eckert IV
		Wagner IV (or Putnins P2)
		Wagner VII (or Hammer-Wagner)
		McBryde-Thomas flat polar quartic
		Eckert VI
		Goode homolosine
		McBryde S3
	Compromise	Natural Earth
		Winkel tripel
		Robinson
		Wagner V
		Patterson (cylindrical)
		Plate carrée (cylindrical)
		Miller cylindrical I
	Equidistant	Azimuthal equidistant
		Two-point equidistant
Hemisphere	Equal area	Lambert azimuthal equal area
	Equidistant	Azimuthal equidistant
Continent, ocean, or smaller area	Equal area	Albers conic
		Cylindrical equal area
		Transverse cylindrical equal area
	Conformal	Stereographic
		Lambert conformal conic
		Mercator
		Transverse Mercator
	Equidistant	Azimuthal equidistant
		Plate carrée
		Equidistant conic

Source: Adapted from Šavrič et al. (2016).

Even when there is one primary property of importance, it is important to consider other types of distortion across the projection. For instance, when an equal area projection is most important, consider that the area preservation comes at the expense of distortion of angular measurements, and that each equal area projection will have a different "look" in terms of the shape of regions on the map. An interesting debate in the cartographic community involves the use of the equal area Gall–Peters projection—while the equal area property is often beneficial for thematic mapping, this particular projection has been criticized for its appearance. One cartographer has even referred to it as having an appearance "somewhat reminiscent of wet, ragged long winter underwear hung out to dry on the Arctic Circle" (Robinson, 1985, p. 104).

Narrowing down a projection choice based on the map property of importance is much easier if there is only *one* spatial relationship that clearly takes precedence over all others. When there are multiple spatial relationships that are important to preserve in your map, the problem is a little more complex and involves finding balance and compromise in the location and severity of distortion across the map, for instance with a map showing multiple attributes such as a choropleth map showing per capita wine imports by country (where preserving area is important; see Chapter 7), with a flow map showing the exports from France to each country (where distance traveled is important; see Chapter 9). In this case, balancing distortion of both area and distance is preferred.

Of course, a key part of identifying the right projection for a large or small geographic area is to consider the type of data interpretation tasks that the user will need to do. We will dig deeper into this question using problem-based examples and specific map types throughout Part II of this book.

Geographic Location, Size, and Shape

Geographic location, region size, and landmass shape (e.g., compact, elongated along the north–south or east–west axes, etc.) are also important factors to selecting the most appropriate projection. To minimize distortion across your mapped area, it is common to adjust the projection class, location of standard lines, and aspect.

With respect to location, the most important consideration is to center and minimize distortion around the primary region of interest. This can be done through adjustment of the projection aspect (e.g., equatorial, oblique, etc.), the placement of the central meridian, and the standard lines. One suggestion from traditional cartographic texts is that the projection's class can be a factor when selecting projections, as a class has specific ways in which distortion can be controlled (e.g., through the location of standard lines) and the resulting distortion pattern. For regions in the tropics (23.5° S to 23.5° N), a cylindrical projection should be used with the standard line located at the equator; if in temperate latitudes (between the tropics and the polar regions) a conic projection should be used with one or two standard lines located in the tropics; and if in a polar region then a planar projection should be used with a standard point located at the pole. While these are reasonable approximate guidelines, there are enough examples for where alternate projection types are equally, or more, successful for mapping with minimal distortion. Through the book, we *do not* adhere to these general guidelines for selecting a projection based on latitude but provide detail on how to assess appropriateness of various projection types for different mapping needs.

With respect to region size, with larger geographic areas (i.e., using small-scale maps), it can be trickier to find balance between the different types of distortion. In general, the process is to identify the geographic areas of most importance (this may be the entire globe, or select regions), and evaluate the pattern of distortion for any projections being considered. To identify type and level of distortion in the geographic regions of interest, it helps to use a tool for visualizing distortion patterns. For most of the distortion maps in this book, we have used Geocart to create distortion surfaces; however, there are other ways to visualize distortion that can be used (see Mulcahy & Clarke, 2001, for a review).

For local-scale mapping (large-scale maps), it should be possible to effectively relegate much of the distortion in a projection to a small region on the map. To quantify the sizing of regions that should be able to be mapped with negligible distortion, Maling (1992) suggests that, with adequate projection parameter selection, regions as large as western Europe can be mapped without area distortion greater than 2% or angular distortion greater than 1°; however, regions as large as, or larger than, Canada would exceed this threshold.

For maps where the location of interest spans the globe, it is a bit trickier to identify the acceptable level of compromise in the projection selection. The decision-making process is driven by the type of map (e.g., map for navigation, or thematic map such as choropleth, dot density, etc.) and specific tasks that map readers are expected to undertake with the map. Additionally, there is the added challenge of every flat map having an interruption at the periphery (e.g., the outer bounding edge of a global-scale mapped area). If it is important for readers of your map to be able to estimate distance between locations of interest, perhaps to identify clusters or estimate travel times, it is critical to minimize measurements that cross the periphery of the map. That is, do not expect your map readers to be able to make accurate assessments of distance between locations if they have to mentally wrap the distance around the interruption on the map. For an example, consider the visual measurement between Sydney, Australia, and Los Angeles, California, on the two maps in Figure 5.1. In Figure 5.1A, the shorter path (shown in red) can easily be visually measured across the map, and in Figure 5.1B, the shorter path crosses the periphery, though most map readers would incorrectly assume the shortest path to be the one connecting across the Prime Meridian instead of wrapping around the periphery. Crossing the periphery makes it more difficult for map readers to estimate actual distance (not to mention that a straight line is rarely the shortest path on a global-scale map!).

With respect to region size, some projections are better for global-scale, while others are better for more local-scale mapping. For example, consider the orthographic projection in Figure 5.2. This projection works well for areas up to a hemisphere (and arguably, with the distortion around the edges, it is really better for a smaller region around the center of the projection) but

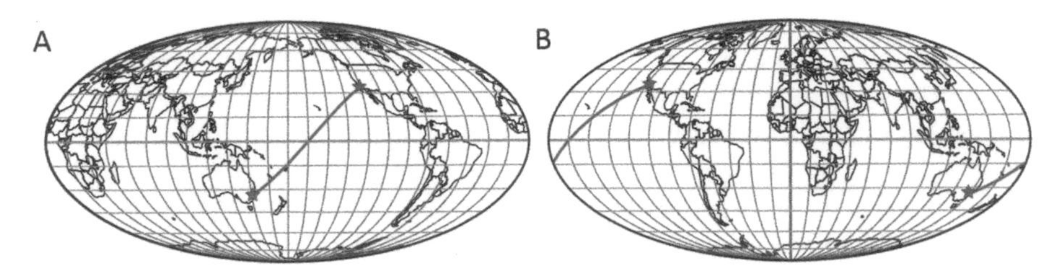

FIGURE 5.1
Mollweide projection (equal area) with Sydney, Australia, and Los Angeles, California, marked with red stars. Map A is a Pacific-centered map with a central meridian of 180°, while map B is centered on the Prime Meridian. The shortest path between Sydney and Los Angeles is shown with a red line on both maps.

FIGURE 5.2
Orthographic projection highlighting South America. The projection is well suited for visualization of regions up to a hemisphere; however, regions near the periphery show enough distortion that this projection is normally used to highlight regions smaller than a hemisphere.

would not be usable for a global-scale map because half of the world cannot be seen.

Or, consider a map using universal transverse Mercator (UTM) as seen in Figure 5.3. UTM zones are optimized for displaying relatively narrow strips of north–south oriented data for a hemisphere. Zone 10 North works well for the Pacific Northwest and portions of California in the United States

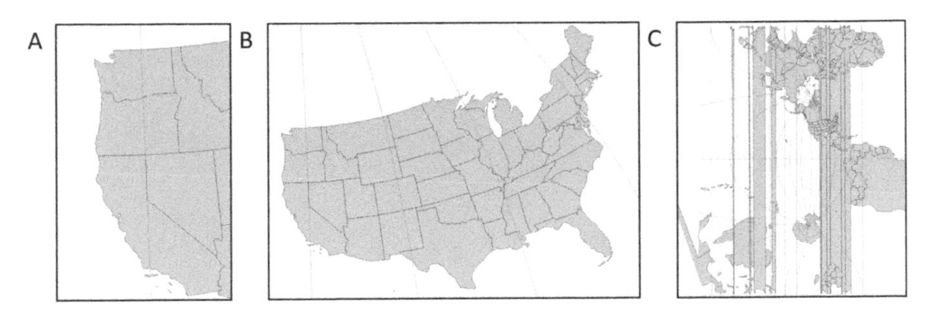

FIGURE 5.3
Universal transverse Mercator (UTM) projection in Zone 10 North, which is suitable for the Pacific Northwest and California (A). Distortion increases rapidly outside of Zone 10 and this projection is not suitable for larger regions (e.g., United States) (B). If one attempts to use this projection for global-scale data, there are many undefined regions and odd geometric transformations that result (C).

(Figure 5.3A; because the projection parameters have been adjusted for that geographic area), but it would not work for a map of the larger United States region (Figure 5.3B), and would result in some very strange geometry (and undefined regions in the projection) if it were used at global-scale (Figure 5.3C).

While projections commonly used for hemisphere or smaller regions may not work well for larger regions, the opposite is often true—global-scale projections can be modified for local-scale mapping *with an appropriate adjustment of the projection parameters*. It is important to note that the default parameters for most global-scale maps will generally not be appropriate for local-scale mapping, so they must be adjusted.

For local-scale mapping (i.e., using large-scale maps), it is easier to minimize distortion through careful adjustment of projection parameters. This is true even with a projection that many would not consider optimized for a specific region. By carefully adjusting the projection parameters you can create a local-scale projection with acceptable levels of distortion for multiple properties (e.g., both angular and areal distortion). For example, most would not consider an equatorial aspect Mercator projection to be appropriate for mapping Greenland. However, adjusting the parameters for the projection can lead to a map with limited distortion for that area (Figure 5.4).

For local-scale mapping, there are many approaches to identifying the best projection. A good first step is to explore whether there is a regional standard in place. For instance, through the United States, many states and counties have standardized around the State Plane Coordinate System (based on either transverse Mercator or the Lambert conformal conic). Many countries also have recommended projections and parameters that are tailored to the specific region. If there is a standardized projection recommended by official government sources for the area in which you are working, it is often a good idea to use that recommended projection unless you have specific data or

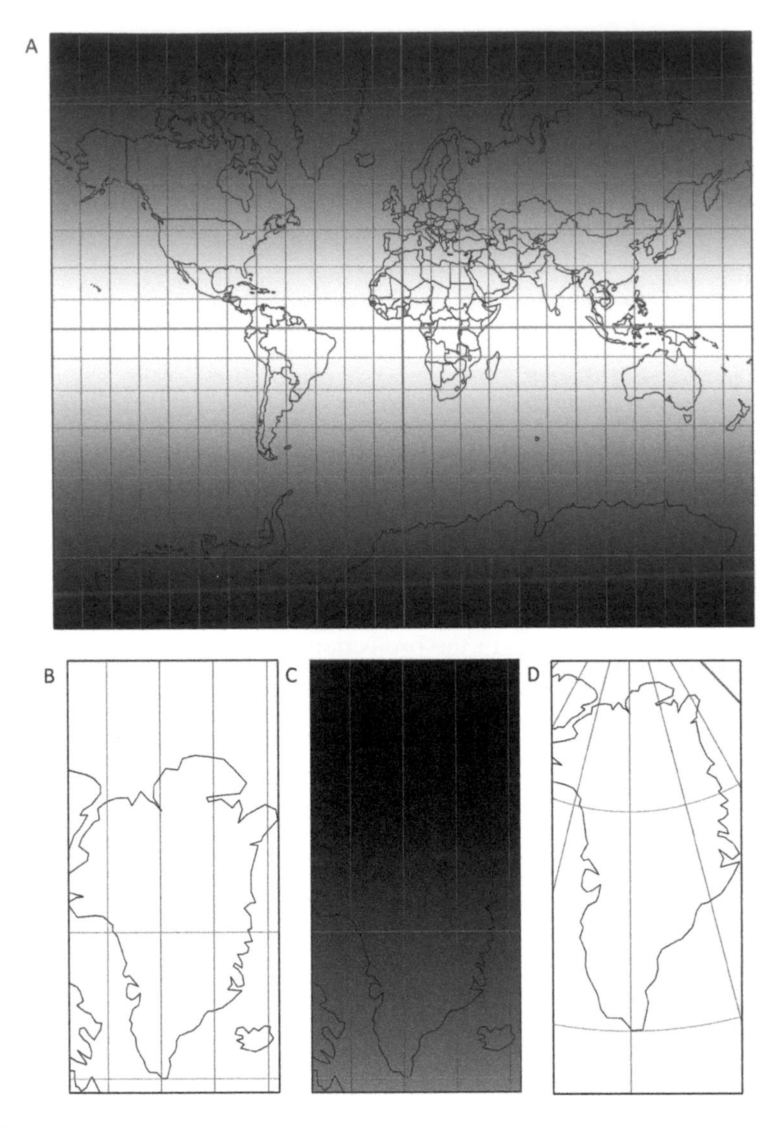

FIGURE 5.4
A global-scale map in Mercator projection shows significant distortion of area away from the poles (A). This is indicated by the green pattern on the map, with darker green indicating greater distortion of area. Greenland is situated in the far Northern Hemisphere. Map B depicts Greenland as rendered on an equatorial aspect Mercator projection. Map C shows the same region as map B but highlights the distortion pattern. This emphasizes the significant distortion of area in this part of the map. Because of this, Greenland as rendered in an equatorial aspect Mercator projection is grossly distorted in area (B, C). Map D shows both the Greenland outline and the distortion pattern for an oblique Mercator projection centered at 70° N. The color scheme for the distortion is the same as in the first map. Note that only a slight hint of green is seen in the distortion visualization, suggesting that the areal distortion is minimal. Through careful selection of parameters in a Mercator projection, we can minimize distortion even for areas that would be greatly distorted in the more common equatorial Mercator projection.

aesthetic concerns. Taking this approach decreases the chance of projection conflicts between local data sources, as they are likely to already be in the locally recommended projection and will ensure that your map has the same "look" as other official maps of the region.

With respect to landmass shape, it is also important to tailor projection choice and parameters to minimize the distortion to the general geographic extent of the region. For regions with a predominant north–south, east–west, or oblique geographic extent, you can minimize distortion by selecting a projection and then modifying its parameters (e.g., center and standard line or lines) to create a zone of lower distortion across the mapped area. Frequently for small regions with a north–south extent, a transverse projection such as transverse Mercator is common; for east–west extent and depending on the latitude of interest, a cylindrical or a conic projection with standard line(s) cutting through the region; for an oblique extent, a projection such as oblique Mercator could be used.

An important element of working with conic projections is in appropriate selection of the standard lines. In Chapter 2, we discussed this in more detail, expanding on the process to adapt a conic projection to the shape of the region being mapped using Kavraisky's constant, K.

Broader Design and Analytic Considerations

For any specific spatial relationships that you strive to maintain, or minimize distortion to, for your map, there will likely be more than one projection available to choose from. There will also be parameters that you can adjust (e.g., standard lines, central meridian, etc.) to tailor the projection for your map purpose. When deciding between multiple projections that have similar characteristics of importance for your region of interest and specific map type, there are many broader design and analytic considerations that can help you narrow down your final selection. This section will discuss the projection's overall shape, graticule arrangement, appearance of the poles, and interpretability.

Projection Shape

Projections frequently fall into one of four traditional shape "families": ovals, epicycloids, rectangles, cones, circles—or variants of these (Figure 5.5). For each of these projection families, there are numerous properties to choose from that are equal area, conformal, compromise, and other projections that are non-equal area and non-conformal.

The physical space allotted for your map can help determine which shape projection will be best for your project. Rectangular projections

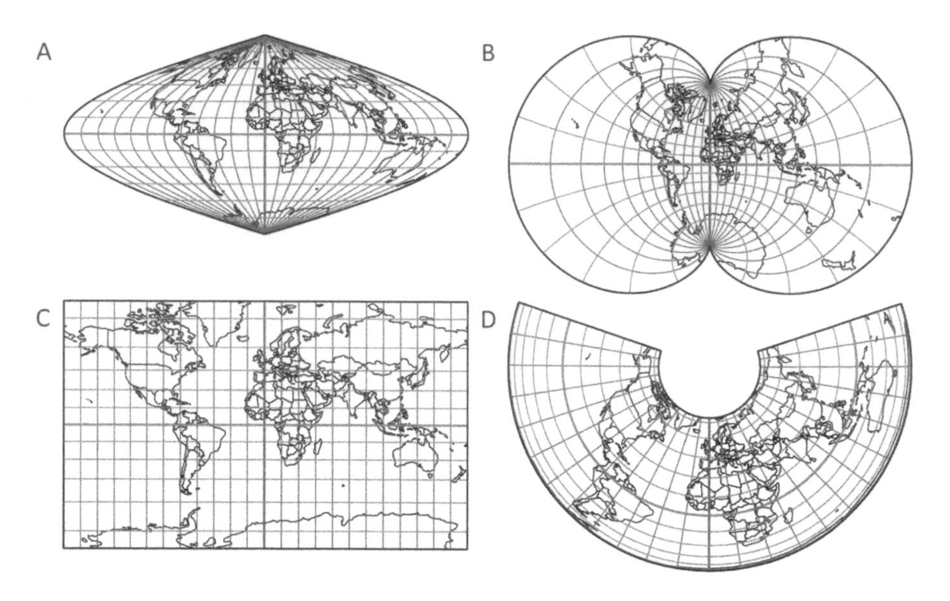

FIGURE 5.5
Sinusoidal projection (A; oval shape, equal area), Eisenlohr (B; epicycloid shape, conformal projection), Miller cylindrical (C; rectangular shape, neither conformal nor equal area); and Albers (D; conic shape; equal area).

often fit well on printed pages or rectangular digital displays, while oval and circular shapes leave more empty space in the corners. While the oval and circular projections can lead to more empty space in your design, this extra space may be beneficial for incorporating other informative or decorative elements to complement your map. For instance, this empty space could be filled with a legend or descriptive text so that it doesn't overlap the map, or small inset maps with detail views or to show complementary spatial patterns, or decorative elements to support the theme of the map (e.g., adding in photos of birds to a global-scale map of migratory patterns).

When working in web mapping environments, you may not have a choice due to the common usage of the web Mercator projection for tiled map services (e.g., Google, Bing, or Yahoo Maps; ArcGIS Online, etc.). The web Mercator projection is an almost-conformal, cylindrical projection and will present significant areal distortion near the poles. For more information on the specific challenges of working in a web Mercator environment, we recommend "Implications of Web Mercator and Its Use in Online Mapping" by Battersby et al. (2014).

If your mapping project is not restricted to use of a tiled map service, there are many online code libraries, such as Proj.4, GDAL, and d3.geo that can help facilitate transformations of your data to appropriate projections for display. See Chapter 11 for more discussion on these and other projection-specific code libraries.

There are also interesting research efforts underway to explore the feasibility of alternate projections for tiled map services, such as the work by Jenny (2012) on "Adaptive Composite Map Projections." This Adaptive Composite Map Projection system for web mapping is designed to adjust the projection and its parameters for the map based on the user's zoom level and geographic location. By doing this, the projection changes from one class and property to another on-the-fly to minimize distortion for the viewing area.

Outside of the four traditional shape families, there are many other special shaped projections that may be of use, though the uniqueness of the projection shape makes many of them less useful for thematic mapping purposes and page space. Some of the more interesting for thematic mapping are projections that are designed to be cut out and turned into physical globe-like shapes. For instance, the Fuller projection (Figure 5.6) can be folded up into an icosahedron (Figure 5.7), a 20-sided 3-D geometric shape.

There are many other creative projections that may add to the aesthetic design of a map, though most of these will not add analytical benefit—and will likely introduce a challenge for map readers trying to interpret spatial patterns. However, they can be quite fun when you are looking for a special design (Figure 5.8).

Arrangement of the Graticule

While inclusion of graticule isn't a requirement for any map, its presence in the map design presents both an analytical and aesthetic consideration for projection choice. On the analytical side of the decision-making process, they contribute important cues to map readers about the distortion in projections and about spatial relationships between locations.

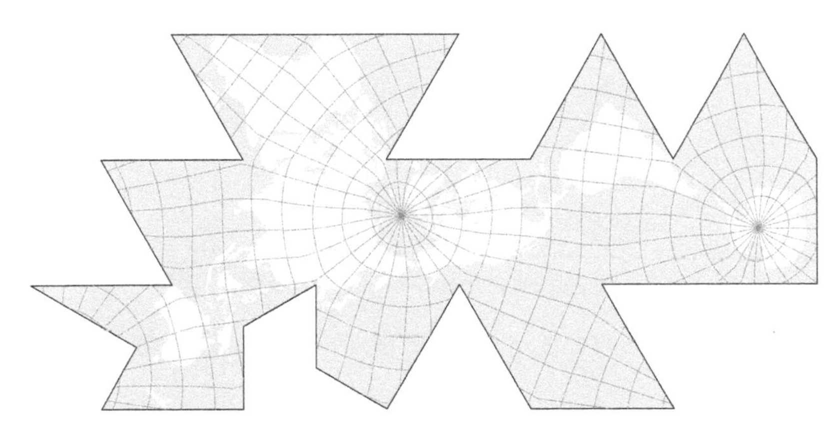

FIGURE 5.6
Fuller projection, unfolded.

FIGURE 5.7
Fuller projection, folded into an icosahedron.

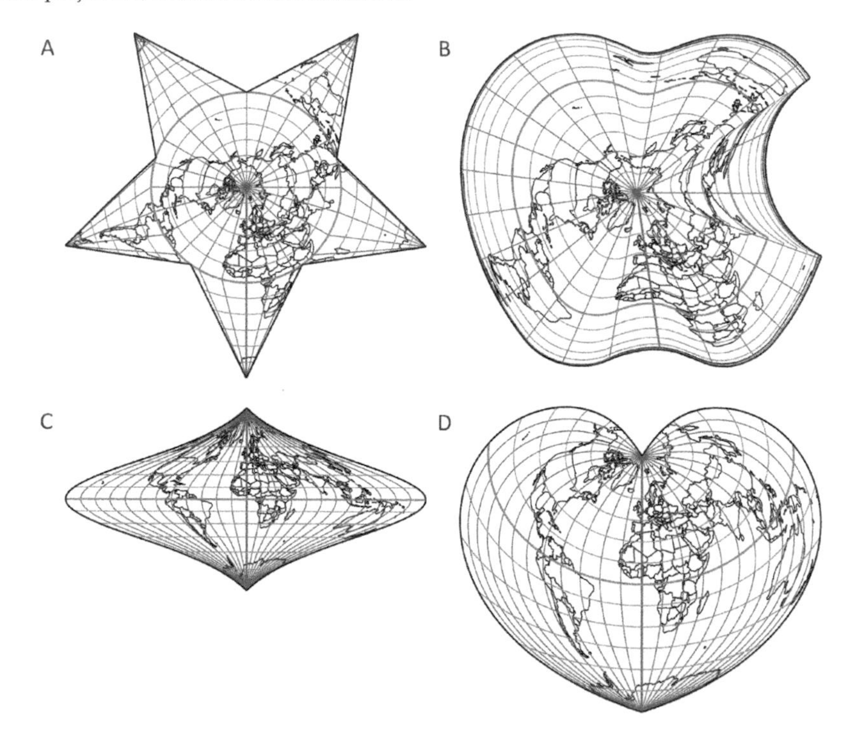

FIGURE 5.8
Creative projection designs, including the Berghaus star (A; equidistant), apple (B; equal area), Foucaut stereographic (C; equal area)—which we think looks a bit like a Christmas tree ornament—and heart (D; equal area).

As described in Chapter 2, the graticule is a network of latitude and longitude lines that are used to locate features on Earth's surface and appears on the map. The graticule shows angular measurements relative to north or south of the equator (latitude), or east or west of the Prime Meridian (longitude). For lines of latitude, or "parallels," the ground distance covered by one degree of latitude will be the same everywhere on the globe. For lines of longitude, or "meridians," however, because the lines meet at the north and south poles, the ground distance for one degree of longitude will vary based on latitude, and will approach 0 at the exact pole; at the equator, the distance of one degree of longitude and one degree of latitude is the same (approximately 111.1 kilometers or 69.1 miles). Parallels and meridians can be drawn either as straight or curved lines. By examining the spacing and shape of the graticule on a map, or the angle at which parallels and meridians meet, a map reader can quickly gauge if the map is likely to be equal area or conformal. If you would like more detail, "Map Projections and the Visual Detective: How to Tell If a Map Is Equal-Area, Conformal, or Neither" by Judy Olson (2006) presents a wealth of information to help you interpret the characteristics of a map based on the arrangement of the graticule. While Olson presents a visual method for approximate evaluation of distortion type, the underlying issue is based on evaluation of scale variation across the surface of the map. Sometimes the scale variation is communicated through use of visualization techniques such as the distortion surface visualizations we use throughout this book, or one of the many other methods that can be used to show how scale varies across a map (see Mulcahy & Clarke, 2001). Another common and informative method for evaluating pattern of distortion across a projection can be the scale bar; for example, see the Mercator projection map in Figure 5.9. (Note that with the web Mercator projection, the reader is typically presented with a single scale bar, but that the bar width and values adjust according to the zoom level and the latitude of the geographic area displayed as the reader pans around the map.)

The graticule also impacts the aesthetic appearance of your map. The graticule is generally considered a supplemental graphical element that should be lower in the visual hierarchy than the thematic data or geographic reference data (e.g., country outlines, points of interest, road networks, etc.). Because of its role as background reference, it is wise to minimize visual complexity, or irregularity, in the graticule where possible. The graticule will either appear as straight or curved lines, depending on the projection selected. Adjustments to the parameters of the projection will impact the arrangement of the graticule.

For instance, with an equatorial aspect (equator and meridian meet in center of the map), the graticule will have an orderly and symmetrical arrangement around straight lines for equator and central meridian (Figure 5.10A–B and Figure 5.11A–B). However, when the aspect is adjusted so that the projection

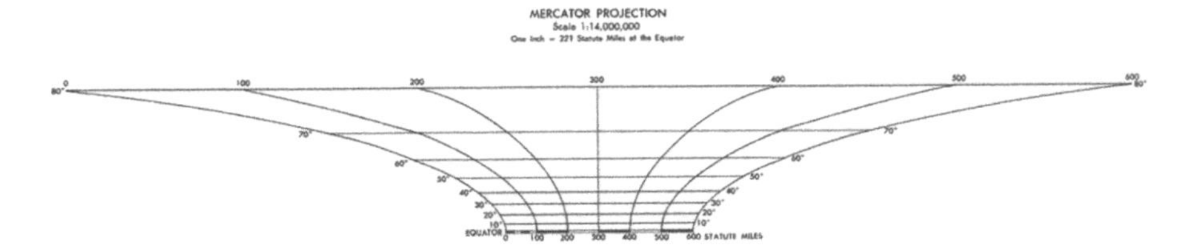

FIGURE 5.9
Mercator projection scale bar indicating the change in scale on the map with increase in latitude. The scale bar here is a scan from a Defense Mapping Agency world map, series 1150. (Image retrieved November 14, 2018, from https://commons.wikimedia.org/wiki/File:World_Scale_from_DMA_Series_1150_map.png.)

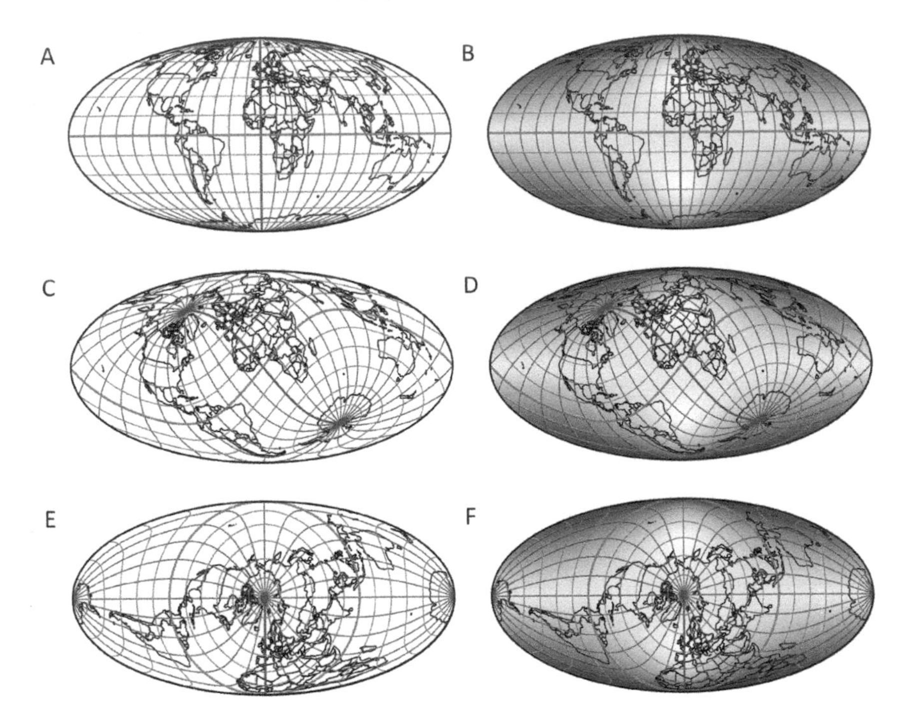

FIGURE 5.10
Mollweide equal area projection shown with an equatorial aspect (A, B) versus oblique aspect 45° N (C, D) versus polar aspect 90° N (E, F). Corresponding patterns of angular distortion in the projection are visualized in the maps on the right (B, D, F). In all maps, the equator and Prime Meridian are shown with a thicker gray line. Note that the distortion pattern remains invariant as the aspect is changed.

does *not* have its horizontal center on the equator, the arrangement of the graticule may appear less orderly (Figure 5.10C–F and Figure 5.11C–F). This is true for cylindrical and noncylindrical projections.

Some projections will show even stranger patterns in the graticule, such as the gnomonic projection (Figure 5.12).

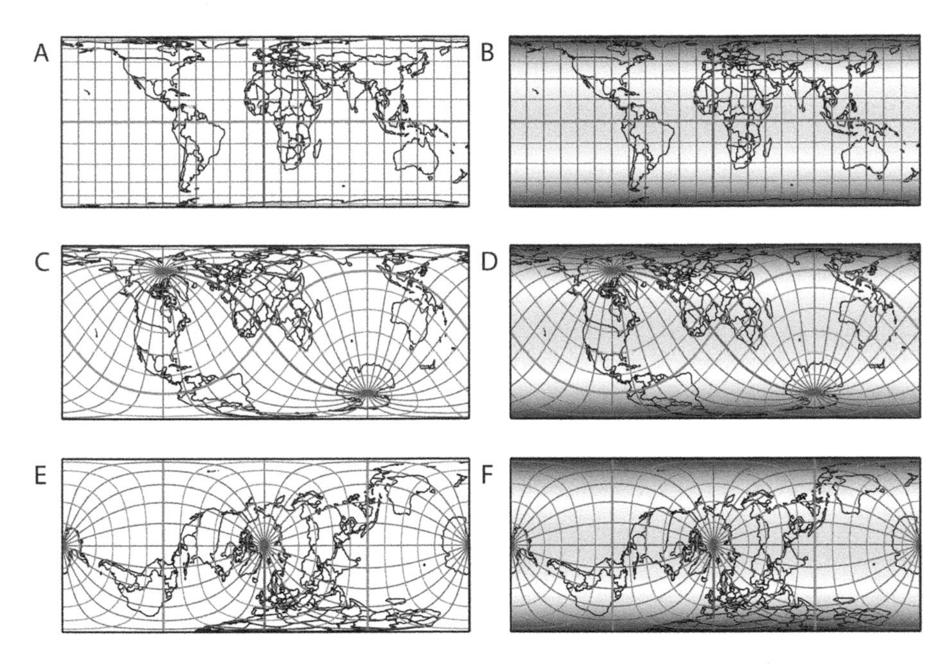

FIGURE 5.11
Cylindrical equal area projection shown with an equatorial aspect (A, B) versus oblique aspect 45° N (C, D) versus polar aspect 90° N (E, F). Corresponding patterns of angular distortion in the projection are visualized in the maps on the right. In all maps, the equator and Prime Meridian are shown with a thicker gray line. Again, note that the distortion pattern remains invariant across all projection aspects.

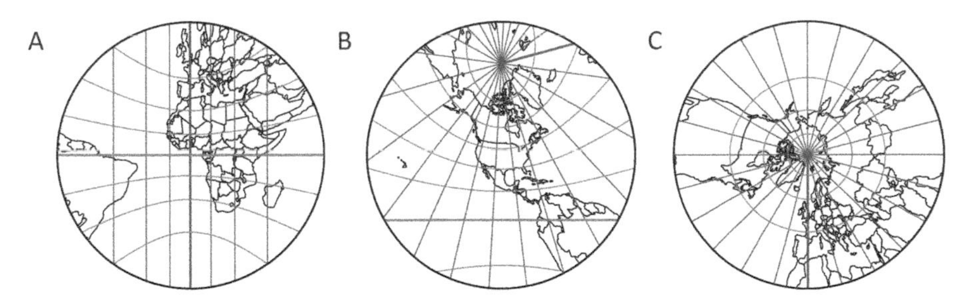

FIGURE 5.12
Gnomonic planar projection with an equatorial (A), oblique (B), and polar (C) aspect .

Appearance of the Poles

In addition to considering projection shape and pattern of the graticule, you must decide how you want to represent the polar regions. On the globe, the poles are represented as *points* where the lines of longitude all meet; on the map, the poles can be represented as points or lines, or not shown at all. The choice of how the poles are represented has an impact on interpretability of data in the high latitudes, as well as the aesthetics of landmasses near the poles.

Rendering the poles as points on a map may seem like the truest representation, since this is how the poles appear on the globe. However, showing the poles as points comes at the expense of horizontal (and/or vertical, depending on the projection) compression of the high-latitude regions. This compression can be seen in the quartic authalic and sinusoidal projections shown in Figure 5.13, where the far northern landmasses become difficult to distinguish. This is a particularly notable challenge when working with datasets where there are important data features (e.g., clusters of points on a dot density map) or regions of interest in high latitudes (e.g., Greenland, Nunavut in northern Canada, etc.).

The polar regions may also be represented as lines, instead of as points. When the poles are represented as lines, the length of the line used to show the pole can vary. In cylindrical projections, the pole is shown with a line that is the same length as all parallels.

In the case of other projections, the poles can't be rendered. For instance, in the Mercator projection, the equations used to generate the projection have no valid value for 90° north or south (if latitude is ±90° than the y value calculated is infinite). A related interesting case is seen with the *web* Mercator projection. No values are shown above 85.051129° north or south, even though valid y coordinates can be calculated. This is the latitude at which the full map extent is represented as a square, so anything above 85.051129° is not shown because it would be beyond the boundary for the smallest-scale tile rendered in web maps.

While the poles are not shown in either Mercator or web Mercator, the high-latitude landmasses are more easily distinguished compared to projections where the poles are represented as points—which can be a nice benefit for comparing values distributed across high-latitude regions. However, the ability to distinguish the polar regions in a global-scale map comes at the expense of exaggeration of areas. In projections like the Mercator, it also leads to potential over-exaggeration of visual importance of these high-latitude regions. We will discuss this in more detail through the remaining chapters of the book, where we provide examples of the visual challenge of interpreting patterns across global-scale maps.

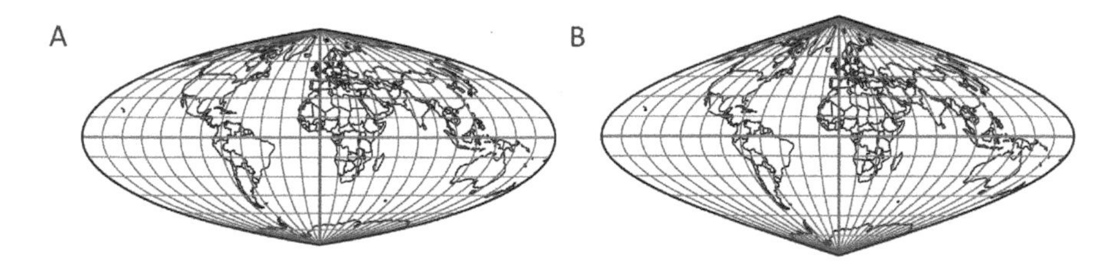

FIGURE 5.13
Quartic authalic (A) and sinusoidal (B) equal area projections. Both represent poles as points, leading to compression of features in the high absolute latitudes.

Interpretability

Interruptions on the map can significantly impact ability to interpret patterns across global-scale regions. While every flat map is interrupted at the periphery, some projections incorporate additional interruptions, where the surface of the map is cut along one or more lines. Interruptions are placed so that they fall in areas that are deemed to be less important to the overall map purpose. For instance, the interruptions on the Goode homolosine interrupted equal area pseudocylindrical projection (Figure 5.14) can be arranged so that they emphasize the land masses (Figure 5.14A) or the oceans (Figure 5.14B). Interrupted projections allow for a more accurate view of the uninterrupted regions, at the expense of accuracy across the interrupted regions—for instance, you cannot easily take a distance measurement between Santiago, Chile, and Johannesburg, South Africa, in the Goode homolosine because it will cross over an interruption.

While we normally think of maps like the Goode homolosine as examples of interrupted projections, technically, *all* global-scale maps are interrupted because the periphery of the projection introduces a discontinuity in the representation of the Earth's surface.

In addition to the challenges of interpreting maps with multiple interruptions, there are interesting visual interpretation challenges that crop up when background cues, such as when the graticule, is not drawn on a map. While the graticule is typically just reference information in the background, and not the primary focus of a reader's attention, they are very important to identifying spatial relationships. Not only can the graticule be used to identify some basics of projection distortion (see Olson 2006), but it also provides a key cue to compare locations across a map. Because maps are flat, and there are many common cylindrical projections that present the lines of latitude and longitude as straight lines meeting at right angles, it is possible that a map reader will intuitively assume that locations along a horizontal or vertical line will have the same latitude or longitude value. This will not always be true for any projection with *curved* lines of latitude and/or longitude (Figure 5.15). Note that in the cylindrical projection used in Figure 5.15A and Figure 5.15C,

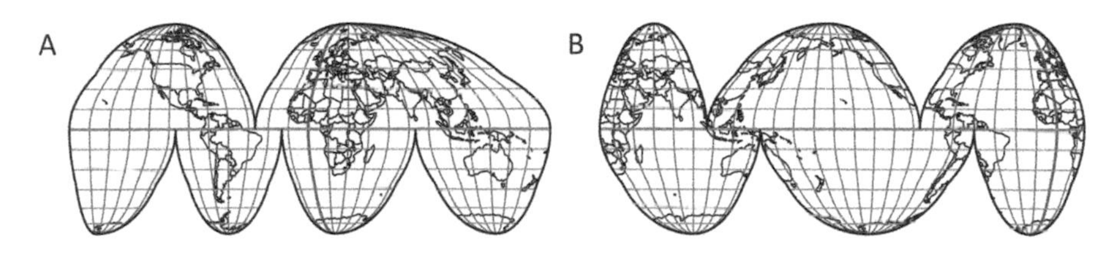

FIGURE 5.14
Goode homolosine interrupted equal area pseudocylindrical emphasizing land areas (A) and water areas (B).

the marked locations shown as red stars appear horizontally inline with each other, at the same latitude. Even with the graticule removed Figure 5.15C) it is fairly apparent that the started locations are located on the same latitude. The conic projection in Figure 5.15B with a graticule for reference shows the same locations shown by red stars are located at the same latitude even though the starred location further east on the map is "higher" up on the map. A reader would only be able to easily interpret these two locations as being at the same latitude if a graticule is provided because the locations do *not* lie along the same horizontal position on the map. This problem is easily seen in Figure 5.15D, where the graticule has been removed. This challenge of deciphering locations without a graticule for a reference can impact interpretation of patterns (e.g., assuming two weather stations are at the same latitude and misinterpreting the general relationship of latitude and temperature), and the formation of cognitive maps of the world (e.g., incorrectly aligning regions along a line of latitude or longitude).

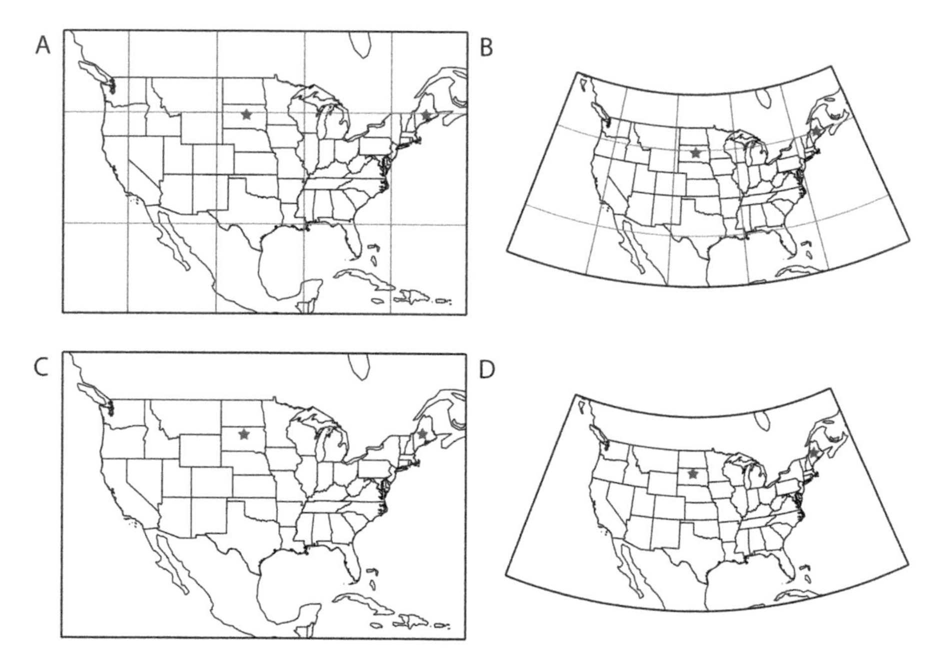

FIGURE 5.15
Which of the marked locations is located farthest north? They are both located at the same latitude: however, due to the projection, the placement on the map may shift one farther "up" than the other. With the graticule, it is clear that the latitude is the same for both cities (A, B). Without graticule (C, D) the map may give the impression that the locations are at a different latitude. For reference, the projection for the maps on the left (A, C) is a web Mercator and the projection for the maps on the right (B, D) is Albers equal area conic.

Conclusion

In this chapter, we addressed different approaches that have been developed to select a projection. Some of these approaches have been brief recommendations that match a specific map purpose to a projection property: Thematic maps should be based on equal area projections. A more detailed projection selection guideline was developed by Snyder (1987). His hierarchical selection guideline recommended a projection based on the geographic extent of the landmasses to be mapped, the projection property, geographic extent of the data, location of constant scale, and other criteria. In the end, his hierarchical guideline presents one or more suitable projections from which to choose. While this nonautomated selection guideline provides some assistance, it still requires the map maker to have a modicum of projection knowledge to work through the hierarchy, make appropriate decisions, and narrow down a single choice from a list of possible projections. Several automated decision-making tools have been developed to assist in the recommendation of a suitable projection. These systems generally replace much of the map maker's decision-making that was involved when using a nonautomated projection guideline. In this chapter, we highlighted the projection decision-making tool by Šavrič et al. (2016) called "Projection Wizard." Projection Wizard aids in selecting projections for small- or large-scale mapping projects. This approach allows the user to interactively experiment with various parameters and controls in order to select an appropriate projection.

Many maps require the preservation of a specific property. Some map purposes focus on determining areas of landmasses or measuring distances. Projection properties allow for these and other metrics to be quantified. In such cases, the projection property should be paramount in guiding the selection process. Despite the importance one can place on preserving the projection property, it is often necessary to consider the influence of distortion across the map and how that distortion will impact other facets of the map (e.g., the aesthetics of the map design). Aside from the desired property to be preserved, selecting a projection also involves considering the geography of the landmass, its location, size, and shape when evaluating the trade-offs of different projections. As a general rule, the region to be mapped should be placed central to the map's extent. This central placement of the mapped landmass generally aligns with the projection's region of lowest distortion. This placement can be controlled by specifying the location of the projection's central meridian, central latitude, and standard lines. The size of the landmass's size can be problematic when selecting a projection. Continental- and global-scale maps experience more distortion than larger-scale maps or maps of countries. To help in refining the projection choice, we recommend using tools to help visualize distortion across the projection's surface. This book used Geocart to create informative visualizations of distortion patterns on many projections illustrated throughout the chapters. Consideration of the landmass's shape and orientation can also be an important factor when selecting a projection. As a solution, the projection's aspect can be altered so that it can be aligned to the predominate extent of the landmass. This alignment usually brings the landmass's shape into the projection's zone where distortion is at a minimum.

There are four additional considerations when selecting projections that were addressed in this chapter. First, the overall shape of the projection (circles, ovals, cones, or rectangles and their variants) should fit into the available space on the page or screen so that the amount of empty space can be avoided. For example, while an oval projection fits into a rectangular frame the "corners" of the frame are empty. In other situations, such as web mapping where tiled maps are required, there may not be flexibility in choosing a projection. We also pointed out that projections come in novel shapes that can give an eye-catching appeal to a map (e.g., stars, hearts, icosahedrals, and so forth). Second, the arrangement of the graticule can be a consideration in the selection of a projection. The graticule will appear as either straight or curved lines. Including the graticule on the map can be helpful to map readers as they try to determine the distortion pattern, preserved property, or measure distances. Third, how the poles are represented on the map may also help guide selecting a projection. Poles can be represented along a continuum, from points to lines the same length as the equator. If the poles are represented as points, then the upper latitudes experience compression. On the opposite end of this continuum, if poles are represented as lines, landmasses in the upper latitudes can be stretched east–west. Collectively, this compression and stretching can negatively impact the appearance of the landmasses and data patterns in the upper latitudes which may be detrimental to map readers carrying out their map reading tasks. Fourth, concerns over interpretability, or the influence that projections have on interpreting patterns in the data should also be considered. Interpretability focused on the advantages and disadvantages associated with interrupted projections and the utility that the graticule provides map readers when relating their cognitive maps to spatial patterns. Interrupted projections can present challenges when trying to carry out specific map reading tasks that cross an interruption. If the map reader is unfamiliar with the geography of the map and data, including the graticule on the map may provide a spatial context to assist in interpreting the map.

Part II

Projection Selection by Map Type

In Part I of this book, we focused on fundamentals of map projections, including the projection process, guidelines for projections selection, and how projections may influence data representation and interpretation. Each of these earlier chapters addressed general principles that lay the groundwork for selecting projections. In Part II of this book, we will demonstrate and apply the principles from Part I in exploring projection selection for specific map types (e.g., choropleth, dot density, isarithmic, etc.).

General Mapping Challenges

While Part I of this book detailed principles of selecting map projections, we wanted to also provide a quick summary and guide for the general challenges that a map maker faces in selecting a map projection. In the subsequent chapters we will dig in in more detail specific to different map types, but we will start here with a set of the general concerns that should be addressed when selecting a projection for any map, regardless of the map type or symbolization method.

1. **Select your projection based on general geographic location and extent.** (See Chapter 2 for review of projection parameters and their selection)

This section includes consideration of how the geographic location (e.g., latitude location) impacts the projection aspect, how the scale (e.g., global- or local-scale) impacts the projection class, and how the extent of the mapped area (e.g., the map is "wide" and runs more along an east—west axis, or is "taller" and covers more of a north—south axis) impacts the location of standard lines and distribution of distortion throughout the map.

2. **Select your projection based on interpretation task and aesthetic needs for the finished product**. (See Chapter 4 for review of how projections can influence data representation and the map reader's interpretation)

   The overall shape that each projection class possesses can impact the landmasses' appearance. Whether a projection's shape is rectangular, oval, circular, or something else, can visually alter the landmasses' appearance and change the location of the symbols used to represent the data. This shape can impact the symbols and the data patterns to appear congested or expanded compared to their true locations on Earth.

   The projection's center can bring a specific region and the symbols representing the data from that region to the map's center. This centering may be advantageous so that an important region is brought to the map reader's attention. But, in so doing, this centering can cause other symbols to be pushed toward the map's periphery where distortion is often greater.

3. **At the global scale, always consider the impact of the map periphery, central meridian, and interruptions on data visualization and interpretation.**

   While some aspects of projections can be controlled through careful selection of projection class and parameters, there are additional challenges that will be true for any map at the global scale regardless of projection selected.

   Map periphery. With any global-scale dataset it is important to consider how selection of projection parameters impacts the placement of regions of interest around the map: what is centered in the map, what is on the periphery, and what is cut-off or interrupted at the periphery and continues on the other side of the map? It is easier for a map reader to make comparisons between regions when they can be viewed in their entirety—as opposed to having to combine two parts of the region on different sides of the map. Figure II.1 shows the impact of several different selections of the central meridian parameter; note how, and by how much, different regions are impacted by the adjustment to projection center. It is also easier to identify patterns when the map symbols are not cut across the

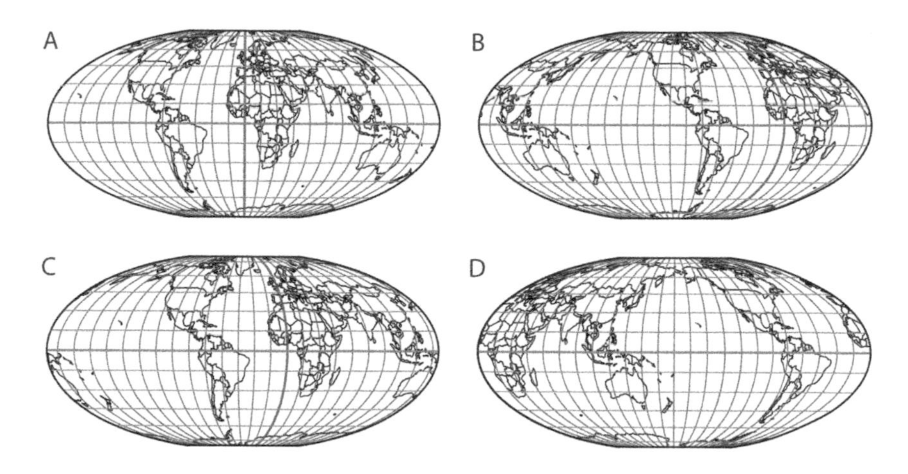

FIGURE II.1
Mollweide equal area pseudocylindrical projection centered on 0° (A), United States-Centered, 100° W (B), Atlantic centered, 40°W (C), and Pacific-centered, 180° E (D). At least one country is split at the periphery with each of these selections of central meridian.

periphery (when possible). While comparisons across the periphery can be a challenge in any map, the impact is particularly notable with a few specific map types. To explore, we provide more detail on the impacts on isarithmic maps in Chapter 7, dot maps in Chapter 8, and flow maps in Chapter 9.

Central meridian. If your map is designed to emphasize comparisons between one, or a few, specific locations, and the remainder of the mapped locations, it is important to prioritize central placement of primary regions of interest and to minimize the potential for their interruption on the periphery of the projection. For instance, if you were mapping GDP per capita for countries around the world and wanted to encourage comparison to the countries in central Africa, you would want to select projection parameters that bring those regions into the center of the view, while balancing the impact on the regions on the periphery. A central meridian of 20°E centers the map on the African continent, but at the expense of Alaska being chopped in two. Instead, a centering at 0° keeps Africa roughly centered while minimizing the impact on the periphery (Figure II.2).

4. **Interruptions.** Interrupted equal area projections, such as Goode homolosine equal area pseudocylindrical (Figure II.3), limit distortion to regions by adding breaks, or interruptions, into the mapped area. This splits the map into several lobes and introduces discontinuities between segments of the map. While the interruptions can be beneficial in minimizing angular distortion and may improve the perceived accuracy of shape for regions, they may also make it more difficult for map readers to assess distance-based spatial

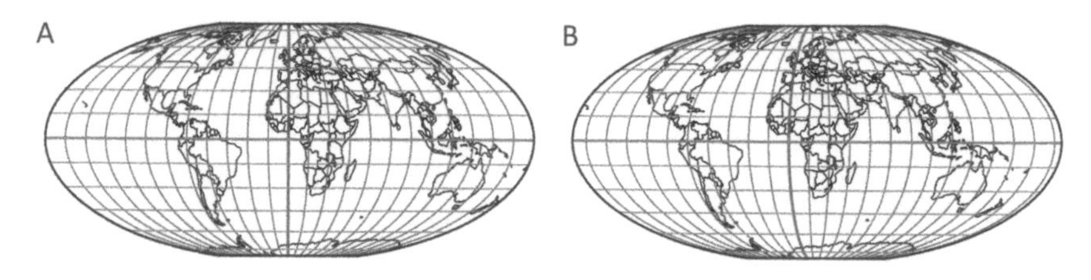

FIGURE II.2
Mollweide projection with a central meridian of 0°(A) and 20°E (B).

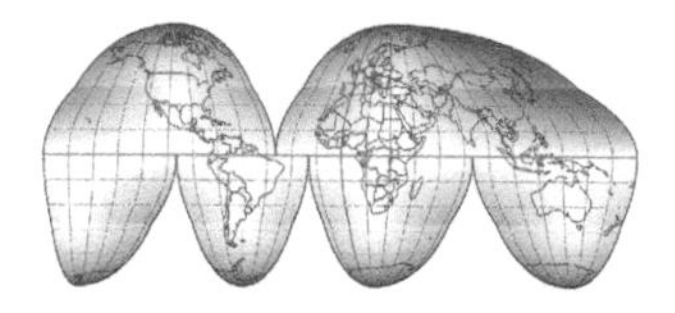

FIGURE II.3
Goode homolosine interrupted equal area pseudocylindrical projection with interruptions placed to minimize distortion in and areas. Angular distortion is shown in magenta; lighter shades = less distortion and darker shades = more distortion.

relationships, such as finding clusters of high or low values or measuring distances between locations, if the regions of interest are separated by an interruption. To explore, we provide more detail on the impacts on isarithmic maps in Chapter 7, dot maps in Chapter 8, and flow maps in Chapter 9.

Book Structure

Because the focus in this book is on map projection issues and not cartographic practice in general, we will not go into detail on the process of creating each map type, or the trade-offs related to specific design decisions—however some of this is covered in the earlier chapter on map design and cartographic principles if you are in need of a refresher (Chapter 3). There is also extensive discussion available throughout the cartographic literature and online should you desire more background information on any of the map types that we discuss.

We have used MacEachren and DiBiase's (1991) model of geographic phenomena to guide the organization of our discussion in Part II of this book—focusing on the different map types that are relevant for phenomena that occur discretely vs. continuously, and that change abruptly vs. smoothly. Using MacEachren and DiBiase's model, we have broken down the chapters to address the following map types:

- Choropleth maps (continuously occurring and abruptly changing)—Chapter 6
- Isarithmic maps (continuously occurring and smoothly changing)—Chapter 7
- Dot maps (discretely occurring and smoothly changing)—Chapter 8
- Proportional point and flow symbol maps (discretely occurring and abruptly changing)—Chapter 9.

Additionally, we include a chapter on "Special Maps" (Chapter 10) where we address projection selection for a variety of map types and map reading tasks that do not fit neatly into the categories defined by MacEachren's model.

In each chapter in this section, we provide a broad overview of a specific map type (or multiple types, in the case of the "Special maps" chapter) and detailed discussion of key topics for selecting appropriate projections for the map type. For each map type, we include an overview of the visual analysis tasks that map readers typically need to accomplish to help guide educated selection of projections that allow the best combination of distortion pattern and intended task. In many cases, there are multiple good projection choices for any given map type, however, the intended task will be the ultimate deciding factor in narrowing down to the right projection for narrowing down to the right projection for *your* map. Through the discussion in each chapter, we provide the tools you need to make educated and confident decisions about projections for your mapping projects.

We assume that most map readers will not go through the chapters sequentially but will target the information more appropriate to the mapping task at hand. To accommodate this reading style, we have designed each chapter so that it can stand by itself as a reference. Because of this, there will be some overlap in the discussion between chapters, for instance, when covering the visual or analytical impact of specific projections or projection classes. Since there are a relatively small number of commonly used map projections, particularly at the global-scale (see the list in the Appendix), our discussion of application to different map types will address the relevance of many of the same projections, however, will have detail tailored to explaining their appropriateness for specific map types and map reading tasks.

6

Continuously Occurring and Abruptly Changing

In this chapter, our focus turns toward selecting map projections for representing continuously occurring and abruptly changing phenomena. These phenomena are typically represented in a choropleth map, which is, arguably, the most commonly used of the map types that we cover in this book. Choropleth maps are made from sets of enumeration units with a single value assigned to each polygon in the set.

The type of dataset used for these maps shows a spatial distribution of phenomena with values occurring continuously across space, but where the values change abruptly at specific boundary locations. For example, if you have data for the gross domestic product (GDP) per capita (Figure 6.1) for countries across the world, the dataset has a single value for every country (polygon), that single value is assumed to be the same value across every location within the country (values occur continuously), and the value shown on the map will only change at the border of each country (abrupt change).

The enumeration units used in a choropleth map are frequently based on political boundaries; however, other measurement areas with set boundaries may be used as well (e.g., sales territories or plant growing zones). This differs from the polygonal boundaries created with filled contour maps, which are used to show phenomena that change smoothly and continuously (e.g., isarithmic mapping; see Chapter 7). In filled contour maps, the boundaries of polygonal regions are defined by the data values, as opposed to the boundaries in a choropleth map, which are set locations defining the data collection area.

Since choropleth maps are designed to visualize patterns and make comparisons across filled polygon areas, the projection distortion impacts are largely based on how well area is preserved in regions of interest. However, the projection choice will ultimately be driven by the map scale and the set of specific visual analysis tasks for which the map is used. The primary considerations are:

- Equal area projections will help facilitate direct visual comparison of area covered in different locations of interest.
- Compromise projections that balance distortion to area and angular relationships are also generally acceptable choices.

- Extreme areal distortion in non-equal area projections can lead to visual overemphasis of regions' importance in the spatial distribution of attributes on the map and hinder ability for map readers to quantify relative and absolute values across enumeration units.
- For local-scale mapping, the recommendation of sticking to equal area projections can be relaxed, provided the projection selected uses appropriate parameters so that areal distortion is minimized across the local-scale mapped area.
- For choropleth maps with expected tasks that involve distance measurements in additional to general comparison of areal patterns, the goal should be to minimize areal distortion while preserving distance from the one or two primary points of interest on the map.

Throughout the chapter, we will discuss these challenges in more detail and provide examples to demonstrate the impact of projection on choropleth mapping.

Visual Analysis Tasks

The predominant visual analysis tasks for continuous and abruptly changing phenomena are to identify values at specific single locations of interest, compare values across multiple locations, and identify broad regional patterns through visual aggregation of groups of neighboring values. In comparing patterns across a choropleth map, a map reader is often looking for regions with visually distinct, differing values (e.g., where are there clusters of relatively high or low values, or values that appear substantially

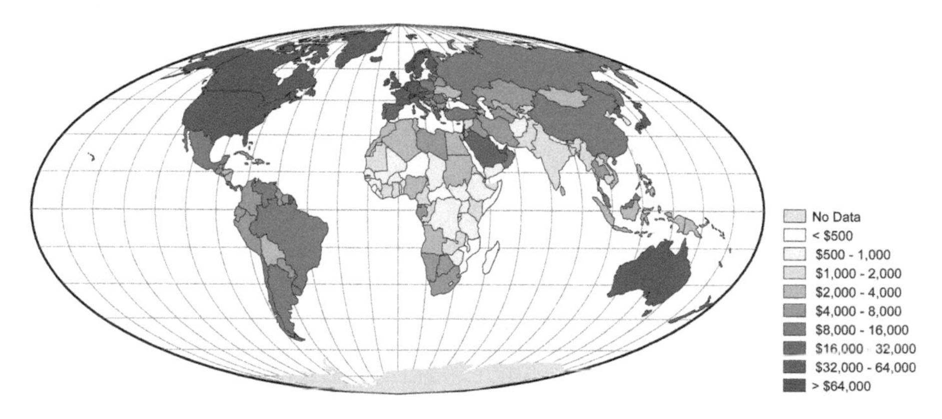

FIGURE 6.1
Per capita nominal GDP for countries (2015). Data source: United Nations Statistics Division.

different than their neighboring regions). Finding the borders that separate these regions may be done at local scale (e.g., find neighbors that differ) all the way up to at the entire-map scale (e.g., visual identification of the border of the "high value" region versus the "low value" region).

While much of the evaluation of choropleth maps is based on visual identification of values for map regions, the underlying task for making sense of the meaning and spatial distribution of the values for these regions is to assess and compare the area that they cover and the distance between the regions (or between areas of interest). For instance, being able to identify that a larger portion of the map is covered by one value compared to another, or that the higher-value regions for an attribute appear to be farther away from the major cities than the lower-value regions.

While it is possible to find choropleth maps with a sole purpose of helping map readers simply identify single values for specific regions (e.g., a doctor using a world map showing the risk potential for being infected by the Zika virus, where they only need to look at specific countries where a patient has visited), the majority of choropleth maps are used for making comparisons across locations in a single map or between two or more maps. In making comparisons between locations on the map, the map reader is assessing variation in the represented value (the color or pattern of fill), as well as the shape and size of regions with values of interest. There are numerous projection-related challenges that can help or hinder interpretation of differences in value across locations in choropleth maps. The next section will discuss these challenges with respect to specific projection characteristics.

Impact of Projection on Choropleth Maps

Equal Area Projections

Interpretation of choropleth maps is largely based on comparison of values for large areas on the map or across multiple regions. To make valid comparisons, the regions need to have the same relative areas (i.e., not grossly enlarged or reduced in area due to projection distortion). Because of this, for choropleth mapping of large areas (medium to small scales) we recommend equal area projections for several reasons.

Minimize differences in visual importance of enumeration units. The enumeration units in a choropleth map are rarely the same shape and size; there is nothing that can be done to change this fact of geography. From a visual perception standpoint, the difference in size means that some enumeration units will be more likely visual targets—the larger the area, the more likely a map reader is to focus on it. Representing regions using an equal area projection ensures that the visual difference in size is not further exacerbated by areal distortion in the projection. Consider the relative size of Russia when using an equal area projection versus a non-equal area projection (Figure 6.2; Table 6.1). In either map, the region appears to be very large compared to other countries—after all, it is ~17 million square kilometers. However, in the non-equal area projection (Mercator; Figure 6.2A) the difference in relative area is visually more noticeable due to the extreme exaggeration of areas in the high latitudes. In the equal area projection (Mollweide; Figure 6.2B), it appears more appropriately proportioned to the other regions on the map, since the projection preserves relative area.

Facilitating direct comparisons between enumeration units, and ensuring accuracy of "backward transformations." A key interpretation task in choropleth maps is to make comparisons across locations. When mapping area-based data (e.g., population density), if area is not preserved on the map, it is impossible to make valid direct comparisons between locations. Consider the hypothetical problem of comparing a density value between

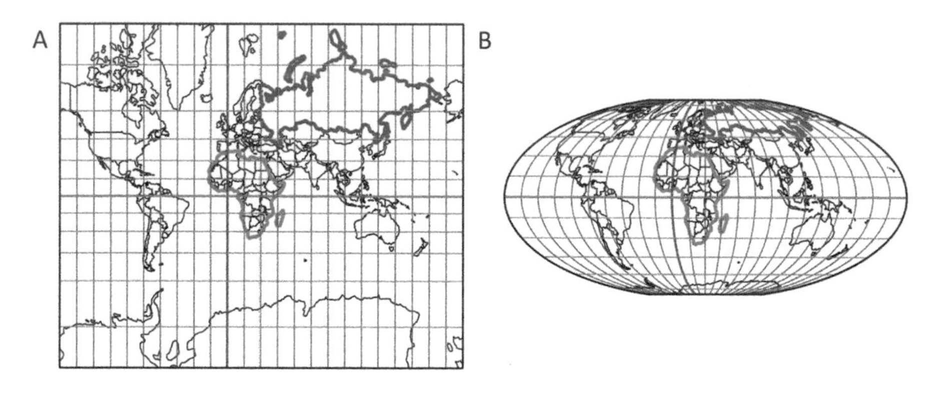

FIGURE 6.2
Visual importance of regions will vary significantly between projections that preserve and distort relative areas. Note the relative size of Russia (outlined in red) versus the continent of Africa (outlined in blue) in the Mercator conformal cylindrical projection (A) compared to the Mollweide equal area (B) projections.

TABLE 6.1

Comparison of Areas for Russia and Africa in Equal Area (Mollweide) and Non-Equal Area (Mercator) Projection

	Area as Measured Using Mercator Projection (km²)	Area as Measured Using Mollweide Projection (km²)
Russia	82,930,996	16,962,483
Africa (not including Madagascar)	32,709,228	29,516,528
Africa as % of Russia	**39.4%**	**174.0%**

(Area calculations based on natural Earth 1:10 m vector Admin 0 (countries) dataset and calculated using Esri's ArcGIS. With the Mercator projection, Russia becomes visually dominant due to areal distortion.)

countries. Figure 6.3 shows simplified choropleth maps to emphasize the challenges that projection choice may introduce for direct comparisons. If the orange shade on the map represented a matching density measure for the number of Tribbles per square kilometer (an entirely made up attribute) in both Sweden and Morocco, would it be possible to identify which country we expect to have more Tribbles? To complete this task, a map reader would compare the density values reported for each location and then visually compare the geographic area for each location to do a mental backward-transformation to identify if one location is expected to have more or less of the attribute. Intuitively, people assume that densities and counts are related directly based on the size of the enumeration unit as shown on the map, not the actual land area. In the case of Sweden and Morocco, the actual land areas are roughly equivalent (both are around 450,000 km^2), but the projected land areas are substantially different. This difference in area due to the projection distortion would lead to widely differing interpretations of the number of Tribbles in each country (Table 6.2).

More accurate for data preparation. With respect to the data on a choropleth map, it is also important to consider the projection used when normalizing data counts by area. If a geographic information system (GIS) is used to generate the areas used in the data normalization, it is imperative that an equal area projection is used to calculate the areas. This ensures that relative land areas are correct. If a non-equal area projection is used, the resulting calculations will be incorrect and, depending on the projection, may be complete nonsense. Imagine what would happen if we

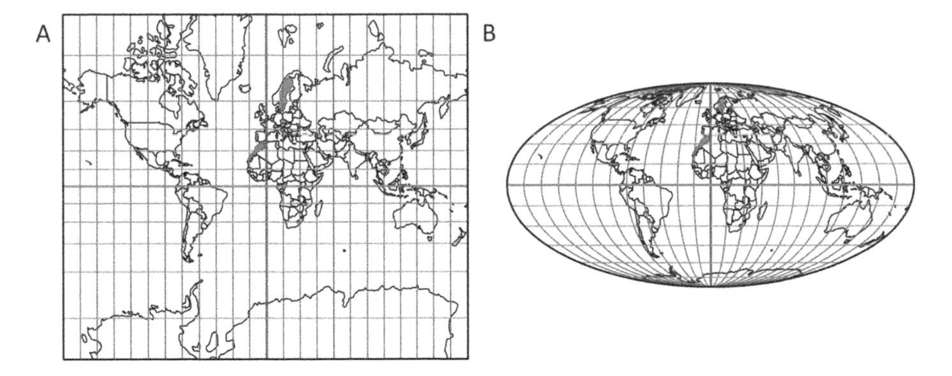

FIGURE 6.3
When comparing values by area, the relative sizes of enumeration units can lead to incorrect interpretations of value. These maps highlight the locations of Sweden and Morocco as an example. These two countries have approximately the same land area. If their values for Tribble density per square kilometer were the same, the non-equal area Mercator projection (A) would lead the map reader to assume that Sweden has over three times as many of the attribute as Morocco. However, if we use the Mollweide equal area projection (B), where the countries are appropriately sized and the map reader can more easily interpret that their areas are similar, the map reader would assume that the count is roughly the same in both countries.

TABLE 6.2

Actual and Projected Land Areas for Morocco and Sweden and Impact on Density vs. Count of a Mapped Phenomenon

	Land Area as Reported in CIA World Factbook (2017)		Projected Land Area (Mercator Projection)	
	Morocco	Sweden	Morocco	Sweden
Land area (km²)	446,550	450,295	790,452	2,187,138
Tribble density per km²	2	2	2	2
Est. count (area × density)	~893 k	~900 k	~1.6 m	~4.3 m
Ratio of est. counts (Morocco/Sweden)	**99.2%**		**37.2%**	

(Actual land area based on CIA World Factbook (2017) measures. Projected land area calculated in Esri's ArcGIS. Note that people are poor at estimating exact areas, so even though we've used fairly exact values for the land areas, the true estimates an individual would be expected to make for area and count of Tribbles would be approximate at best.)

used the projected land areas for Sweden and Morocco in Figure 6.3 to calculate density values; if the countries both had the same count of any given attribute, the density for Sweden would be *much lower* than that of Morocco if the Mercator area (Figure 6.3A) were used in the calculation, because Sweden's land area in the Mercator projection is significantly larger than Morocco's.

With larger-scale mapping projects, for instance, at the individual country- or state-level, the impact of area distortion as part of a calculated field is less significant. If you need to create your own density measure and the attribute table for your dataset doesn't already have an area field (with presumed accurate values), you will need to calculate area yourself. Most, if not all, GIS software packages or services have this capability. However, you must be careful to ensure that the calculation is based on an appropriate projection. The impact of projection on density calculations is shown in Table 6.3.

TABLE 6.3

Densities as Calculated Based on Actual and Projected Land Areas for Morocco and Sweden

	Land Area as Reported in CIA World Factbook (2017)		Projected Land Area (Mercator Projection)	
	Morocco	Sweden	Morocco	Sweden
Land area (km²)	446,550	450,295	790,452	2,187,138
Count of Tribbles	1 mil	1 mil	1 mil	1 mil
Density (count/area)	**2.2 per km²**	**2.2 per km²**	**1.3 per km²**	**0.5 per km²**

(The areal distortion in Mercator leads to widely differing density calculations.)

The unit for the coordinate system is an additional challenge to consider for calculating area-based measures. Consider calculating area for countries of the world using a dataset represented in what Esri's ArcGIS refers to as "Geographic Coordinate System" (GCS). The coordinates in this "projection" represent *degrees*. What would the area measure be for a region using these coordinates? Square decimal degrees! Fortunately, many GIS packages are smart enough to not allow for calculation of area using decimal degrees as the unit, but ultimately it is your responsibility to pick an appropriate projection and to know the geographic unit (miles, kilometers, meters, etc.) for those results!

Conformal Projections

Conformal projections preserve local angles at the expense of relative area, so for global-scale choropleth maps this means that it will be difficult, if not impossible, to make accurate comparisons of area covered by any given attribute. The impact of not preserving areas means that we see a "spreading" or "contracting" effect where relative distance between points increases or decreases in locations across the projection. As we noted in the earlier section discussing the benefits of equal area projections, and demonstrate in the previous section, this distortion will also lead to errors in visually assessing area-based density measures and counts for attributes.

Unfortunately, the near-conformal web Mercator projection is the projection of choice for many web mapping services. It is important to consider the impact that use of web Mercator may have on how map readers will interpret patterns in your maps (e.g., making invalid estimation of counts or densities, over- or under-estimation of amount of area on the map covered by a particular attribute, etc.).

Equidistant Projections

Map readers rely on distance between locations to help identify clusters of similar value. While identifying these clusters is an important part of interpreting patterns on choropleth maps, it is not the top priority to consider when selecting a projection for a choropleth map. Because an equidistant projection will only preserve distances from one or two specific locations, it will still not provide a valid distance measurement between all locations that map readers are likely to compare. Equidistant projections will also introduce areal distortion, which is more likely to be detrimental to the map reader than the distortion of distance.

Compromise Projections

For small-scale choropleth maps, the next best alternative to an equal area projection is a "compromise" projection. These projections preserve no specific property, but often are designed for a balance between distortion and aesthetics. For instance, Figure 6.4 shows the distortion in two compromise projections, Winkel tripel modified azimuthal (Figure 6.4A) and Robinson pseudocylindrical (Figure 6.4B), which have been used frequently by National Geographic for small-scale mapping. Because these projections are designed to limit, but not eliminate, both area and angular distortion, they often present minimal areal distortion that would hinder interpretation of patterns in a small-scale choropleth map. In both instances, the areas with extreme angular and areal distortion (the darkest regions on each projection) are isolated to the polar regions.

Large-Scale Projections

Even though we recommend specifically working with equal area or conformal projections that are near-to-equal area for choropleth maps, as is the case with many projection problems, if you are working with a small enough area (local-scale mapping), the impact of the distortion can be mitigated through selection of appropriate parameters (e.g., standard lines and central meridian) appropriate for the region. For instance, in mapping the conterminous United States, an Albers equal area conic or a Lambert conformal conic will have a similar appearance and present minimal areal and angular distortion across the main region of interest (Figure 6.5). While the Lambert conformal conic (Figure 6.5B) shows notable distortion in the southern portions of the map, the areal distortion is not excessive and would sufficiently minimize areal distortion to be used for choropleth mapping. For context on the level of distortion, compare the distortion visualization in Figure 6.5B for the Lambert conformal conic to the level of distortion in the global-scale version of the projection (Figure 6.6).

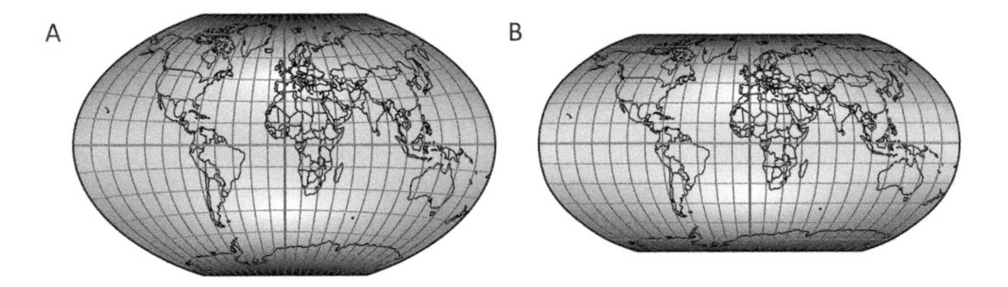

FIGURE 6.4
Distortion pattern for the Winkel tripel modified azimuthal (A) and Robinson pseudocylindrical (B) projections. In both projections, angular distortion is shown in magenta, while areal distortion is shown in green. In both cases, lighter shades represent less distortion and darker shades represent more distortion.

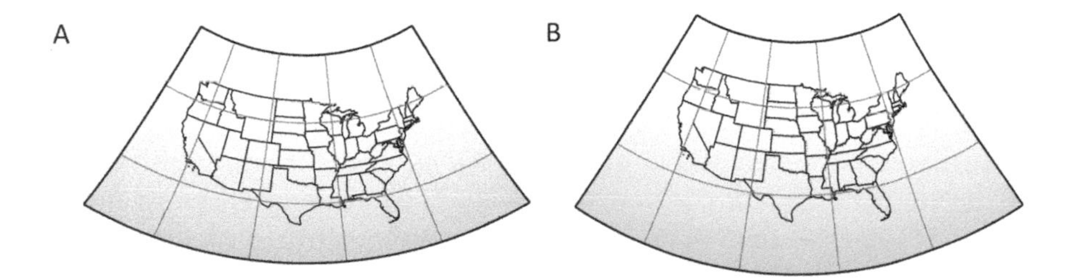

FIGURE 6.5
Distortion for the Albers equal area (A) and Lambert conformal conic (B) projections tailored for the continental United States. The Albers equal area conic projection uses two standard lines at 29.5° and 45.5°. The Lambert conformal conic uses standard lines at 33° and 45°. Both have a central meridian at −96°, and a latitude of origin of 23°. Angular distortion is shown in magenta, while areal distortion is shown in green. In both cases, lighter shades represent less distortion and darker shades represent more distortion.

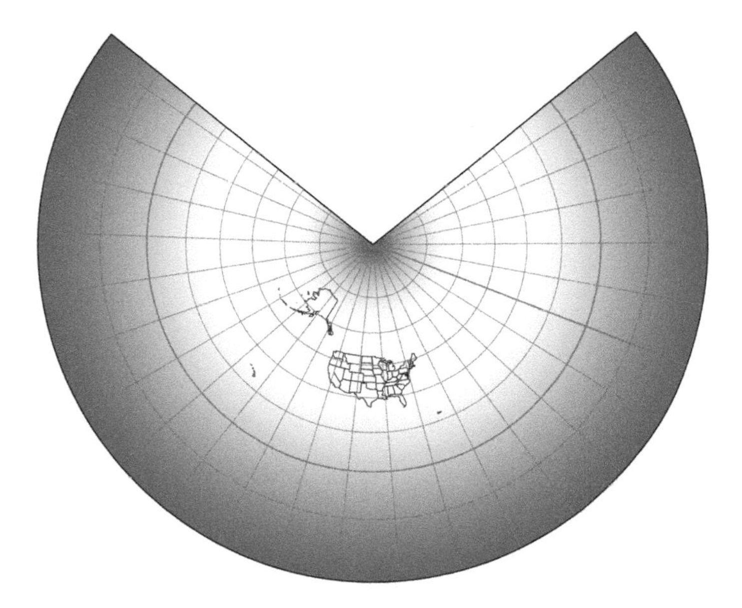

FIGURE 6.6
Lambert conformal conic projection tailored for the conterminous United States. Areal distortion is shown for the entire world. Areal distortion is shown in green; lighter shades represent less distortion and darker shades represent more distortion.

Conclusion

In this chapter, we have examined the implications of projection on choropleth maps (a map type for depicting continuously occurring and abruptly changing phenomena). To address the most common visual analytical tasks supported by choropleth maps, equal area projections are generally a good choice. An equal area projection will help facilitate direct visual comparison of area covered in different locations of interest. Compromise projections can also be acceptable choices for global-scale mapping projects, as they balance distortion to both area and angular relationships. For local-scale mapping, our recommendation of equal area projections can be relaxed, provided the projection selected uses appropriate parameters so that areal distortion is minimized across the local-scale mapped area.

7

Continuously Occurring and Smoothly Changing

In this chapter, our focus turns toward selecting projections for phenomena that are considered continuous and smooth. These phenomena exist at every point location, which makes the data continuous in nature. The variation between the phenomenon values changes smoothly across space. For example, sea surface temperature could be considered continuous and smooth, as there is a discrete temperature value everywhere on the surface of Earth's oceans, yet specific temperature values change smoothly from place to place. Because it is impossible to collect data for every possible location, data for continuous and smooth phenomena are collected by sampling a finite set of values and recording each coordinate location. A spatial interpolation method is then applied to the data values, creating a surface representation of the underlying phenomenon, which can then be symbolized by one of several methods. It is the values in this interpolated surface that are then mapped. The interpolated surface representation is referred to as an isarithmic map. Isarithmic maps symbolize data with isolines, which are lines of equal value. The regions between isolines are considered to vary in value, although within a region the values are bounded by the isolines (the lines of equal value) on each side. These isolines might be referred to with terms specific to the map—for example an isoline showing time is an isochrone, showing temperature is an isotherm, and showing rainfall is an isohyet. In some maps, hues or color values are added to the isolines or the areas between them to help distinguish the general trend in the data, as shown in Figure 7.1. Maps with shading between the isolines are sometimes referred to as "filled contour" maps.

Isarithmic maps may be based on either true or conceptual point locations for the phenomena they represent. An example of a *true point location* is a weather-recording station. At each station, a specific, measurable data value (e.g., barometric pressure) is collected along with the exact coordinates of the station. Isarithmic maps made with isolines that connect this kind of data are called *isometric*. An example of a *conceptual point location* is the centroid of an enumeration unit, such as from a county or state. At each centroid, a single data value represents the aggregation of some attribute, such as population, measured across an enumeration unit. Isolines are then drawn through the centroid of each enumeration unit, creating a surface of the aggregated population data. Maps that are created using this kind of conceptual data are called *isopleth*.

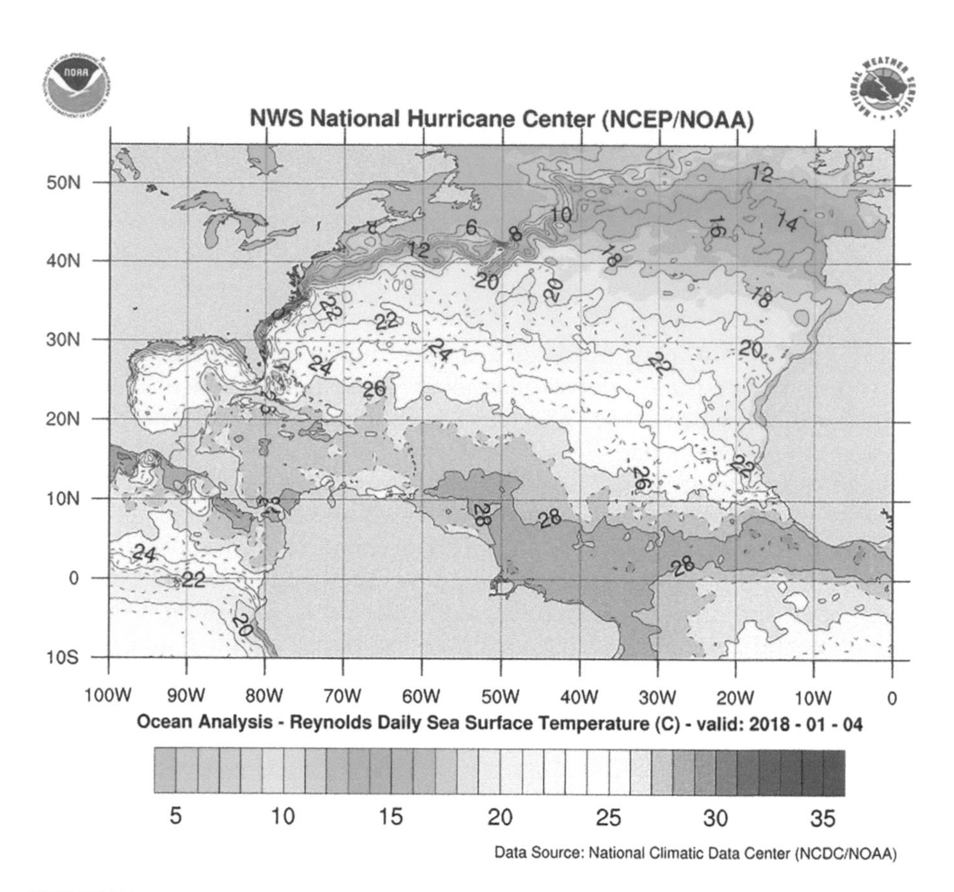

FIGURE 7.1
Isarithmic map of ocean sea surface temperatures. (Reprinted from National Oceanic and Atmospheric Administration and National Weather Service, in National Hurricane Center Reynolds SST Analysis, n.d. Retrieved December 28, 2018, from www.nhc.noaa.gov/tafb/sst_loop/14_atl.png.)

In this chapter, we will use two different datasets (based on true point locations) to demonstrate how projection choice can influence the creation, appearance, and interpretation of isarithmic maps. The first dataset is a collection of 13,241 oil wells drilled in the North Sea. This dataset will be used to show the impacts that scale distortion in a projection has on the *distance calculations* associated with spatial interpolation methods. The second dataset is global sea surface temperature, which we use to illustrate the impact of projection on the *visual tasks* associated with reading isarithmic maps.

As we see in other examples throughout this book, distortion due to projection will vary depending on specific characteristics of the projection as well as the map scale (e.g., local scale, global scale, etc.). Therefore, the choice of projection should inform the choice of symbolization method. With isarithmic maps, there are a few key projection concepts to consider:

- Many interpolation methods rely on distance measurements to estimate data values at point locations. These estimates are then used to create the isolines that represent the continuous smooth surface. Distance measurements on a projection will be distorted in some places and thus will reduce the accuracy of the estimated data values. Unfortunately, no projection preserves distances throughout a map. While equidistant projections, for example, preserve distances, they do so only in limited ways (e.g., from one or two set locations or along meridians).
- To avoid the consequence of projection distortion when measuring distances, spatial interpolation methods should measure distances on a reference ellipsoid instead of a projected surface (see Chapter 2 for discussion of reference ellipsoids). Reference ellipsoids are accurate models of Earth's surface and are accessible for use in interpolation methods built into most GIS environments.
- Despite the advantages of measuring distances on a reference ellipsoid, it is not advisable to use one when displaying the results of an interpolated surface. The reference ellipsoid only expresses latitude and longitude values rather than projected Cartesian coordinates. We recommend projecting your data for visualization.
- A primary goal of isarithmic maps is to show how an interpolated surface changes across a mapped area using, for example, the spacing of isolines. Projection distortion can alter the visual distance between isolines. This can lead to misinterpretation by the map reader of how the mapped values change, and possibly faulty estimation of data values. Therefore, the choice of a projection should be made in accordance with the anticipated goal of the individuals reading the map.

Visual Analysis Tasks

The predominant visual analysis tasks for continuous and smooth changing phenomena are estimating values at specific locations on, or between, isolines on the map, understanding the dataset's overall spatial trend with respect to how quickly the phenomena change across space, and comparing the coverage and pattern of values across different regions on the map.

For estimating approximate values for locations on isarithmic maps, the map reader can simply look at the "band" in which the value falls. For instance, a point located in the light green area in Figure 7.1 would have a temperature value between 16° and 18°C. For more precise estimates, the map reader would need to perform a visual interpolation process of estimating values based on distance between two isolines. Projection-related distance distortion would impact this ability to approximate true value based on the distance from an isoline.

When using isarithmic maps to evaluate the overall spatial trend, the spacing of the isolines is used to interpret how quickly the phenomenon being mapped changes across the mapped area. Closely spaced isolines suggest a quicker change (e.g., greater change of value over a shorter geographic distance) to the data values than more widely spaced isolines, which suggest a slower change.

Isarithmic maps may also be used to estimate how much of Earth's surface area is encompassed by a specific value. For example, to compare how much of the map is covered by regions estimated with higher than average yearly precipitation totals, and how much is covered by regions with lower than average precipitation. Or, a map reader may wish to do this across multiple isarithmic maps of the same region to show temperatures at different time periods. A map reader may want to compare the area covered by the highest temperature isotherms across the maps to identify change (e.g., the growth of an area impacted by a heat wave).

A final potential task is in interpreting distance between two values of interest. For instance, determining how far a specific amount of pollution has spread from its origin by noting the distance between a point on the isoline of interest and the source of the pollution. Or, a map may show the extent of a pollution plume across an area of interest where the pollution amount is symbolized by isolines.

Spatial Interpolation: An Initial Step in Isarithmic Mapping

To better frame the challenges when working with and selecting projections for isarithmic maps, we will separately discuss the impact of projection on the *creation* of the interpolated dataset and on the *interpretation* of the map data.

To really understand the impact of projection on isarithmic mapping, we think that it helps to start with the initial step of turning discrete data into a surface through spatial interpolation. Spatial interpolation starts with a point dataset of a smooth and continuous phenomena. Then, one of several interpolation methods can be selected for use. Inverse distance weighting (IDW) and kriging are two common ones. Generally speaking, spatial interpolation, regardless of specific method selected, uses a mathematical formula to estimate a data value at an unknown location. As a simple example, Figure 7.2 shows four points where known data values exist (as shown in the parentheses). A simple way to estimate an unknown value at a given location (indicated by "?") is to average all four of the surrounding values; doing so

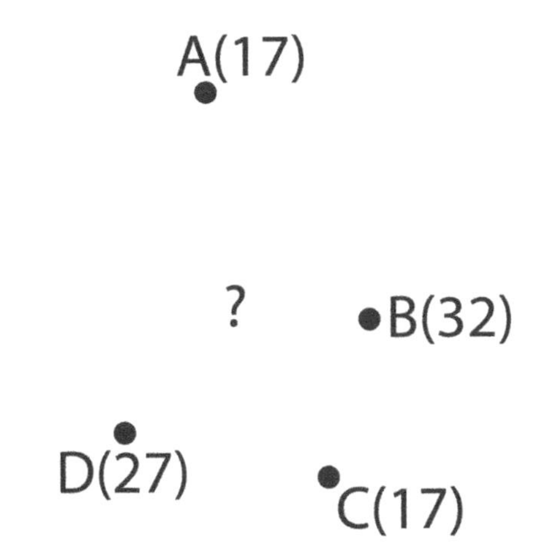

FIGURE 7.2
Four individual point locations, their associated data values, and a point, "?", whose value is to be estimated.

results in a calculated value of 23.25. However, by using averaging as our calculation method, each of the four data points is contributing equally to estimating the data value at "?" which may or may not be very realistic across geographic space. This idea is reflective of Tobler's (1970, p. 236) First Law of Geography where "everything is related to everything else, but near things are more related than distant things." Considering Tobler's assertion, a different approach is suggested where we can more heavily weight the closer values in the calculation.

Different interpolation methods can incorporate different parameters into their equations, such as varying distance between points. Figure 7.3 shows the same four points as Figure 7.2, but also includes the distances between the four known points and the point to be estimated. The underlying assumption is that points closer to "?" will contribute more to the estimated value than points that are farther away. This idea is central to the inverse distance weighted (IDW) spatial interpolation method shown in Equation 7.1.

$$Z_n = \frac{\sum_{i=1}^{n} Z_i / d_i^k}{\sum_{i=1}^{n} 1 / d_i^k} \tag{7.1}$$

Where

Z_n = the estimated value to be determined
Z_i = the data value at the point used in making the estimation
d_i = the distance between the points
k = the exponent to which the distance is raised
n = the number of points used to estimate the point in question

Inserting the distances from Figure 7.3 into Equation 7.1 with $n=4$ points and exponent $k=1$ produces the following equation:

$$Z_n = \frac{\left(\frac{17}{2.0}\right)+\left(\frac{32}{1.0}\right)+\left(\frac{17}{1.45}\right)+\left(\frac{27}{1.25}\right)}{\left(\frac{1}{2.0}\right)+\left(\frac{1}{1.0}\right)+\left(\frac{1}{1.45}\right)+\left(\frac{1}{1.25}\right)}$$

Note that point B (32) is geographically closer to "?" than point A (17) and, therefore, influences the estimated value to a greater extent. Finishing the calculation, we obtain a value of $Z_n=24.69$, which is more reflective of the underlying surface where distance has an influence on the estimation process.

Remember, however, that projections may distort distance measurements. Thus, when performing spatial interpolation that relies on distance measurements, it is important to choose an appropriate projection that minimizes the distortion for the dataset being used.

Impact of Projections on Calculating Values for Isarithmic Maps

There are five general caveats to the impact that a projection can have on a distance-based spatial interpolation method.

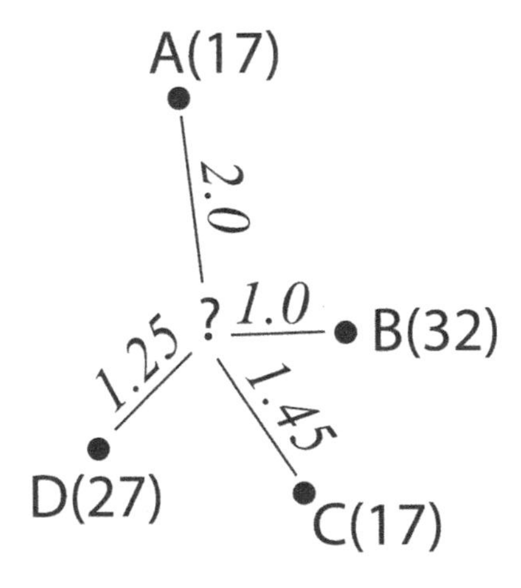

FIGURE 7.3
Showing the distances between each point (in italics) and the point to be estimated at "?".

- Spatial interpolation is a method that uses several parameters to estimate a value. When using any spatial interpolation method, it is unlikely that you will ever know what the true value is at a point location that is to be estimated. Even though we used distances to calculate estimates for, even in a distortion-minimized projection we can never truly know which estimate is correct. While you can ground-truth the values at some locations, this will still only provide an estimate of the accuracy of your interpolation, as it is impossible to collect values for every location in your area of interest.
- To ensure that interpolated estimates are as accurate as possible, an appropriate datum (see Chapter 2 discussion on reference ellipsoids), or the azimuthal equidistant projection (ellipsoidal form) should be used. Using either approach will reduce errors in the distance calculations and produce estimates closer to the true value. Recall that calculated distances on a reference ellipsoid are more accurate than those carried out on a spherical Earth model.
- Before deciding on a suitable projection for spatial interpolation, it is important to fully understand the way that any given projection handles scale distortion and note what parameters need to be set in order to minimize distortion across the mapped area. For instance, placing the center of the azimuthal equidistant projection directly over the location to be estimated allows distances to be accurately measured from the unknown point to any other point and, thus, improve the accuracy of the estimated value. Had the projection been centered elsewhere, the measured distances would not have been accurate. Ideally, the center of the azimuthal equidistant projection should be automatically moved to each location with a value to be estimated.
- Another important consideration is the geographic extent of the dataset. Some datasets have a narrow longitudinal extent but cover a considerably greater range of latitudes, and vice versa. Some datasets are more compact than others (related to the ratio of area of the shape to area of circle with same perimeter; circles are considered most compact). Regardless of the geographic extent of the dataset, you should be careful to set the projection parameters so that distance distortion is minimized. As the extent of the geographic area increases, the ability to control distortion for a specific purpose, such as measuring distances, becomes more challenging.
- After the interpolation is completed, you will still need to select a projection that appropriately illustrates the interpolated surface or fulfills a specific map purpose.

Distance Measurements

Scale, extent, and location become factors when selecting a projection, because projection affects distance calculations for spatial interpolation. To more deeply examine these impacts with interpolation, we will walk through an example using a point dataset of oil wells drilled in the North Sea. The dataset used in this example was sourced from the United Kingdom's Oil and Gas Authority.

Figure 7.4 shows the geographic extent of the dataset. When assessing data extent to select an appropriate projection, consider the spatial aspects of the data:

- Latitudinal extent: In the case of the oil well dataset, we have a latitude range of approximately 14.6° (47° 52' 35" N and 62° 28' 46" N) but the geographic location is in the upper latitudes.
- Longitudinal extent: Here, the longitude extent is rather narrow of approximately 7° (11° 54' 6" W and 4° 21' 18" W).
- Property of interest: Distance is an important consideration in spatial interpolation, making equidistant projections a good place to start. However, equidistant projections are still limited in that distance measures cannot be preserved from all locations to all other locations.

Using our oil well dataset, we will create an isarithmic map of surface depth to oil. We will use IDW as the spatial interpolation method to estimate unknown depth values at additional point locations. To simplify and explore the impact of different projection choices, we will focus on a small sample of the data (Figure 7.5) to estimate the value for a single point location (symbolized by "?") based on three of the closest surrounding points. The value for each point in Figure 7.5 includes the depth, in meters, from the oil drilling platform to the underground oil bed. Coordinates for each point are also provided.

As previously discussed, many spatial interpolation methods calculate distances between points in the estimation process. Since no projection can preserve distances throughout a map, this makes it difficult to select an ideal projection. However, we will discuss several approaches that you can use to minimize the distance errors during spatial interpolation. In the next sections we will discuss the results of distance calculations using a reference ellipsoid (International Ellipsoid, 1924[1]), grid system (UTM Zone 30N), two equidistant projections (azimuthal equidistant planar and plate carrée cylindrical projection), and one near-conformal projection (web Mercator). To provide context for the following discussion, Table 7.1 shows the measured distances between "?" and A, B, and C on the reference ellipsoid, grid system, and projections to be discussed.

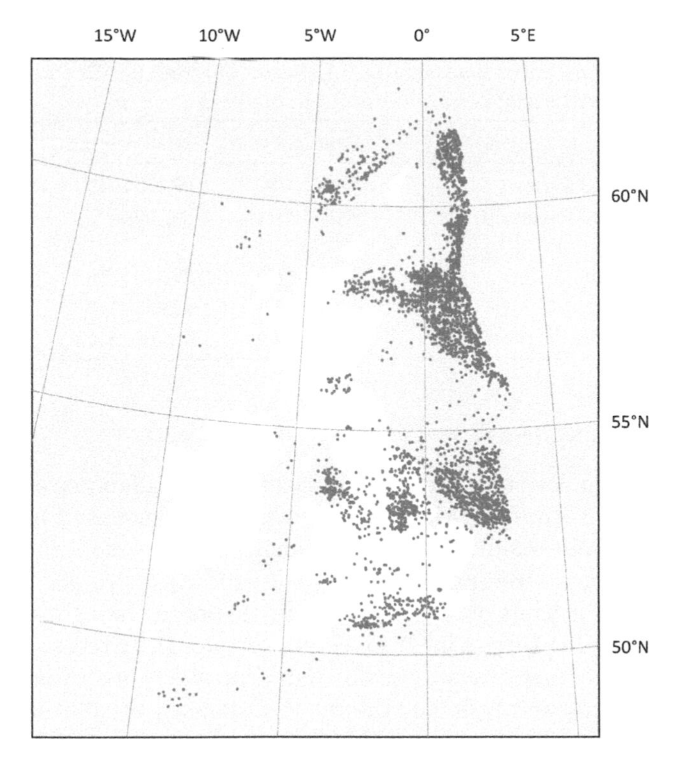

FIGURE 7.4
The spatial distributions of oil wells in the North Sea. Data source: United Kingdom's Oil and Gas Authority (www.gov.uk/guidance/oil-and-gas-offshore-maps-and-gis-shapefiles#offshore-gis-data).

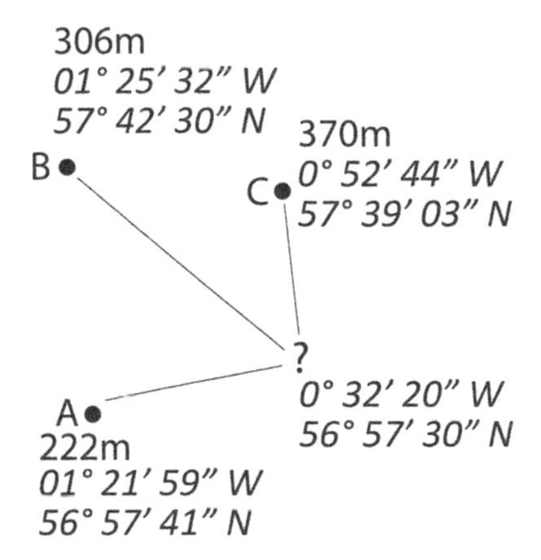

FIGURE 7.5
Estimating the depth of oil for a location where no sampled value is available ("?"). To estimate the value, we use three of the closest surrounding points (A, B, C) for which the depth values are known. The coordinate values of all points are shown in italics.

TABLE 7.1

A Comparison of Distances between Each Lettered Oil Well (A–C) and the Point to Be Estimated "?" (All distances are reported in meters.)

Name	Distance to A	Distance to B	Distance to C
International Ellipsoid 1924	50,342.7980	99,130.7710	82,041.0508
Universal Transverse Mercator (UTM) Zone 30 N	50,332.3591	99,109.4475	82,030.5159
Azimuthal equidistant	50,342.7980	99,130.7710	82,041.0508
Plate carrée	92,114.7875	72,453.4177	82,032.5996
Web Mercator	92,116.1793	126,696.6919	151,614.5827

Datum and Grid Systems

Before continuing, let's consider the utility of using a reference ellipsoid and grid system for minimizing distance distortion. As discussed in Chapter 2, a reference ellipsoid is a model of Earth's shape and size, and measurements carried out using a reference ellipsoid will produce more accurate results than using other models or a projection (e.g., a sphere). In a similar manner, grid systems like the Universal Transverse Mercator (UTM) were developed, in part, to create maps where scale distortion was low providing accurate distance measurements. For the oil well dataset, we will use both the International Ellipsoid 1924 and UTM Zone 30 North grid system as specific examples for the Earth model and grid system, respectively.

The International Ellipsoid 1924 is commonly used for data collected on and off shore of western Europe. This ellipsoid is a mathematically defined smooth model of the Earth's surface. Coordinates cast on this and other reference ellipsoids are not subjected to distortion resulting from the sphere-to-plane conversion process that is common with projections, so they can be considered "true" to the distance as it would be measured directly on Earth's surface. For our example here, the International Ellipsoid 1924 is relevant to the location; however, it is good to know that other reference ellipsoids are defined for many regions in the world and Earth in its entirety,[2] so there should be one relevant for wherever your data is located.

The UTM Zone 30 North grid system is tailored to the geographic region where the oil wells are located. Note that this grid system is based on the transverse Mercator conformal cylindrical projection and divides the world into a series of zones that are each 6° wide. Since the underlying Mercator projection is conformal, it preserves angles and ensures that scale is constant but not accurate in every direction about each point. However, scale will vary from location to location according to the latitude. The UTM system, referred to simply as UTM, was designed for local-scale mapping, and is advantageous with predominantly north–south oriented datasets that are contained within any individual UTM zone. If your area of interest is predominately east-to-west and is greater than 6° of longitude or your area is split between two UTM zones, then the UTM may not be the best choice for your mapping purpose. For our example here, this grid system seems like a good choice as the point locations conveniently fit near the center of UTM Zone 30 where distortion is minimized (less than 1% in the area of the oil wells) and will allow greater accuracy in distance measurements.

In addition to looking at the impact of reference ellipsoids and grid systems on distance calculations, we also consider the impact of projections and their specific properties: equidistant, conformal, and equal area. For the equidistant property, we look at the azimuthal equidistant and plate carrée projections, and for the conformal property, we look at web Mercator. While equal area projections have utility in thematic mapping, they do not preserve distance, making them inappropriate for this task, so we will not consider them further in this discussion on isarithmic mapping.

Equidistant Projections

To explore the impact on interpolation, we examine two equidistant projections: azimuthal equidistant and plate carrée. The azimuthal equidistant projection has a tangent point that specifies the projection's center. This tangent point can be assigned to any pair of latitude and longitude values. From that central location, a measured straight line to any other point on the map produces an accurate distance (Figure 7.6).

As seen from Table 7.1, all measurements from "?" to points A, B, and C using the azimuthal equidistant projection result in identical distances as calculated from the International Ellipsoid 1924. The reason the distance measurements between the azimuthal equidistant and the International Ellipsoid 1924 are identical is due to the appropriateness of the parameters defined for the azimuthal equidistant projection. Had the selected tangent point been located other than at the point to be interpolate ("?"), the distance values would have not matched those measured using the International Ellipsoid 1924. The placement of the tangent point is particularly important to note here—in order for the azimuthal equidistant projection to accurately be applied to the entire oil well dataset, the projection's center would have to be centered on each unknown point to be calculated, in essence creating a new projection aspect for every point. At the present time, we are unaware of any mapping software that allows for this automatic centering at every point from which a distance is to be measured.

To show that all equidistant projections are not created equal, let's explore the plate carrée (Figure 7.7). The equidistant property on the plate carrée preserves accurate distances only along the meridians—as opposed to the azimuthal equidistant projection's preservation of distance from the point of tangency to any other point on the map. On the plate carrée projection, scale distortion is constant along any line of latitude, but scale distortion varies with latitude. This means that when a distance is measured exactly east-to-west on this projection, the scale distortion is constant (e.g., 3%), but that

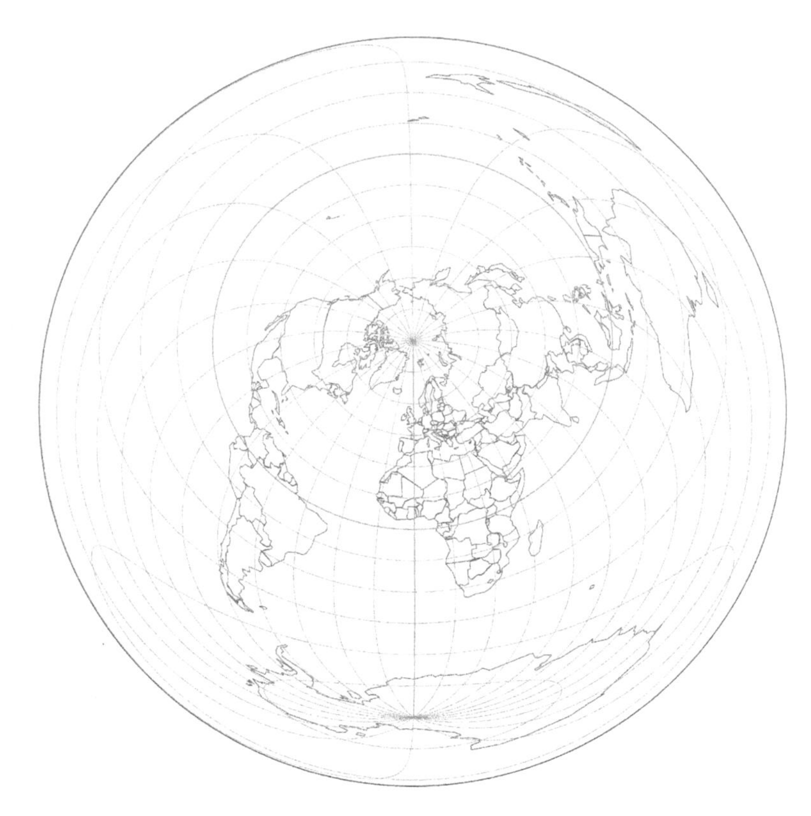

FIGURE 7.6
The azimuthal equidistant projection centered on "?," the point to be interpolated (0° 32' 20" W and 56° 57' 30" N).

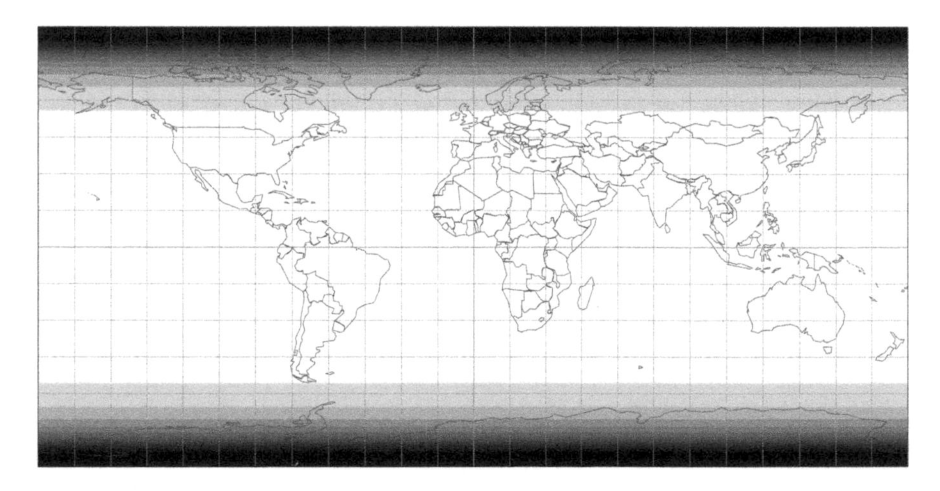

FIGURE 7.7
Scale distortion on the plate carrée projection. Darker shades of gray represent more scale distortion.

value will be different for other latitudes. Other projections exhibit scale distortion along a line of latitude or longitude and changes can be rather abrupt. For example, the scale distortion along the central meridian of the Robinson projection changes from roughly –9% (scale is compressed relative to what is on Earth's surface) at the equator to over 11,000% (scale is stretched relative to what is on Earth's surface) at the poles. Thus, it is important to understand where your dataset is geographically located, where the zone of low-scale distortion is, what parameters control that zone, and how it can be brought into alignment with the spatial extent of the data.

In Figure 7.7, note that the lines of latitude are equally spaced along every meridian, which suggests that north–south distances are true. As distance measurements depart from a strictly north–south alignment, or are measured away from the equator, distortion becomes more of a factor. This distortion is noticeable in Table 7.1. While the distance from "?" to point C is very close to the distance found on the International Ellipsoid 1924 ("?" and C are nearly along a north–south line), the distance between "?" and point A (along an east–west line) is closer to the exaggerated distance measured with web Mercator projection. The distance between "?" and point B (along a diagonal line) is exaggerated but not as severely as that measured with web Mercator. The difference between the distance measurements on these two cylindrical projections is due to their underlying properties, their inherent distortion patterns, and the resulting graticule spacing.

Conformal Projections

The web Mercator projection has been popularized through its application in various online mapping services such as Google Maps and Bing Maps. Web Mercator is *nearly* conformal, which means that this projection does not preserve angular relationships, nor does it ensure constant scale around each point as found on the Mercator projection. Since both Mercator and web Mercator are cylindric, they share many of the same distortion characteristics. As Figure 7.8 illustrates, scale distortion on web Mercator increases poleward away from, and in parallel bands, to the equator. The latitude extent of the oil well dataset is located between 47° 52' 35" N and 62° 28' 46" N. This high-latitude location coincides in areas with higher scale distortion than is found along the equatorial region, which results in exaggerated distance measurements. These exaggerated measurements can be seen in Table 7.1, where the measurements on web Mercator differ from the International Ellipsoid 1924 by quite a bit—about 42,000 meters, 27,000 meters, and 70,000 meters, respectively!

If you inspect the graticule on web Mercator, it is easy to see why these measurements differ by so much. In Figure 7.8, the spacing of the parallels increases along a meridian away from the equator and are stretched to the same length as the equator. The greatest measured difference rests in the distance between "?" and C (located approximately at 57° N in Figure 7.5).

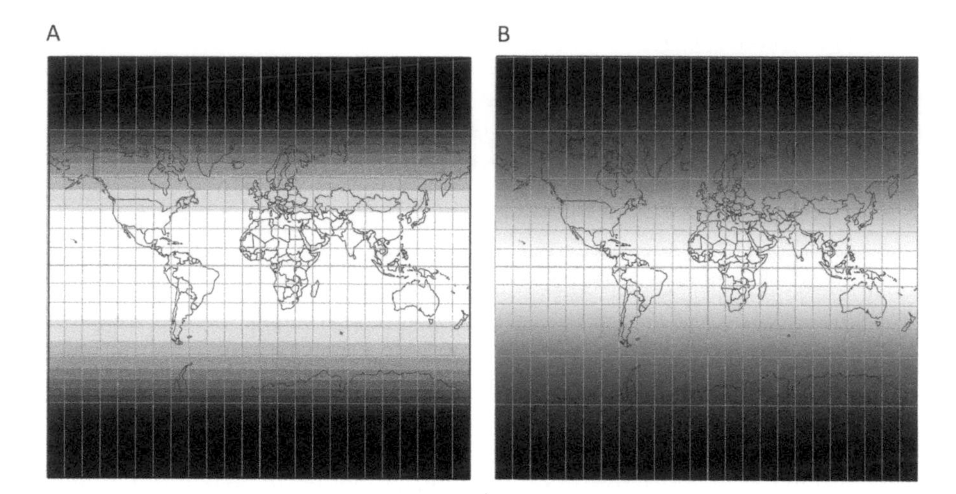

FIGURE 7.8
The web Mercator projection centered along the Prime Meridian 0°. Scale distortion (A) is shown in gray; lighter shades represent less distortion and darker shades represent more distortion. Overall distortion (B) is shown as a mix of green and magenta; angular distortion is shown in magenta, while areal distortion is shown in green. In both cases, lighter shades represent less distortion and darker shades represent more distortion.

At 60° N the scale distortion on this projection is approximately 2.0 (or an 100% exaggeration). Since the scale distortion is *nearly* the same in all direction on this projection, the measured distance from "?" to A is also exaggerated about 2.0 times (or an 100% exaggeration).

From exploring each of the projections, reference ellipsoid, and grid system in this section, hopefully it is a little clearer how the alignment of the projection distortion pattern and points used in a calculation can have a significant impact on the accuracy of any measurements made on the map.

Spatial Interpolation Results

In the last section, we explored some general projection-related issues in distance calculations. Now, let's look at the overall impact of projection distortion on the results of the interpolation process.

Table 7.2 reports the results of IDW interpolation for calculating the value at "?." The following parameters were used for the IDW equation: $n=3$ and $k=2$ (see Equation 7.1 for a refresher on these values). The impact of projection can be quite substantial!

Note from Table 7.2 that the interpolated values associated with the International Ellipsoid 1924 and the azimuthal equidistant projection are the same. Measuring distances on an ellipsoid produces a true distance. Since the azimuthal equidistant projection preserves distance measurements from the center (the "?" in Figure 7.5), we can assume that the scale distortion does not have a negative consequence on the interpolated value for this projection.

TABLE 7.2

The Results of Estimating the Value at "?" Using the IDW Spatial Interpolation Method (All Estimates Are in Meters.)

Name	Interpolated Value at "?"
International Ellipsoid 1924	331.9024
Universal Transverse Mercator (UTM) Zone 30 N	331.9031
Azimuthal equidistant	331.9024
Plate carrée	287.4883
Web Mercator	316.326

On the other hand, the value at "?" on web Mercator and plate carrée projections are both underestimated. Both of these projections are cylindrical, and we have established that locations in the upper latitudes (where the oil well data are located) suffer greater distortion than if they were in the equatorial region. Note also that the interpolated value using UTM is close to those calculated using the International Ellipsoid 1924 and the azimuthal equidistant projection. This close result should not be too surprising as the transverse nature of the UTM zone aligns the area of low-scale distortion to the area where the oil wells are located.

Impact of Projection on Design for Isarithmic Maps

So far, this chapter has discussed the impact of projection distortion on calculation of the interpolated data values represented in isarithmic maps. These distance measurements are important to the calculations associated with spatial interpolation methods. However, once data is interpolated and you are ready to design the map and symbolize the data, there are additional factors to consider. In this section, we examine how projection impacts design choices so that the final map aligns with the visual analysis tasks that map readers typically perform. The choice of projection can have a significant impact, including:

- Some projections represent the poles as points while others use lines of various lengths. When the poles are represented using points the polar regions tend to be compressed. This compression affects the shapes of the landmasses but can also cause congestion of the isolines. When the poles are represented using lines, the polar regions are stretched out, giving more visual space to show the landmasses and isolines without congestion.
- While interrupted projections offer lower distortion over the individual lobes or gores, the interruptions can be problematic for isoline

display. For example, isolines that symbolize an ocean phenomenon (e.g., sea surface temperatures) should appear continuous, and so projection interruptions should be placed on land. Likewise, for isolines that symbolize a land-based phenomenon (e.g., nitrogen levels in soil), the interruptions should coincide with the oceans.

- The geographic scale at which the final map is to be produced also factors into the selection of a projection for isarithmic maps. At larger scales, the impact of the projection on the appearance of the isolines can be minimized. As scales become smaller, the projection can have greater impact on map appearance. At global scale, it is particularly important to consider the visual appearance along the periphery of the map.
- Some projections have considerable distortion along the periphery that introduce substantial distortion to landmasses' appearance. Any isolines that extend into these peripheral areas are also subject to that same distortion. At small scales, projections with curved meridians may make it difficult for map readers to mentally connect isolines from opposite sides of the projection. This difficulty is especially true in the online environment where maps are served as tiles. Cylindrical projections, with their straight parallels and meridians, alleviate the difficulty in mentally connecting isolines from opposite sides of the map.

To illustrate these points, a dataset showing global sea surface temperatures for July 30, 2017, will be used. The dataset was sourced from the National Weather Service's Climate Prediction Center.*

The dataset has undergone spatial interpolation to create isotherms. In evaluating the impact of projection on this dataset, we will use five common projections (Figure 7.9) with different properties to help explain the impact that each projection has on the resulting global sea surface temperature dataset: plate carrée (equidistant), web Mercator (nearly conformal), Goode homolosine and sinusoidal (equal area), and Winkel tripel (compromise). Plate carrée is commonly used to illustrate global climate-based phenomena. Web Mercator is commonly used in the online maps. Goode homolosine interrupted equal area pseudocylindrical, sinusoidal equal area pseudocylindrical, and Winkel tripel modified azimuthal projections are commonly used in printed atlases to show thematic data such as sea surface temperatures. For our discussion, we highlight two specific isotherms that have been filled to help visually compare their areal extents. Consider the two areas labeled **A** and **B** filled in with orange representing a specific isotherm highlighted in Figure 7.9. Area **A** is located in the Gulf of Alaska while area **B** is located along the equator east of the Philippines. Table 7.3 shows the area

* Available at www.cpc.ncep.noaa.gov/products/GIS/GIS_DATA/sst_oiv2/index.php.

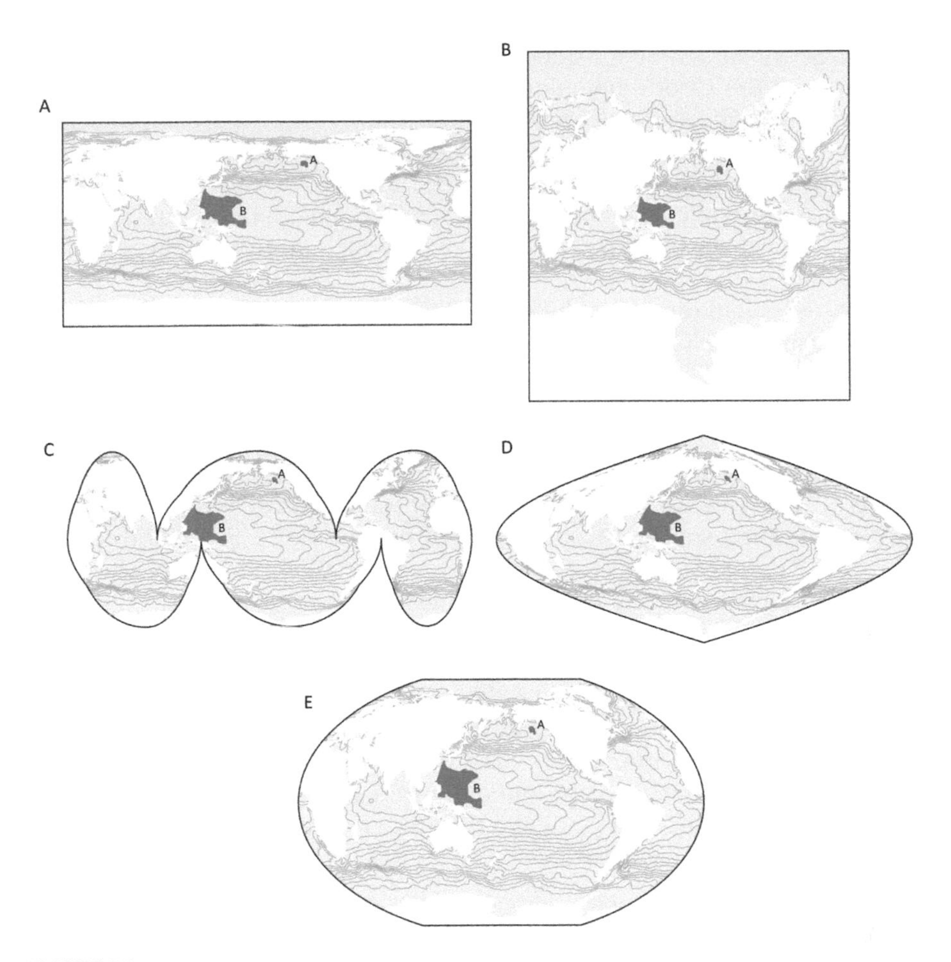

FIGURE 7.9
The plate carrée equidistant (A), web Mercator conformal (B), Goode homolosine interrupted equal area (C), sinusoidal equal area (D), and Winkel tripel compromise (E) projections. For reference in our discussion about projection distortion throughout this section, the areas of two isotherms (A and B) are highlighted in each map.

TABLE 7.3

Area Measurements for the Two Locations Shown in Figure 7.9 according to the Five Projections Listed (Plus the WGS84 Ellipsoid)

Projection	Area A (square kilometers)	Area B (square kilometers)
WGS84 ellipsoid	8,990,484	240,133
Web Mercator	9,581,360	566,014
Plate carreé	9,267,920	371,498
Goode homolosine	8,972,940	244,041
Sinusoidal	8,920,020	244,291
Winkel tripel	7,558,030	256,697

measurement for the two isotherms on the five projections. From Table 7.3 we see that the Goode homolosine and sinusoidal projections report the approximate same area values for **A** and **B**, the Winkel tripel compresses the area for location **A** while exaggerating the area for location **B**, the plate carreé exaggerates the areas of both locations, and the Mercator projection considerably exaggerates the areas of both locations. We will refer to Figure 7.9 and Table 7.2 as we discuss the details of how each projection alters the area calculation.

The area calculations for the projections shown in Table 7.3 were determined using the Calculate Geometry tool in ArcMap 10.6. All projections were based on their ellipsoidal form using the WGS84 ellipsoid. Determining the area for locations **A** and **B** directly on the WGS84 ellipsoid was accomplished through an online tool that was developed to calculate the surface area of a polygon on the WGS84 ellipsoid.* Interestingly, from Table 7.3, the area values for locations **A** and **B** determined on the Goode homolosine and sinusoidal are close but not the same. Since both projections are equal area, intuitively, their area calculations should be the same. Since we don't know the exact mathematical manner in which the areas were calculated, we can't speculate as to the discrepancies. Not surprising, their reported area values for locations **A** and **B** are different than what is reported by the online tool. The difference between the results obtained through ArcMap 10.6 and the online tool is likely to be the mathematical approach to solving the surface area on an ellipsoid. The bottom line is that before you blindly accept one measurement (whatever that measurement may be) be sure that you have an independent method to validate the returned value from software.

Equidistant Projections

Figure 7.10 shows the global sea surface temperature isotherms displayed on the plate carrée cylindrical equidistant projection. The plate carrée has one standard line that coincides with the equator and, as such, the polar areas experience the greatest amount of distortion. This distortion can result in compression, exaggeration, or shearing of landmasses or symbols. Note that in Figure 7.10, the isotherms in the polar areas are stretched out in an east–west fashion much like the landmasses. While there is notable stretching, this cylindrical projection offers a distinct design advantage for online display. Cylindrical projections are characterized by straight and orthogonal parallels and meridians. For example, in the online environment, web maps in a mapping service typically use a tile mapping service (sometimes just called "tiles"). Displaying online maps as tiles allows for smooth and seamless panning and zooming across the projection, and also allows areas of the map to be loaded quickly and held in memory as the map is panned and

---

* Available at https://geographiclib.sourceforge.io/cgi-bin/Planimeter.

FIGURE 7.10
Sea surface isotherms for July 30, 2017, on the plate carrée projection.

zoomed. To facilitate the panning process, the graticule should also be easily divided into square tiles. Projections possessing curved meridians or parallels make it difficult to create map tiles, since the tiles would not be square.

As far as visual analysis tasks are concerned, cylindrical projections offer some notable benefits. For example, since plate carrée is equidistant, the spacing of the parallels is the same as they would appear on Earth's surface. This is important when representing global-scale isotherms, especially with regard to how temperature changes with latitude. The temperature gradient (represented by the spacing between the isotherms) is reflective of that spacing as it would appear on Earth. If the mapped dataset varies by latitude, as temperatures generally do, a projection showing equally spaced parallels offers advantages to the map reader in relating a phenomenon to changes in latitude.

Conformal Projections

Figure 7.11 shows the isotherms displayed in the web Mercator projection. Though web Mercator is close to conformal, it is not truly conformal and preserves no particular property. On this projection, the equator is the standard line. Distortion increases slowly at first but more rapidly when approaching the polar regions (Figure 7.8B). This rapid change in distortion is especially noticeable in the exaggeration of size for Greenland and Antarctica. Note that we shifted the central meridian in the projection in Figure 7.11 to the 180th meridian so that the Pacific Ocean and the two highlighted areas **A** and **B** are brought into focus.

From a design standpoint, it is difficult for us to recommend web Mercator for small-scale maps. This is especially true when mapping any data that varies with latitude. The distortion in the polar regions is excessive and can give an erroneous impression of spatial patterns in the data. At larger map scales, depending on the location mapped, the visual effect may not be as dramatic as shown in Figure 7.11. With web Mercator, regions located approximately 15° north or south of the equator are mapped with relatively low distortion.

While we stated earlier that it was difficult for us to recommend the web Mercator projection for global-scale isarithmic mapping, it is important to acknowledge the wide use of web Mercator in the online environment as the basis for mapping services. Due to this prevalence, web Mercator may be the only projection option available. If you opt to use this projection, particularly for small-scale maps, you need to be aware that the map reader may make erroneous conclusions based on the distortion shown by the projection.

The web Mercator presents both advantages and disadvantages for visual analysis. Although scale distortion increases toward the poles, it remains constant along any parallel, and this projection nearly preserves scale around any given point. Given this distortion pattern, web Mercator

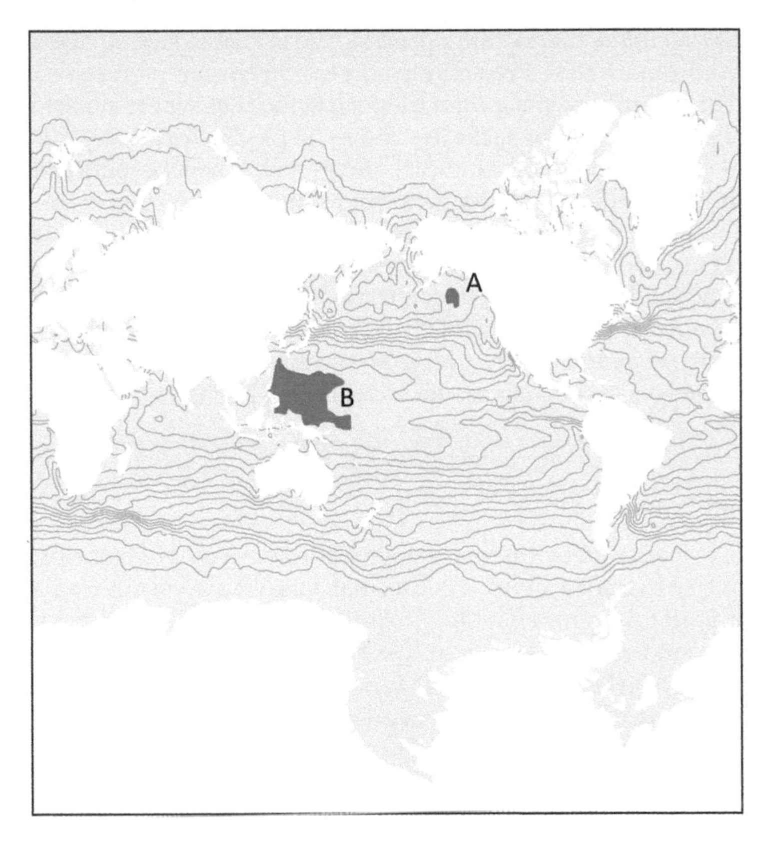

FIGURE 7.11
Sea surface isotherms for July 30, 2017, on the web Mercator projection.

can be useful for specific measurement tasks, for example, if a map reader wanted to measure the distance over which a phenomenon varied between two isotherms, or along the same isotherm. As with any measurement task on a projection, however, caution is needed. At very large map scales, the ability to measure distance is relatively well preserved since the ground areas covered is sufficiently small (i.e., the variation in distance distortion is minimal over the small region). If the phenomenon of interest trends along a parallel, the ability to measure distances can also be carried out rather accurately, provided that the map reader is equipped with the appropriate scale bar that reflects true scale at the latitude where measurements are to be taken. In most mapping services, for example, the scale bar's length will change according to the zoom level. However, as the map scale becomes smaller or varies with latitude, the ability to measure distances becomes more inaccurate—and only one scale bar is provided for the map view, without providing clear indication as to at what latitude, exactly, the scale bar is correct.

Due to the distortion patterns, web Mercator is not well suited to all visual analysis tasks. For example, compare the spread of the isolines around Svalbard in the Arctic Ocean between Figure 7.10 and Figure 7.11 (or you can see both at the same time in Figure 7.9). Since plate carrée is equidistant, Figure 7.10 shows the isotherms appearing close together as they should, indicating a rather rapid change in temperature over a short latitudinal distance. In Figure 7.11, web Mercator is nearly conformal and shows increased spacing between lines of latitude toward the polar regions. This increased spacing is evident as the isotherms appear spread out, more so than in Figure 7.10. This gives the impression that the area encompassed by a range of temperatures is greater compared to reality. In addition, the spread-out appearance suggests that the change in temperatures is not as rapid as shown in Figure 7.10.

Looking at Figure 7.11, we see the same bounded isotherms appearing on web Mercator. Note that the isotherm labeled **A** in Figure 7.11 appears to encompass a larger areal extent than shown in Figure 7.15. The isotherms bounded by **B** in Figure 7.15 and Figure 7.11 are both near the equator and both appear to have approximately the same areal extent. Because web Mercator is conformal, it does not preserve area, and because of this, it presents extreme distortion in the upper latitudes. This may lead the map reader to draw erroneous conclusions about the areal extent of the isotherm bounded by **A** in Figure 7.11.

Equal Area Projections

Equal area projections have long been the projection of choice for thematic mapping of global-scale phenomena. Figure 7.12 shows the distortion pattern on Goode homolosine pseudocylindrical projection, which is equal area. This projection uses interruptions (see Chapter 2 for discussion of interrupted projections) so that areas (such as the polar regions) are not as distorted as in many other projections. Figure 7.12 shows an ocean-focused Goode homolosine projection, where the angular distortion has been minimized over the oceans (lighter shades) at the expense of increasing distortion over the land areas (darker shades). Since the projection is equal area, there is no distortion to area depicted.

Figure 7.13 shows the global sea surface isotherms on the Goode homolosine projection. From a design perspective, this projection is eye-catching, in part due to the interruptions providing a unique look compared to many other projections. Typically, Goode homolosine is created as either

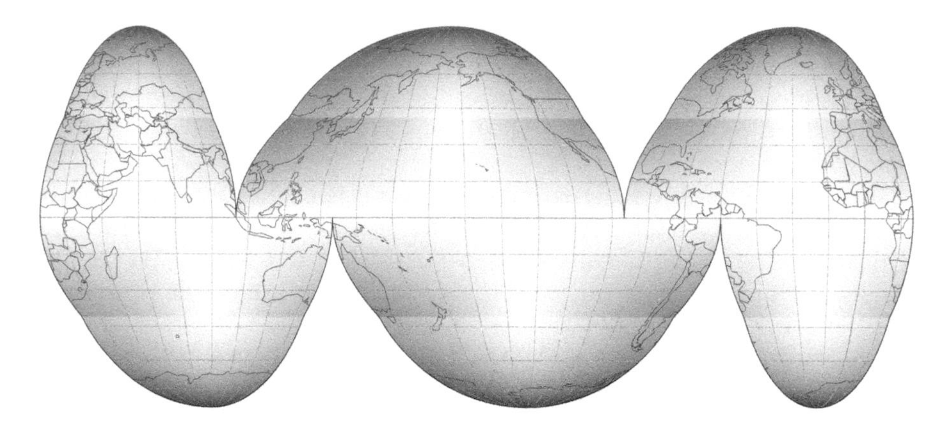

FIGURE 7.12
Overall distortion on the Goode homolosine projection. Angular distortion is shown for the entire world. Angular distortion is shown in magenta; lighter shades represent less distortion and darker shades represent more distortion.

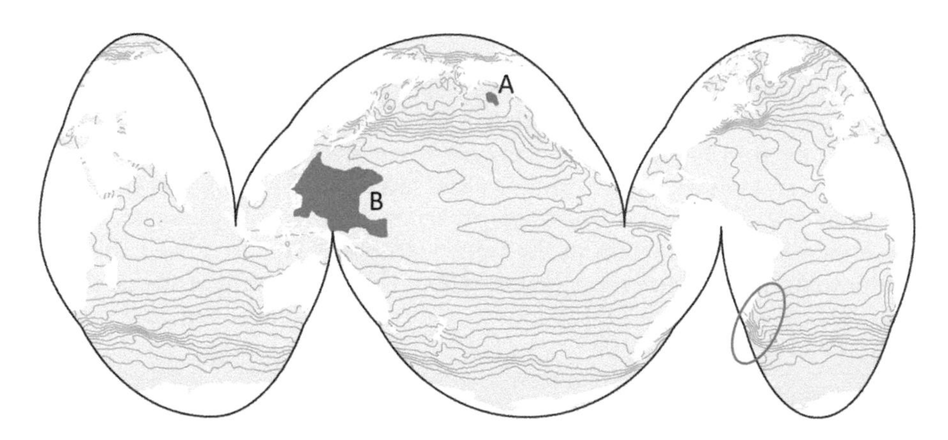

FIGURE 7.13
Sea surface isotherms for July 30, 2017, on the Goode homolosine pseudocylindrical equal area projection. The orange-circled area on the map demonstrates the congested isotherms that can result near the interruptions.

a land-centric or ocean-centric map. To highlight the isotherms across the different ocean basins, this map uses the ocean-centric version to focus the view on isotherms across the ocean basins. This highlights the sea surface temperatures in the Pacific, Atlantic, and Indian Oceans to a greater extent than on either the plate carrée or web Mercator. The interruptions also redistribute distortion so that it is minimized across the ocean regions; this is advantageous in displaying sea surface temperature.

Despite the visual appeal this projection presents, there are two major concerns with its use. First, due to the interruptions, some of the isotherms are split. This splitting of the isotherms can make it more difficult for a map reader to interpret continuity of the data, especially in the mid-latitudes and polar regions. The interruptions also may have unintended consequences of pushing data of interest into the map's periphery where distortion is greater, making it more difficult to visualize patterns in the data. For instance, the area circled in orange southeast of South America in Figure 7.13 shows a group of isotherms that become congested due to the distortion in the projection.

Second, criticism of Goode homolosine centers on the possible difficulty of understanding them due to the interruptions. For example, even though a landmass is split, the split represents a single line of longitude and not two separate lines. Moreover, some map readers may be confused that the area between the split represents extra land or water, which would lead to an overestimation of area covered by any isotherm. It can also be difficult for map readers to mentally connect isotherms across the split, particularly in areas where there is fairly rapid change in value and the isotherms are spaced close to one another.

Of course, there are many equal area projections that are not interrupted like the Goode homolosine. Figure 7.14 shows the distortion pattern on the sinusoidal projection. This projection is also equal area and there is no scale distortion along any parallel; however, this characteristic does not imply equidistance. The projection shows lows distortion along the central meridian and the equator and represents both poles as points.

Figure 7.15 shows the same set of isotherms on the sinusoidal projection. The central meridian has been moved to the 180th meridian so that the Pacific Ocean is centered on the map. From a design point of view, sinusoidal is advantageous for mapping phenomena along the equator, but not so advantageous for mapping the polar regions, where the poles are represented as points and compress the isotherms until they become almost unrecognizable (Figure 7.15). The sinusoidal projection is equal area, so areas shown on the map correspond to the same areas as on Earth's surface. This is an important consideration for visual analysis tasks, especially if (in this case) map readers examine a region bounded by a specific isotherm to get a sense of the expanse of an area coincides with a specific a temperature value.

Comparing the patterns on sinusoidal and Goode homolosine projections demonstrates an important point, even when the basic property of the

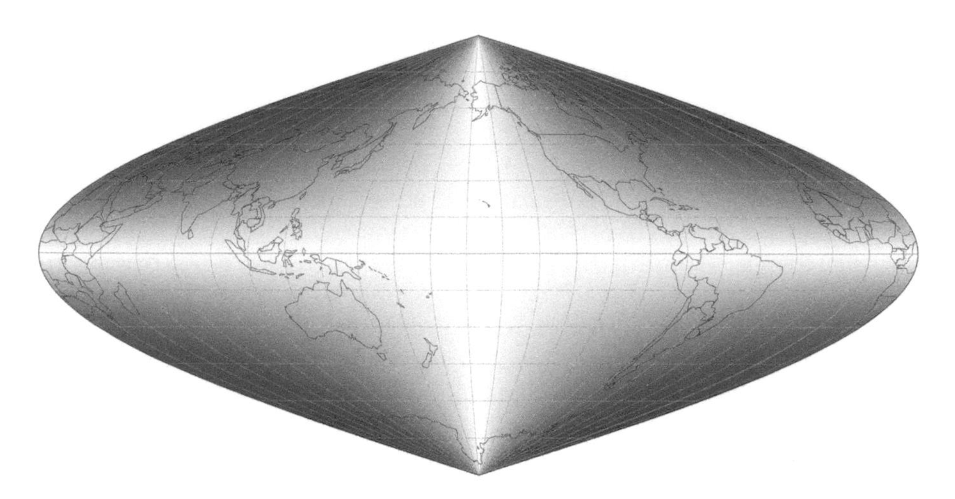

FIGURE 7.14
Overall distortion on the sinusoidal projection. Angular distortion is shown for the entire world. Angular distortion is shown in magenta; lighter shades represent less distortion and darker shades represent more distortion.

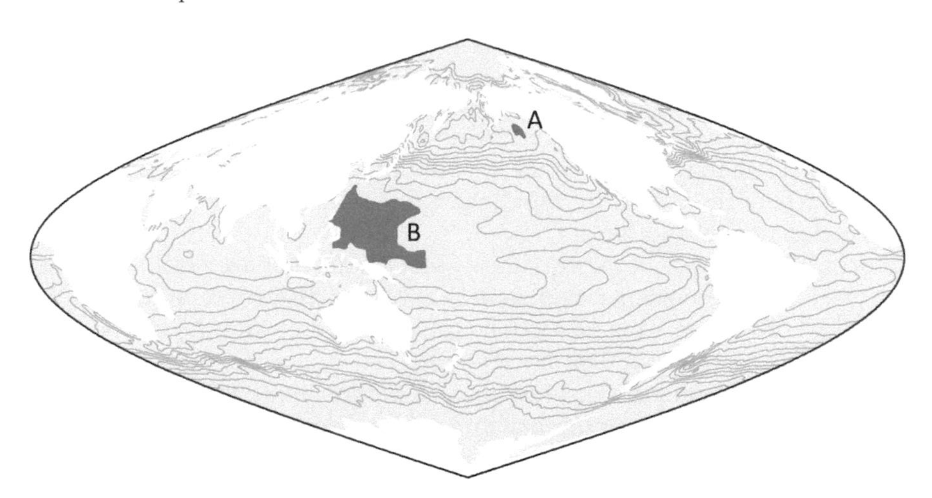

FIGURE 7.15
Sea surface isotherms for July 30, 2017, on the sinusoidal projection.

projections is the same, it is important to examine the geographic location of the more important or interesting data values and assess how the projection alters the appearance of data in this location. With the sinusoidal projection, if the important or interesting data values are located in the polar regions, the poles represented as points would hinder the map reader in identifying those data values. Generally, unless the designer wishes to emphasize isotherms that exist near the equator, the sinusoidal projection would not be a suitable candidate for global-scale phenomena.

Compromise Projections

Figure 7.16 shows the overall distortion pattern for the Winkel tripel projection. This modified azimuthal projection is considered a compromise projection that preserves no specific property. There are no standard lines on the Winkel tripel; rather, there are two zones of low distortion roughly corresponding to 45° North and South. Spacing between parallels tends to increase slightly from the equator to the poles. The poles on Winkel tripel are represented as lines 0.389 times as long as the equator, rather than the same length as the equator as shown on the plate carrée or web Mercator. Because poles are represented as lines rather than as points like on the sinusoidal projection, there is some stretching of landmasses east–west in the polar regions. But this stretching is viewed as a compromise between the severe compression on the sinusoidal and the stretching on plate carrée and web Mercator.

Figure 7.17 shows the sea surface temperature isotherms on the Winkel tripel projection. This compromise projection offers some advantages regarding the visual tasks. When represented as lines, mid-latitude and polar regions are not as compressed, which allows easier reading and interpretation of data in those areas (see the isotherms that bound **A** and **B**). Compared to sinusoidal, the meridians are gentler curves and suggest a spherical Earth. The curvature of the meridians allows the isotherms to appear more spread out, especially near the periphery of the projection and in the polar areas where

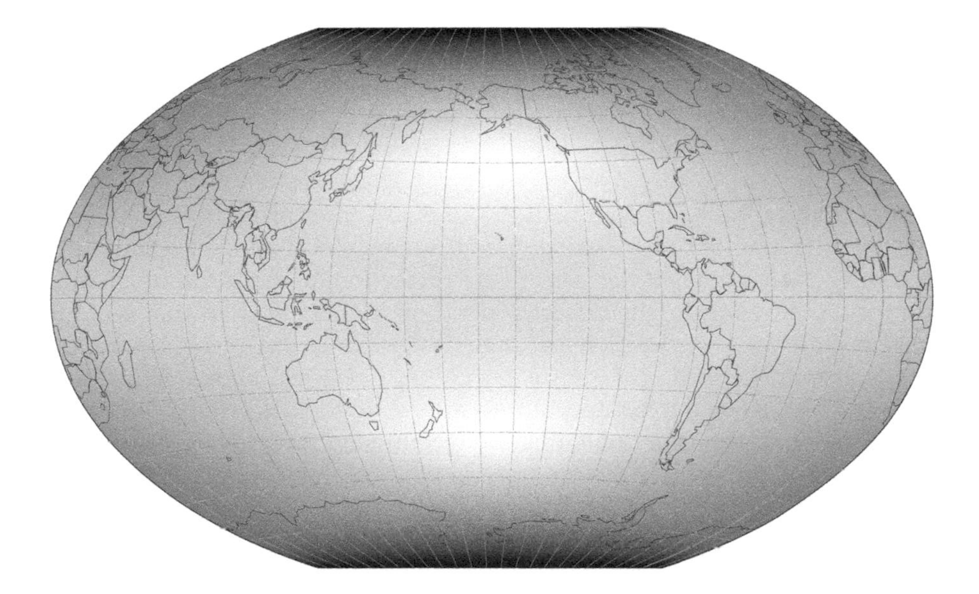

FIGURE 7.16
Areal (shades of green) and angular (shades of magenta) distortion on the Winkel tripel projection. Angular distortion is shown in magenta, while areal distortion is shown in green. In both cases, lighter shades represent less distortion and darker shades represent more distortion.

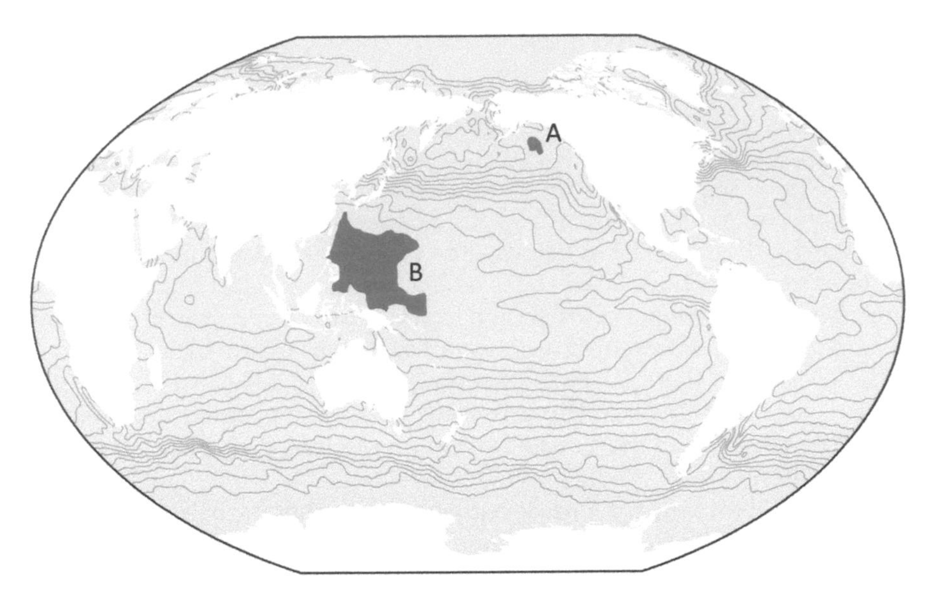

FIGURE 7.17
Sea surface isotherms for July 30, 2017, on the Winkel tripel projection.

the greatest levels of distortion are found. Map readers may have an easier time carrying out other visual tasks such as visualizing the overall trend in the temperature patterns, and interpolating values between isotherms.

Conclusion

In this chapter, we have examined the implication of projection on isarithmic maps (a map type for depicting continuously occurring and smoothly changing phenomena). When selecting a projection for an isarithmic map, it is important to consider the impact of the projection on the two phases of the mapping project: interpolation and map design. Since the interpolation process is generally based on distance calculations between known point locations in your dataset, it is important to minimize the distortion of distance across the projection. For doing this, we recommend interpolation always be done using an appropriate datum (see Chapter 2 for discussion on reference ellipsoids), or using the azimuthal equidistant projection (ellipsoidal form). When interpolating distances over a large area, keep in mind that equidistant projections cannot preserve distance measurements between all locations across the map, and typically only preserve distances from one or two set locations, or along a meridian.

When considering projections for designing isarithmic maps using a dataset that has already gone through the interpolation process, there is no specific projection type that is best for all potential tasks on an isarithmic map. Ideally, the projection will be selected based on the dominant tasks expected from map readers. For instance, if the map will primarily be used for estimating values on or between isolines, or for estimating how quickly a phenomenon changes over space (e.g., how close or far apart the isolines are), preservation of distance is the most important characteristic. If the primary task is to compare the distribution and size of regions on the isarithmic map (e.g., how much area will have an estimated sea surface temperature change of +2°C in 2020 vs. 2040), the key property to preserve is area. This is the only way to allow valid comparison of the size of the impacted regions across the maps.

Outside of the general projection properties of importance for different map reading tasks, interpretation of isarithmic maps is particularly sensitive to the overall "shape" of the projection—for instance whether the polar regions are presented as points or lines or any interruptions introduced in the projection (e.g., with Goode homolosine). We recommend trying out several different projections and their centers to evaluate how well the isolines can be interpreted across the entire mapped area (e.g., are they compressed to the point of illegibility in the polar regions due to representing the poles as points).

At the local scale, the guidelines we present for global-scale isarithmic mapping can be relaxed, as it is easier to tailor the projection parameters to find compromise between the different distortions inherent in map projections.

Notes

1. The International Ellipsoid 1924 is part of the European Datum, 1950.
2. Prior to the International Ellipsoid of 1924's development, reference ellipsoids were modeled specifically to "fit" a given region or county. The International Ellipsoid of 1924's was an early attempt to model the entirety of Earth's size and shape.

8

Discretely Occurring and Smoothly Changing

In this chapter, our focus turns toward mapping phenomena that occur discretely (i.e., at distinct locations) and have patterns that change smoothly over space, without being constrained by the boundaries of enumeration units. A common form for representing this type of phenomena is in a dot map (Figure 8.1). Because dot maps rely on discrete point symbols to visually encode values at specific point locations, as opposed to choropleth maps encoding using single values across enumeration units, it is possible to depict a more realistic, non-uniform pattern to represent the spatial distribution. Often, but not always, the point marks on the map each represent a count of greater than one of the phenomena. An example of this would be a map showing population distribution where each mark is equal to 1,000 people. The "dots" in a dot map are often circular marks; however, there is no cartographic rule to suggest that the mark cannot be another mark shape, such as a square, particularly when the mark value is 1:1.

Dot maps may represent either true or conceptual location for the phenomena they represent. These maps may also be colloquially referred to as "dot density" maps since the map *shading* resulting from the dot placement implies a relative change in density of a phenomenon. Through this chapter we use the dot map as an exemplar representation for discretely occurring and smoothly changing phenomena.

With a true location dot map, where the dot value is 1:1, each mark represents a single discrete entity at the location where it was measured (e.g., address for a house, touchdown location for a tornado, etc.; Figure 8.2A). In the instance of multiple entities being measured at the exact same location, the mark may either be sized based on count of entities at that location (see Chapter 9 for more discussion on this type of proportional symbol), or may simply appear as a single mark, which masks the true count at that location due to overprinting, where the point marks all lie exactly in the same place. On the other hand, with conceptual location dot maps, each individual mark represents a *proxy* of more than one discrete entity (e.g., one mark equals 10,000 sheep). Conceptual location dot maps are created through random or semi-random placement of a predetermined number of points representing disaggregated data for a region, such as conversion of a count of total population for a county into an appropriate number of points. This type of point

FIGURE 8.1
Dot density map representing population distribution in the United States; each dot represents 2,500 people.

mark is not indicative of the true location of an individual phenomenon; however, cartographers attempt to place point locations to represent where the phenomenon is more likely to occur (Figure 8.2B).

As we have seen in other examples throughout this book, distortion due to projection will vary depending on specific characteristics of the projection as well as the map scale (e.g., local scale, global scale, etc.). With dot maps, there are a few key projection points to consider:

- The data being mapped are discrete entities (point marks), and the primary visual interpretation task is estimating density using relative distance between points. Projection distortion that alters the visual distance between the points can lead to misinterpretation of densities, and will lead to faulty estimation of data values across the map.
- Equal area projections are preferable as the marks-per-area ratio should stay consistent across the map.
- While equal area projections should make it easier to interpret densities, angular distortion across the projection will change the shape of point clusters, which may negatively impact a map reader's ability to estimate density.
- Exercise caution with global-scale dot maps where the dot clusters may be split across an interruption in the map (e.g., with Goode homolosine equal area pseudocylindrical) or at the periphery of the map.
- Global-scale maps usually have significant distortion at the periphery of the map. Dot clusters located at the periphery of the map may

experience noticeable distortion and negatively impact a map reader's ability to estimate density.

- For local-scale maps it is important to carefully adjust the projection parameters to minimize distortion in the area of interest. The smaller the region being mapped, the easier it is to adjust any projection to show an acceptable level of distortion.

Throughout the chapter, we will discuss each of these points in more detail and provide examples to demonstrate the impact of the projection on dot mapping.

Because the focus in this book is on projection issues and not cartographic practice in general, we will not go into detail on the process of creating dot maps, or the trade-offs related to specific design decisions such as aggregation level, mark size, and method of mark placement. For more information on these aspects of the map design, there is extensive discussion available throughout the cartographic literature (see Chapter 3 for more detail on basic cartographic data/symbolization methods and relevant cartographic references).

Visual Analysis Task

The general visual analysis task associated with interpreting smoothly changing phenomena is to identify differences in approximate value or quantity of phenomena across one or more regions of interest. With dot maps, this is based on variation in *apparent density* of marks across individual regions of interest or across the entire mapped area. Regions of interest may be defined with set boundaries on a map (e.g., a specific country), cognitive regions defined by general geographic knowledge (e.g., "the northeast"), or patterns seen in the map (e.g., a cluster of point marks). This visual analysis task is based on assessment of spacing between individual point marks in order to visually approximate a measure of density. The densest areas are the locations where the point marks coalesce and spacing between marks cannot be measured. Density approximation is an estimate of relative area covered by point marks versus that which is not covered, the coalescence of the marks, and a non-quantitative approximation of distance between the set of marks. When marks are particularly close together, they coalesce to form visual clusters, with the edge of the cluster identified as a transition region between higher and lower density. Since recognition of clusters is dependent on interpretation of relative density, changes in area or distance between points due to the distortion patterns in a projection may have significant impact on the map reader's interpretation.

A

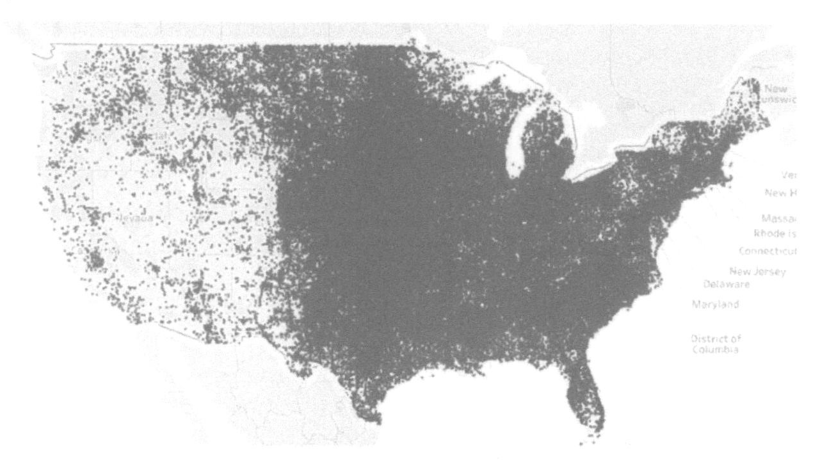

B

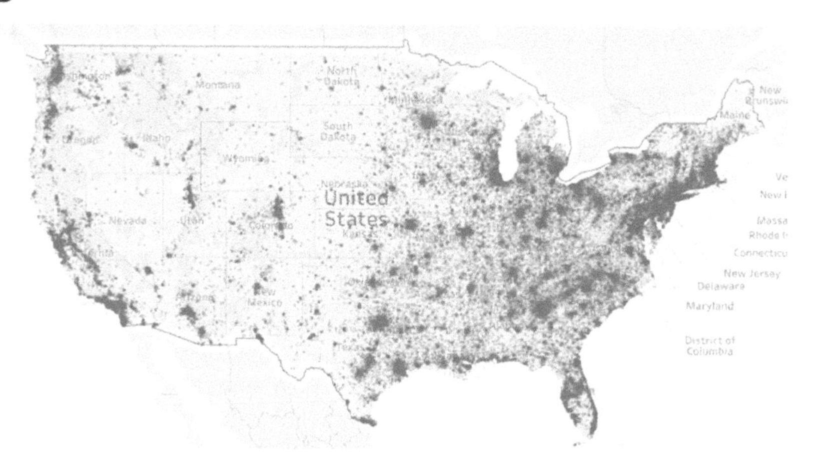

FIGURE 8.2
Example maps of true (A; hailstorm origin locations; each point represents one storm) and conceptual (B; population distribution in the United States; created with population counts at Census Tract level with 1 dot = 2,500 people) location dot maps.

Impact of Projection on Dot Maps

Equal Area Projections

The goal with a dot map is to facilitate the map reader in making direct visual estimates of, *and comparisons between,* densities at any location on the map (see Chapter 3). The reliance on density measures—the count of points divided by area—emphasizes the importance of preserving area in the projection. For most dot density maps, an equal area projection is the ideal choice. However, as we have discussed throughout this book, there are trade-offs with preserving relative areas on a map. One of these trade-offs is that the preservation of area in a projection comes at the cost of angular distortion. With respect to dot maps, preservation of area means that point clusters on the map will be directly comparable to one another with respect to the density of point marks in any given area; however, the form or shape of the cluster will vary.

Of course, while the area that different point clusters cover will be directly comparable across the entirety of an equal area projection, it would be remiss of us to not also include a warning about how the variation in *shape* of the cluster may impact a map reader's ability to estimate areas accurately. Studies have shown that accuracy of area estimates for regions varies based on the shape and level of stretching or elongation of the shapes (see Krider et al., 2001; Gescheider, 2013), so even though an equal area projection will *present* the areas correctly the reader may still *interpret* them incorrectly. Inaccuracy in estimating areas would then impact a map reader's ability to estimate the area covered by marks and mentally calculate the relative density. Unfortunately, there is nothing that you can do about this problem other than be aware of it. Preserving area in your projection is still the best option for dot maps, even with potential for map readers to under- or over-estimate areas due to angular distortion.

Equal area projections will always present some level of angular distortion, and individual projections distribute angular distortion in different ways. Depending on the extent of the data shown on your map, the distribution of angular distortion may make it advantageous for you to select one specific equal area projection over another so that the angular distortion is minimized in the areas of greatest interest (or with the greatest volume of data) on your map. To explore this problem in a little more detail, let's map the same dataset on a few equal area projections and visually compare the dot patterns. The dataset we're using consists of four point clusters placed in varying locations around the globe (centered around (0°, 0°), (180°, 0°), (140° E, 70° N), and (60° S, 30° W)) (Figure 8.3). Each of the separate point clusters shown has exactly the same spatial distribution on the globe, but the appearance on the map will vary depending on the distortion pattern on each projection. We have arranged the point clusters in a deliberate manner in order to facilitate comparison between regions with differing levels of angular distortion, or locations that we would consider "problem regions" worth highlighting specifically on each map.

In the projections shown in Figure 8.4, pay close attention to the shape and apparent size of the point clusters. For each projection we will show the pattern of angular distortion in magenta, with darker magenta indicating greater angular distortion and lighter magenta or white indicating lower levels of angular distortion. We have used blue point marks to highlight the data.

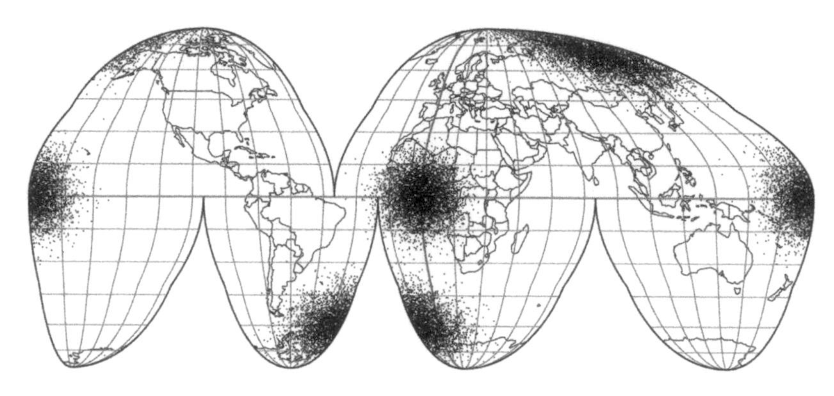

FIGURE 8.3
Location of the four point clusters used; shown on Goode homolosine interrupted equal area pseudocylindrical projection.

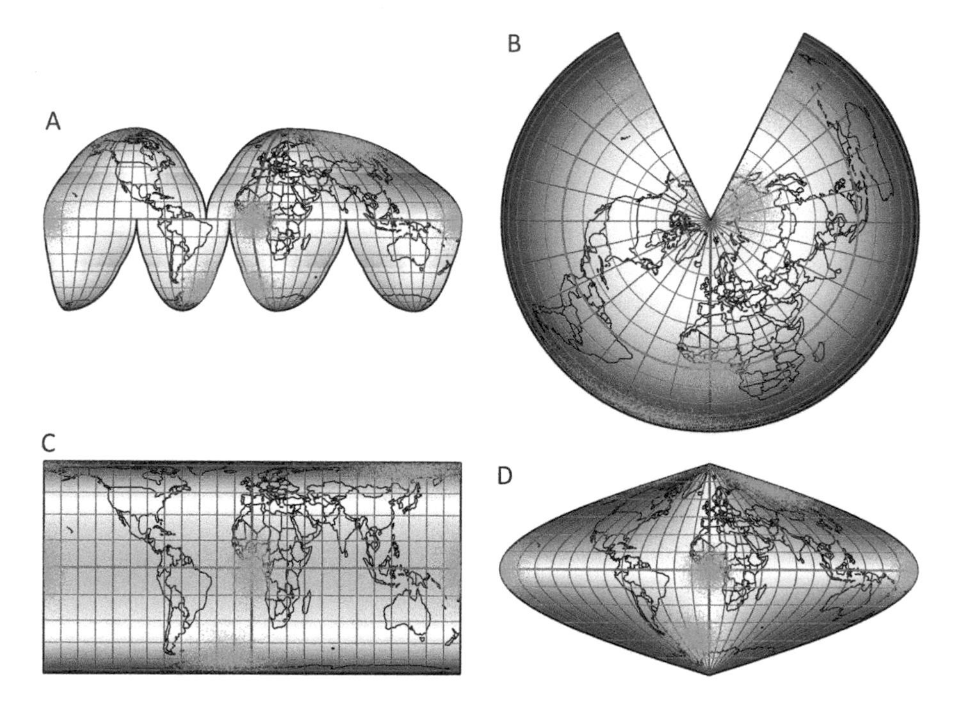

FIGURE 8.4
Four point clusters (same spatial distribution on the globe for each cluster), as shown on four equal area projections: Goode homolosine (A), Lambert equal area conic (B), Craster equal area cylindrical (C), and sinusoidal (D). Angular distortion is shown for the entire world. Angular distortion is shown in magenta; lighter shades represent less distortion and darker shades represent more distortion.

Though each of these projections is equal area and the point clusters take up the same relative area on each map, note the extreme angular distortion (see Figure 8.5 for a more detailed example).

While all of the examples in Figure 8.4 use equatorial aspect projections with a central meridian of 0°, as we have shown throughout the book, the parameters for any projection can be adjusted to tailor the projection to specific locations or for specific datasets. Adjusting the parameters for a Lambert equal area conic angular distortion will have less of an impact on visual interpretation of the clusters (Figure 8.6). In Figure 8.6, we adjust the central meridian so that the projection is centered at 105° W with a standard line at 45° N. These parameters were selected to roughly center the projection on the conterminous United States. With this adjustment, we see far less "skewing" (warping of the shape) on most of the point clusters because of how the angular distortion is distributed across the projection differently than with the previous parameters. While adjusting the parameters for this global-scale example minimized many of the problem areas for the demonstration data, there was still one cluster of data in the southern polar region that is located in an area with significant angular distortion. This situation is common with global-scale datasets. Careful selection of projection parameters can minimize the distortion, but often cannot effectively remove *all* potential problems for visual interpretation.

For larger-scale mapping (more local areas), however, the angular distortion in equal area projections can be more effectively minimized by selecting parameters appropriate to the location. For instance, you can center the projection on the location of interest and ensure that any standard line, or lines, pass through the area. In Figure 8.7, we show how a Lambert *azimuthal* equal area projection (note that this is a different projection than the Lambert equal area *conic* projection used above) can be re-centered to present minimal angular distortion across any one particular region of interest for local-scale

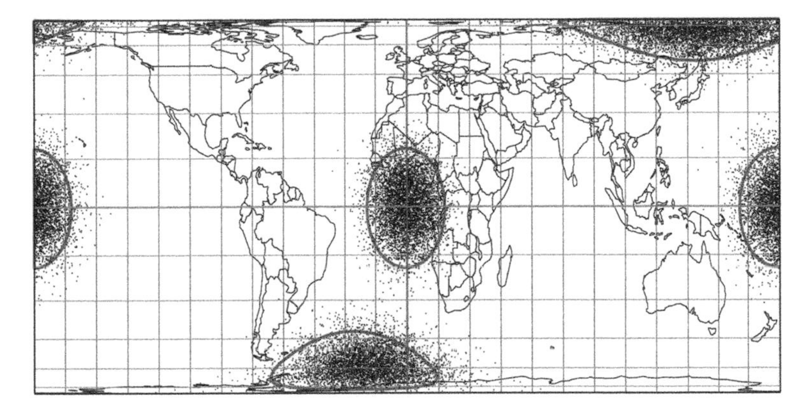

FIGURE 8.5
Point clusters shown on Craster projection. To highlight the angular distortion in the projection, red outlines have been drawn around the central cluster of dots.

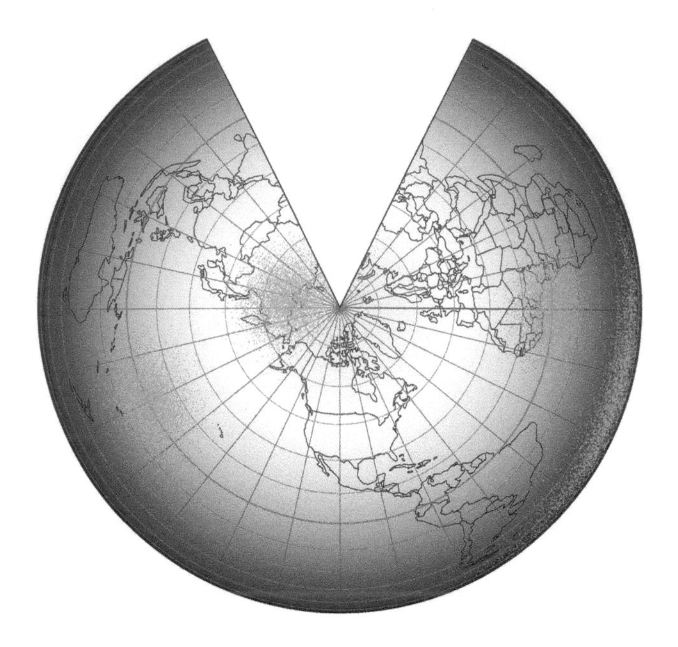

FIGURE 8.6
The same cluster of points shown above but placed in various locations on a Lambert equal area conic projection with parameters that shift much of the significant angular distortion to the southern latitudes. Angular distortion is shown for the entire world. Angular distortion is shown in magenta; lighter shades represent less distortion and darker shades represent more distortion.

mapping. Each example uses one of the four example dot clusters (Figure 8.3). Note that beyond that immediate region of focus in the projection, the angular distortion increases rapidly. The point here is that if we only needed to focus the map on a smallish region—for instance, if we only had to map one of these point clusters—we could define projection parameters that allow us to minimize both area and angular distortion across that small region.

Conformal Projections

To dig deeper into the logic behind area-preserving projections: in general, being more suited to visualization of dot maps, it helps to consider the impact of *not* preserving area. Conformal projections preserve local angles at the expense of relative area. The impact of not preserving areas means that we see a "spreading" or "contracting" effect where relative distance between points increases or decreases in locations across the projection. This inflation or deflation of area increases the possibility that map readers will have difficulty visually assessing density. For small-scale mapping, this is a particularly acute problem. For instance, compare the spatial distribution of points in the Mercator as seen in Figure 8.8. While we noted in the last section that map readers may have challenges interpreting areas correctly on an equal

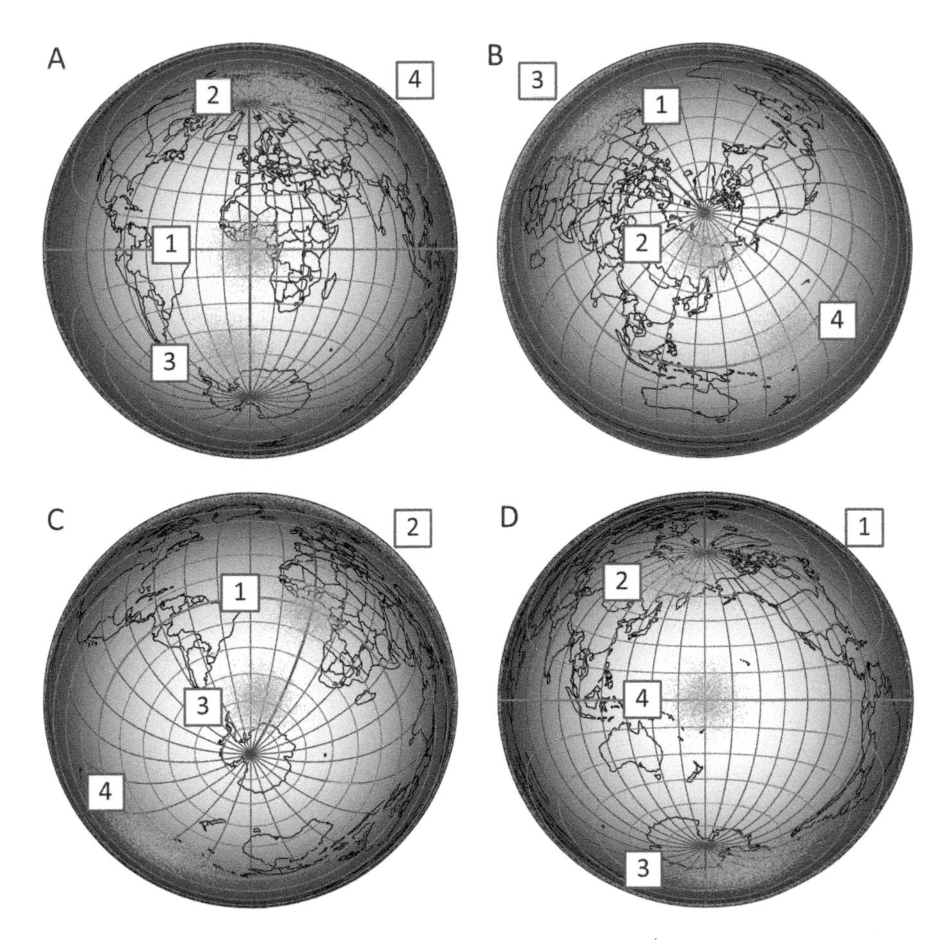

FIGURE 8.7
Lambert azimuthal equal area projection centered on each of the four clusters. Angular distortion is shown for the entire world. Angular distortion is shown in magenta; lighter shades represent less distortion and darker shades represent more distortion. Cluster 1 is centered on (0°, 0°), cluster 2: (140° E, 70° N), cluster 3: (60° S, 30° W), and cluster 4: (180°, 0°). For this projection, adjusting the parameters minimizes distortion in the center of the projection, with increase in angular distortion radiating away from the center. For reference, the clusters have been numbered 1–4 so that it is easier to track them across the four projections. When the label is placed on the periphery of the projection, the point cluster is spread out along the periphery and may not be readily visible due to the extreme angular distortion at the periphery.

area projection due to the angular distortion, at least they are starting with a base map showing correct areas. With small-scale conformal maps, the base map is *not* showing areas correctly *and* there are still potential issues with map readers incorrectly estimating areas for regions—it's doubly bad from an accuracy standpoint. To emphasize this, Table 8.1 lists the areas and relative densities for the central region in each of these clusters of dots. Recall that these clusters are all the same spatial distribution on the globe, so the relative visual densities should be the same.

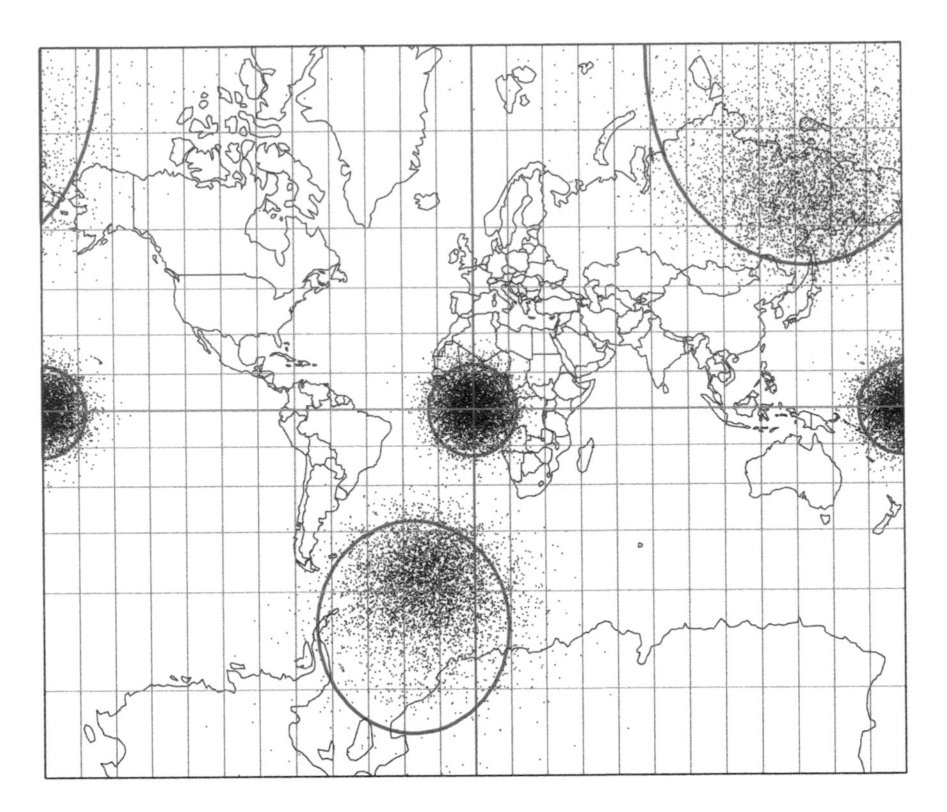

FIGURE 8.8
The same spatial distribution of points shown centrally located, at the periphery, and at a high latitude (where the areal distortion is the greatest) on the Mercator projection. Red circles indicate the region deemed to be the "core of the cluster."

Note that not all conformal projections will show areal distortion as extreme as what we see in the high latitudes of Mercator (Figure 8.8). One example is the Lambert conformal conic projection. With this particular projection we see lower amounts of areal distortion for three of the point clusters with a selection of standard lines of 30° and 60° N. Selection of these

TABLE 8.1
Area for Each Cluster Circled in Figure 8.8 and Numbered in Figure 8.7 in Projected Units, as the Clusters Are Seen in Web Mercator Projection (Nearly Conformal)

Cluster/Location	Cluster Core Area (km^2)	Cluster Core # Points	Cluster Core Density
1–(0°, 0°)	13,693,138	4451	0.000325 points/km^2
2–(180°, 0°)	13,693,138	4451	0.000325 points/km^2
3–(140° E, 70° N)	229,623,586	4451	0.0000194 points/km^2
4–(60° S, 30° W)	69,121,146	4451	0.0000644 points/km^2

(While the clusters each have the same spatial distribution, areal distortion in web Mercator leads to a visual decrease in density of the clusters across the projection.)

parallels for the projection minimize areal distortion across the Northern Hemisphere, where three of the point clusters are located. However, this is at the expense of the fourth point cluster near Antarctica, which is cut off entirely. The Antarctic region is not defined for this projection.

For another example of how characteristics of specific projections within any given class might impact the visual representation, compare the map in Figure 8.9 to an elliptic conformal projection (Figure 8.10) where the areal distortion increases rapidly from the center of the projection, and is not isolated to the high latitudes as in the Mercator projection. In the elliptic projection, the spatial distribution of points on the periphery of the map (the cluster centered at approximately (180°, 0°)) is so heavily impacted by the areal distortion that it loses the same visual appearance that the other denser clusters present (or is seen as an enormous, but sparser, cluster).

Equidistant Projections

On first glance, the name "equidistant" makes it seem like this would be a logical projection choice for a map type where distance between points is a primary factor in interpreting the pattern. However, equidistant projections are selective in the way that distance is preserved. For instance, distances may be correct only along meridians, as in the cylindrical equidistant

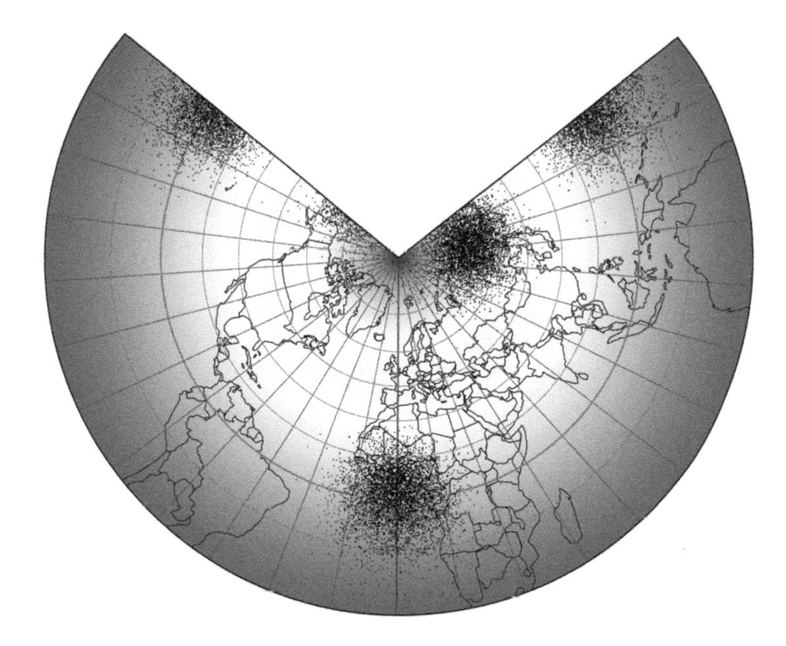

FIGURE 8.9
Lambert conformal conic projection with standard lines at 30° N and 60° N. Areal distortion is shown for the entire world. Areal distortion is shown in green; lighter shades represent less distortion and darker shades represent more distortion.

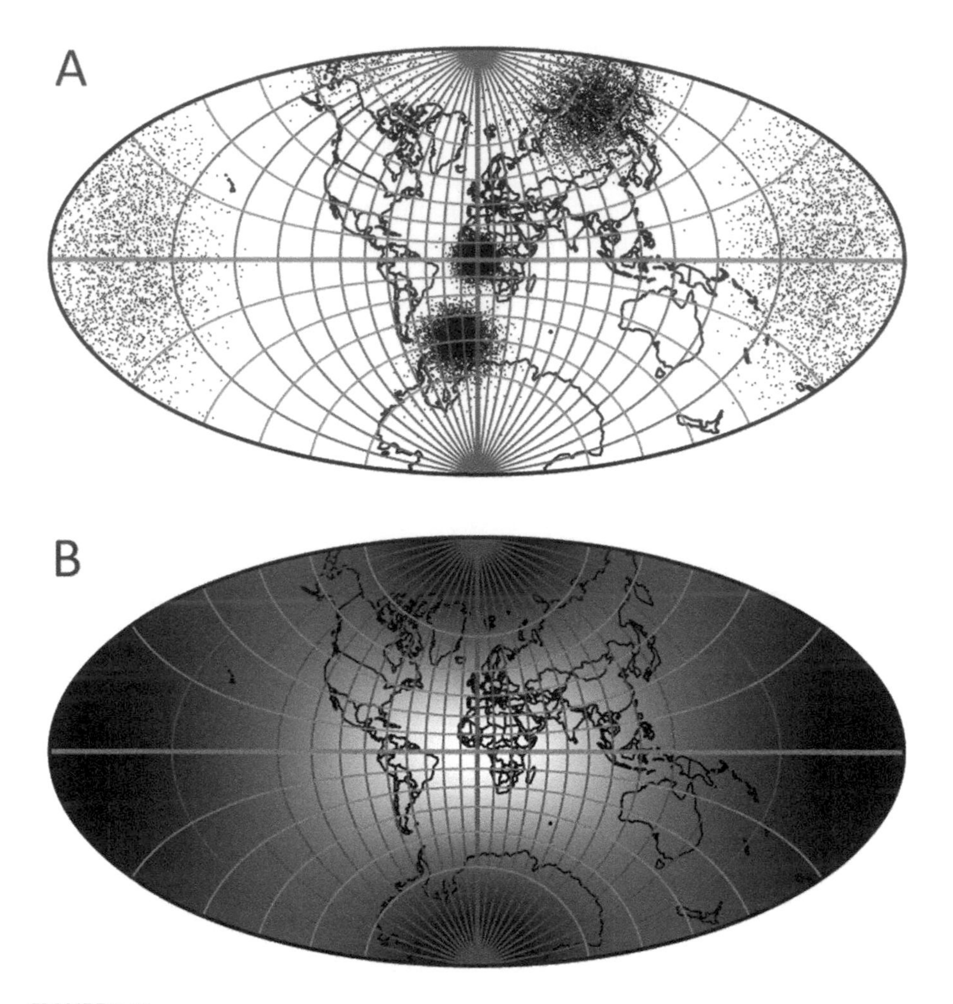

FIGURE 8.10
Elliptic conformal projection centered at (0°, 0°) showing the point separately (A) and with areal distortion pattern (B). Areal distortion is shown for the entire world. Areal distortion is shown in green; lighter shades represent less distortion and darker shades represent more distortion.

projection (Figure 8.11), or distances may be correct between *one or two* points on the map to any other point on the map, as in the two-point equidistant projection (Figure 8.12); distances between all combinations of points *cannot* be preserved on a projected map. While some distance measures may be correct, this comes at the expense of distortion to both relative angles (Figure 8.11A and Figure 8.12A) and areas (Figure 8.11B and Figure 8.12B) across the projection. For small-scale mapping, the distortions will be more apparent and will have a more substantial impact on visual interpretation of dot densities.

A

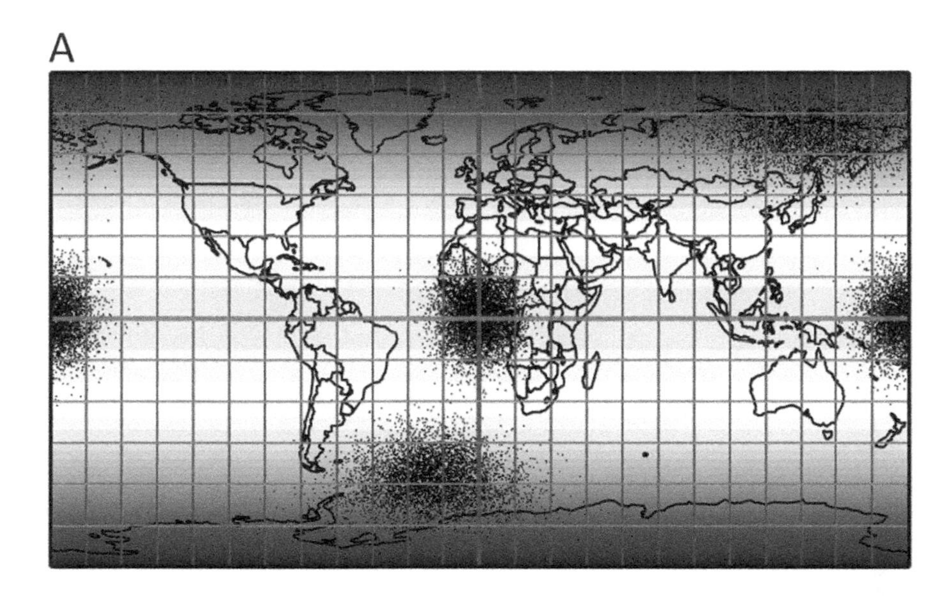

B

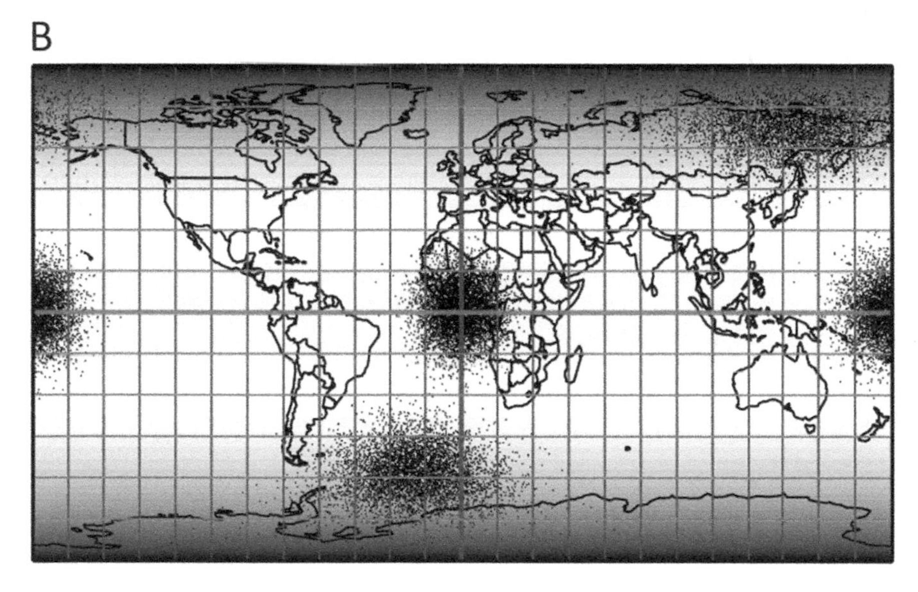

FIGURE 8.11
Cylindrical equidistant projection showing patterns of angular distortion (A) and areal distortion (B). Angular distortion is shown in magenta, while areal distortion is shown in green. In both cases, lighter shades represent less distortion and darker shades represent more distortion.

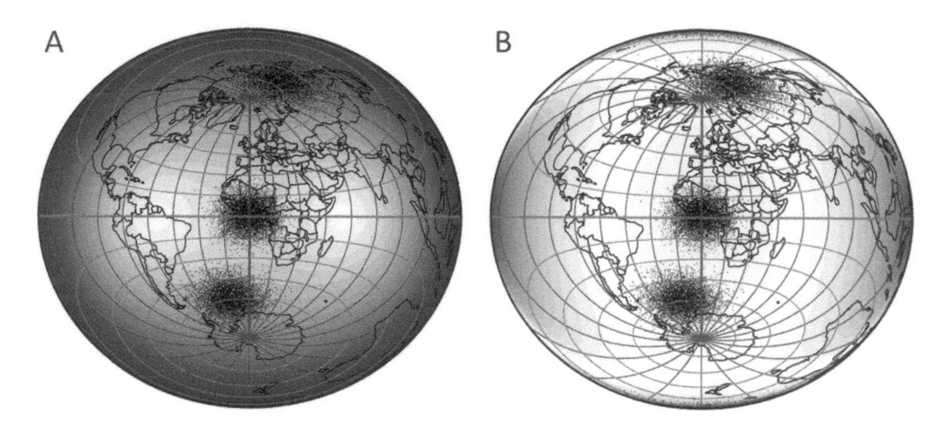

FIGURE 8.12
Two-point doubly equidistant projection centered at (0°, 0°) showing patterns of angular (A) and areal (B) distortion. Angular distortion is shown in magenta, while areal distortion is shown in green. In both cases, lighter shades represent less distortion and darker shades represent more distortion.

Large-Scale Projections

Even though we have suggested that equal area projections are optimal for dot mapping, as is the case with many projection problems, if you are working with a small enough area (local-scale mapping), the impact of the distortion can be mitigated through selection of appropriate parameters (e.g., standard lines and central meridian) appropriate for the region. For instance, the parameters of the two-point doubly equidistant projection discussed earlier can easily be adjusted so that it is centered off the southeastern tip of South America (Figure 8.13), thus minimizing the distortion around the southernmost cluster of points.

General Challenges

While some aspects of projections can be controlled through careful selection of projection class and parameters, there are some additional challenges that will be true for dot maps at the global-scale regardless of projection selected.

Interruptions and the map periphery. Interrupted projections present a special challenge for dot mapping; they introduce discontinuities, or interruptions, along select lines in the map, creating a projection with multiple segments, often referred to as lobes or gores. The locations of the interruptions are selected so that they fall in areas deemed less important to the map's purpose (e.g., through the oceans in a land-based Goode homolosine

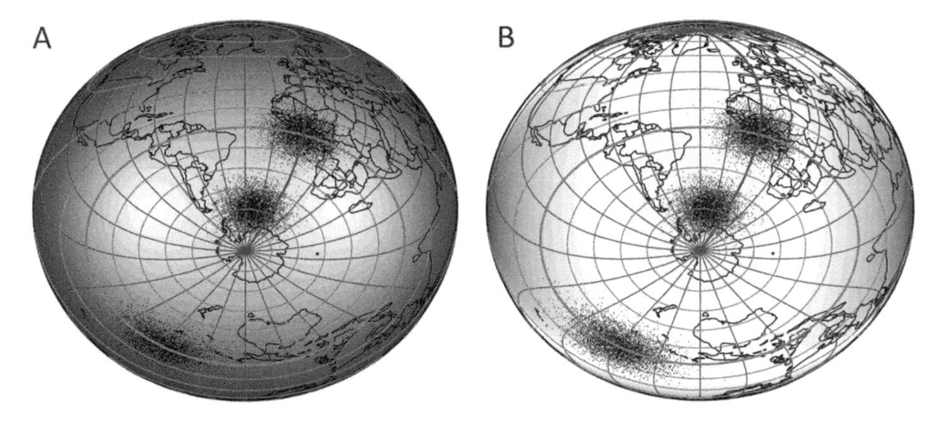

FIGURE 8.13
Two-point doubly equidistant projection with a projection center at 30° W, 60° S. The images show the angular (A) and areal (B) distortion for the entire world with this projection and parameter selection. Angular distortion is shown in magenta, while areal distortion is shown in green. In both cases, lighter shades represent less distortion and darker shades represent more distortion. For small areas the distortion can be made negligible through careful parameter selection. This can be seen in these two top images, where there is low distortion of angles for the central point cluster, and no visible distortion to angles—but just for a relatively small region in the middle of the projection.

projection, such as in Figure 8.14). Interruptions in projections are primarily used with pseudocylindrical projections as a method to help control the overall distortion. With respect to dot mapping, the impact of interruptions is of particular interest due to the impact of the spatial continuity of patterns on the map.

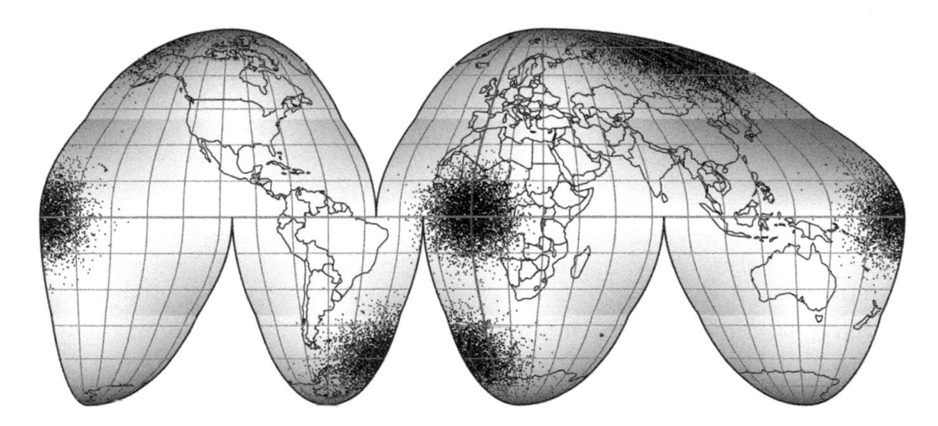

FIGURE 8.14
The four matching dot clusters used for examples throughout this chapter as shown on a Goode homolosine projection. Angular distortion is shown for the entire world. Angular distortion is shown in magenta; lighter shades represent less distortion and darker shades represent more distortion.

For example, consider Figure 8.14 displaying several dot clusters across the Goode homolosine projection. In Figure 8.14, the dot distributions are the same (on the globe), but present notably different patterns when placed at the equator, in the Southern Hemisphere at an interruption, or on the periphery. When the spatial distribution of dots crosses over an interruption it is difficult for a map reader to mentally stitch these together to assess the true pattern. While we are not aware of perceptual studies on this exact topic, it is a logical assumption that people would be more likely to interpret data crossing an interruption as separate point clusters as opposed to one single group.

In addition to the occasional concerns with interruptions in projections, there is *always* an issue with mapping patterns across the periphery on a map. Regardless of whether you are using an interrupted projection like the Goode homolosine, or any other projection—there is always a border on a global-scale map that interrupts the continuous surface of the globe (see the dot cluster wrapping across the International Date Line in Figure 8.14). It is understandably challenging for many map readers to mentally stitch together patterns from one edge of the map to another, and where possible you should adjust the projection's central meridian and aspect so that you minimize the data distribution being split on the periphery.

Zooming in dynamic maps. For dynamic maps, you should also be aware of how the map changes due to user interaction, such as zooming in and out on regions of interest. As the user zooms in or out the map scale changes and the visual distance between marks will increase or decrease accordingly. Depending on how the map has been created, the mark size may or may not adjust when the scale changes; on many maps the marks will re-size so that they remain a fairly constant (though not necessarily exact) size on the screen. The impact of change in scale on the map is most important when making comparisons of density across differing regions; if the reader is exploring portions of the map at different scales, the densities will not be comparable. While this isn't due to a distortion issue with the projection itself, it is sufficiently important to note here.

Comparisons between locations. Because distortion across a map will vary depending on the specific projection and the parameters (e.g., standard line(s) and central meridian), you should also consider how these variations may hinder comparison between patterns in different locations on the map. For large- to medium-scale mapping, projection parameters can often be selected to sufficiently minimize the relative distortion across the mapped area, however; for smaller-scale mapping the distortion will play a significant role. As we have seen in other chapters, as the area being mapped increases in size, the impact of projection becomes more substantial and requires careful consideration regarding how the spatial distribution of distortion will alter the visual pattern of the data.

Conclusion

In this chapter, we have examined the implication of projection on dot maps (a map type for depicting discretely occurring and smoothly changing phenomena). Since the interpretation of dot maps is largely based on identification of clusters of dots and/or calculation of the density of dots covering any particular area, an equal area or compromise map projection is the safest choice. While it may seem like equidistant projections are ideal to evaluate spacing between dots across the map, since equidistant projections only preserve distance from one or two specific locations on the map, or for special cases like preserving distance along meridians, they are not effective for improving the overall ability to interpret distance between points in any combination of locations across the map.

Additionally, it is important to evaluate how the clusters of dots are impacted by any interruptions in the projection or by the placement of the central meridian (e.g., are dot clusters split across the periphery so that the same cluster is split between two opposite sides of the map), or the way in which the polar regions are represented (e.g., are dot clusters "squished" near the poles due to polar regions being represented as points). To avoid these issues, you may wish to select a general type of projection that is best suited to the tasks for which your map will be used, and then consider the impact of the overall "shape" of different projections with the properties of most importance to your map (which will likely be equal area or compromise for dot maps).

For local-scale mapping, our recommendation of equal area or compromise projections can be relaxed, provided the projection selected uses appropriate parameters so that areal distortion is minimized across the local-scale mapped area.

9

Discretely Occurring and Abruptly Changing

In this chapter, our focus turns toward selecting projections for data that are considered discrete and abrupt. This kind of data represents a spatial distribution of phenomena existing at discrete locations while abruptly changing value between those locations. For example, assume you have a dataset that reports the count of people in each county in the United States. In this case, the location of the population is assigned to the centroid of each county and is then considered discrete as the count for the entire region is assigned to a single point location within each county. Although the population does exist across each entire county region, we are only considering one point centrally positioned in each county as a representative location. In addition, the population changes abruptly from county to county. To represent the differing values at discrete locations, the data is often symbolized using the same symbol (e.g., a circle) but the symbol varies in size (Figure 9.1). Mapping discrete and abrupt data allows the map reader to visually compare differently sized symbols across a geographic area of interest revealing a pattern. In addition, those symbol sizes can be compared to obtain estimates of data quantities and calculate differences of values at specific locations.

Discrete and abrupt data are usually collected from point locations. These point locations can be true or conceptual. As a few examples, a true point location can be a weather-recording station, an oil well, or an air quality–monitoring station. Figure 9.2 shows an example of true point locations of weather-recording stations throughout West Virginia. To be a true point location, each point has a single associated real-world coordinate value. On the other hand, a conceptual point location can be thought of as an area that has been reduced to a point. This reduction is sometimes necessary when dealing with scale changes and how features are represented on maps. For example, on a local-scale map a city could be represented as an area while on a smaller-scale map that area would collapse to a point.

Proportional symbols can also represent flows between locations, where the flow path is usually shown with a line illustrating movement of data from point to point, and the width of the line is proportionally scaled to represent quantity of flow. Some of the data mapped with flow lines include, for example, goods, people, or information. Figure 9.3 shows proportional flow lines representing the total tonnage of intermodal freight volume on U.S. railroads for 2017. Note that, in addition to showing the data quantities in Figure 9.3, the flow lines show the railroad route over which the flows took place. Flow lines can also show the distance and direction of the flows.

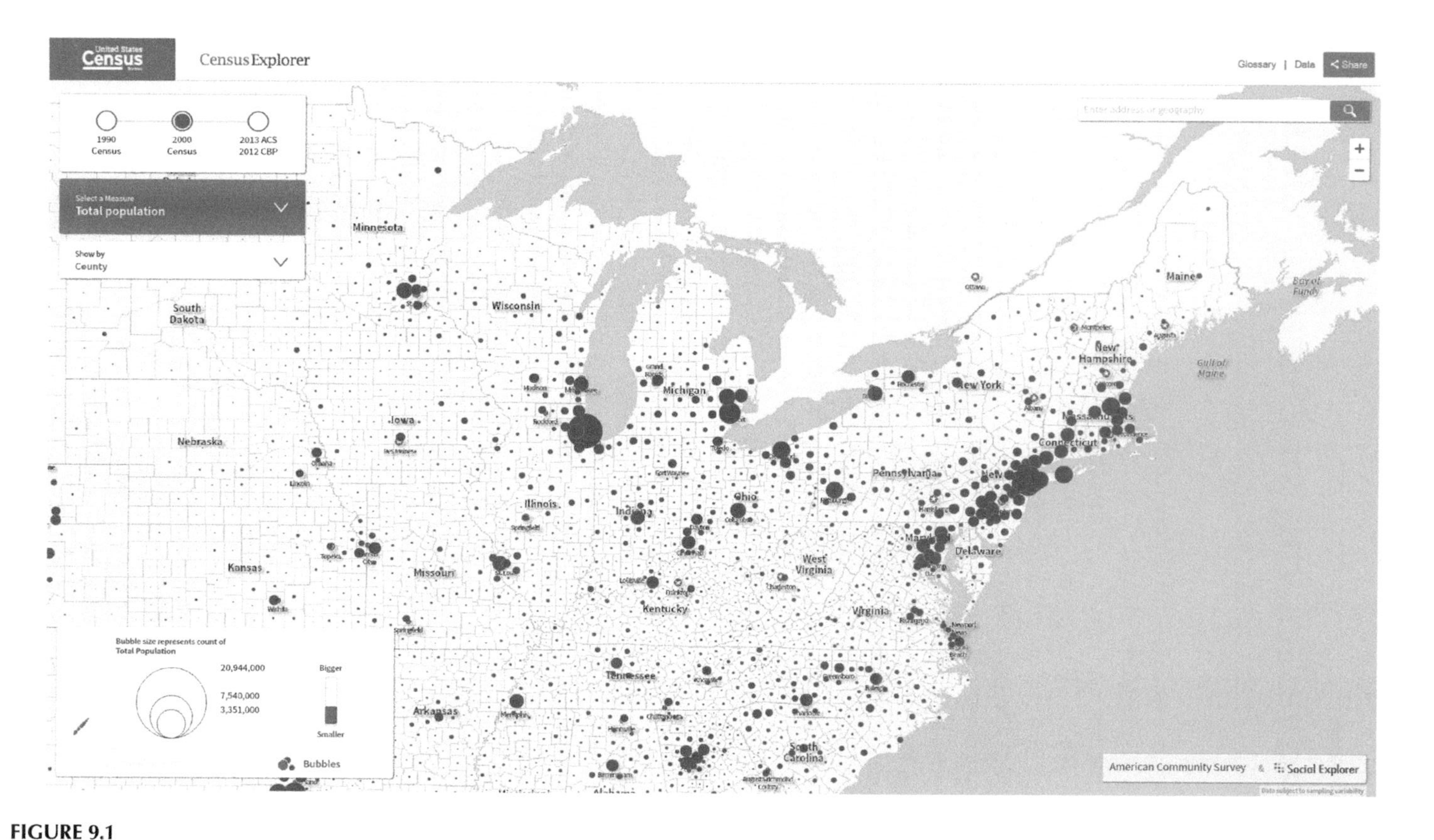

FIGURE 9.1
Map using proportional circles to show total population by county in the United States. Map generated with the U.S. Census Social Explorer: https://census.socialexplorer.com/.

FIGURE 9.2
Locations of weather-recording stations throughout West Virginia. Data from the West Virginia GIS Technical Center (http://wvgis.wvu.edu).

In this chapter, two datasets will be used to illustrate how the selection of a projection can impact mapping discrete and abrupt data using proportional symbols. The first dataset uses the 2050 estimated population of large urban areas throughout the world.* This dataset is mapped in Figure 9.4 using proportional circles. Figure 9.4 shows that even though the data for each urban area represent an areal extent, for symbolization purposes, the data are conceptualized as points (the centroid of each urban area). The second dataset that we will use includes the fictitious tonnage of Tribbles carried on ships traveling from Amsterdam, Netherlands, to various locations (Figure 9.5). This ship tonnage dataset will be used to illustrate how the choice of the projection is a design variable for mapping proportional flow lines.

In addition to the normal projection-related considerations of geographic location, scale, extent, overall shape, and projection center, there are specific concerns related to projection used to represent each of the specific types of proportional symbol (point or line) that should be considered.

* Available at Hugo Ahlenius, Nordpil: UN Population Division and World Urbanization Prospects, 2007 Revision https://nordpil.com/resources/world-database-of-large-cities/.

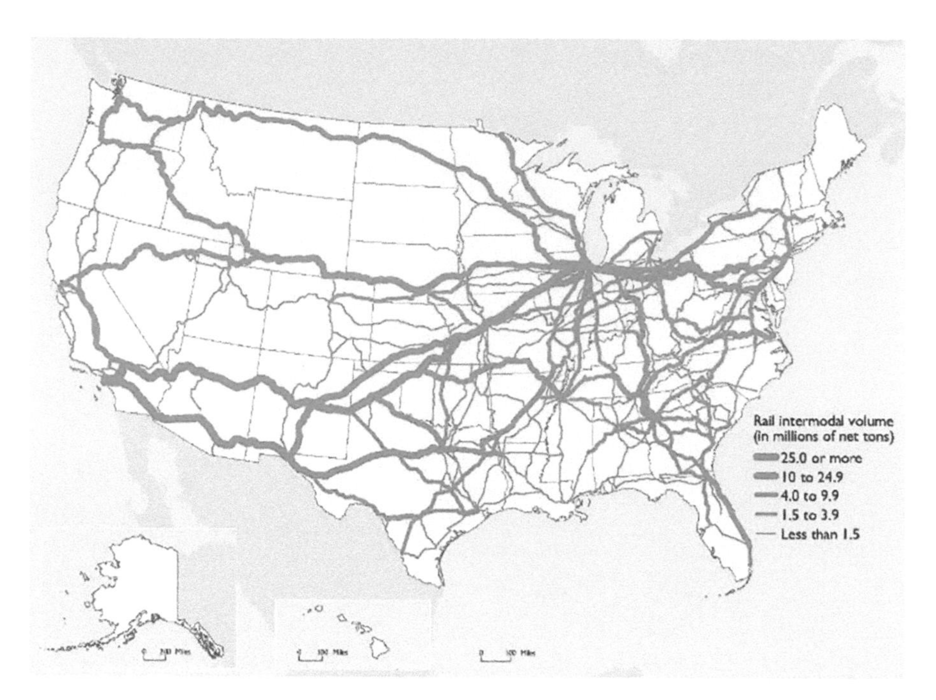

FIGURE 9.3
The total volume of intermodal freight movement of U.S. railroads across the United States in 2017. (Reprinted from Freight Facts and Figures, in Chapter 3 The Freight Transportation System, n.d., from U.S. Department of Transportation Statistics. Retrieved December 28, 2018, from www.bts.gov/bts-publications/freight-facts-and-figures/freight-facts-figures-2017-chapter-3-freight.)

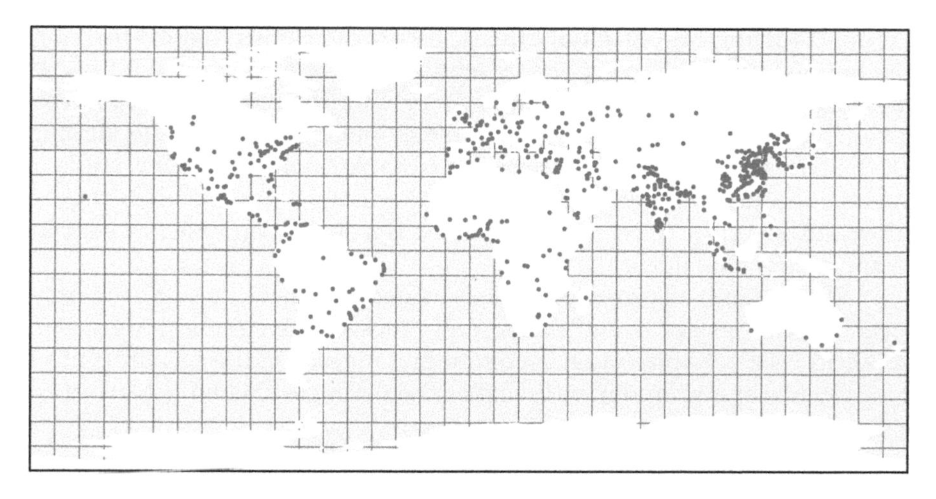

FIGURE 9.4
Location of the larger urban areas.

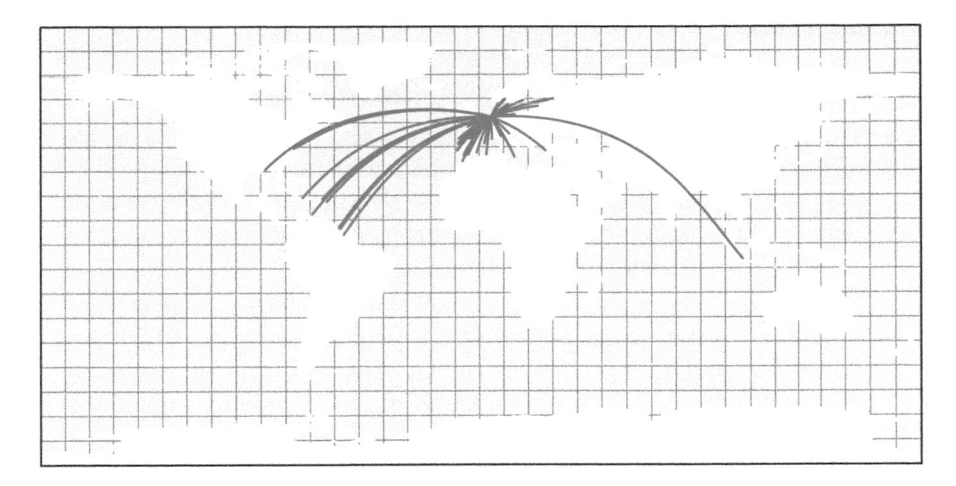

FIGURE 9.5
Flow lines originating from Amsterdam, Netherlands.

Proportional Point Symbols

- Proportional symbols are based on data values existing at discrete point locations. As such, they do not rely upon the preservation of the underlying geographic area as discussed in Chapter 6 and Chapter 8 with respect to continuous and abrupt data or discrete and smooth data, respectively. Thus, the need to select an equal area projection (which is emphasized when using choropleth and dot mapping) for proportional symbol maps can be relaxed in favor of other projection properties such as compromise.
- The pattern exhibited by the proportional symbols is tied to the geographic point locations from where the data was collected (e.g., cities). Due to the inherent distortion that is present with projection properties, landmasses can be stretched, compressed, or experience some combination compared to their appearance on a globe. These changes to the appearance of the landmasses alters the relative location of the points from which the true or conceptual point data were collected and can also change the appearance of the underlying pattern in the data.
- The mapped pattern will also change according to the projection class. With any projection class its overall "shape" will change, leading to symbols appearing more closely or distantly spaced. Depending on the spatial distribution of the true or conceptual point locations for the data on your map, consideration of the projection class may be necessary.

Proportional Line Symbols

- The choice of the projection should be based on the specific goals of the map purpose and how the characteristics and properties of the projection can enhance those goals. For instance, to show overall pattern of the data movement, support measuring distances and/or directions, or a combination of the two.
- Selecting projections for proportional flow lines also involves consideration of the visual appeal of the flow lines. Figure 9.6 shows flow lines appearing on the Van der Grinten compromise projection that represents U.S. exports and imports in a visually appealing, intriguing manner.

Visual Analysis Tasks

Like all thematic maps, the symbolization method chosen to represent the data on a proportional symbol map assists the map reader in carrying out these visual tasks.

- Map readers use proportional symbol maps to understand the data's overall pattern. For example, evaluating for clustering or dispersion of high and low values. The symbols' relative locations appearing on the map can help in communicating this pattern.
- Identifying values for specific areas by associating individual symbols to geographic area. Landmasses that have been subjected to serious compression and exaggeration due to the projection could affect the map reader's ability to recognize the landmasses.
- Map readers use proportional flow line maps for general visualization of paths of connectivity, but also for more analytical evaluation of distance and/or directions between origin and destination pairs.

Impact of Projections on Proportional Symbol Maps

Equal Area Projections

A basic visual task associated with proportional symbol maps is similar to what is expected with other symbolization methods: to present to the map reader an overall visual impression of the data's pattern. Equal area

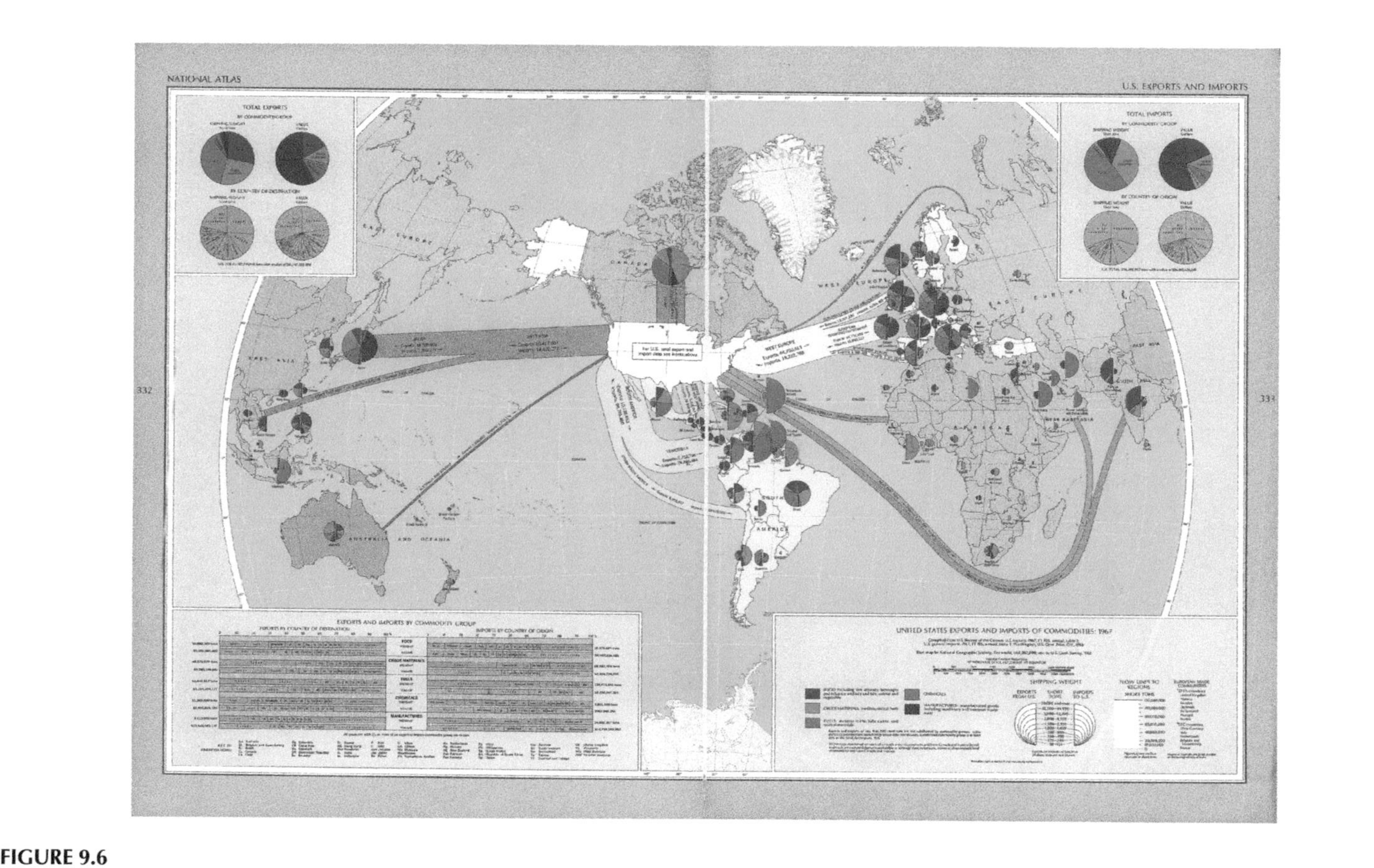

FIGURE 9.6
A map combining flow lines and proportional circles showing U.S. exports and imports 1967 on the Van der Grinten compromise projection. (Reprinted from U.S. Exports and Imports, in United States National Atlas, 1970. Retrieved December 28, 2018, from www.loc.gov/resource/g3701gm.gct00013/?st=gallery.)

projections have a logical association with some forms of symbolization. For example, with choropleth maps, the enumeration unit becomes the symbol. With dot maps, the enumeration unit contains the dots and therefore controls the apparent density of the dot distribution. When considering both symbolization methods, it is important to maintain areal relations of the enumeration units as the size of the enumeration unit can impact the visual appearance of the symbols. Therefore, equal area projections are suitable for choropleth and dot symbolization methods. However, the proportional symbol is a fundamentally different symbolization method. Here, the geometric or mnemonic symbols are scaled in proportion to the data value at point locations. Since the symbol is not dependent upon the enumeration unit's size, the requirement to use an equal area projection can be relaxed in favor of, for example, compromise projections that do not preserve any specific property. Equal area or conformal projections can alter the pattern appearing on the map. Equal area projections warp and shear landmasses and may also alter the point locations, giving the impression of a clustered dataset. Or, using conformal projections that stretch landmasses and the associated point locations, giving the idea of a more dispersed pattern than reality would suggest.

As discussed in greater detail in Chapter 6, equal area projections can contain considerable amounts of distortion, which can greatly alter the visual appearance of landmasses. This distortion is due in part to the necessity of sacrificing angular relations in order to preserve areal relations. Generally speaking, the distortion inherent with equal area projections tends to warp the overall shape of landmasses (especially those lying near the map periphery). For instance, the Lambert cylindrical projection (Figure 9.7) is equal area and has the standard line located at the equator. In Figure 9.7, the area highlighted by the red ellipse shows that the landmasses are compressed in a north–south but stretched in an east-to-west fashion. This compression of landmasses in the upper latitudes is characteristic of equatorial-centered equal area projections. Additionally, the east-to-west stretching is characteristic of cylindrical projections as every line of latitude is represented the

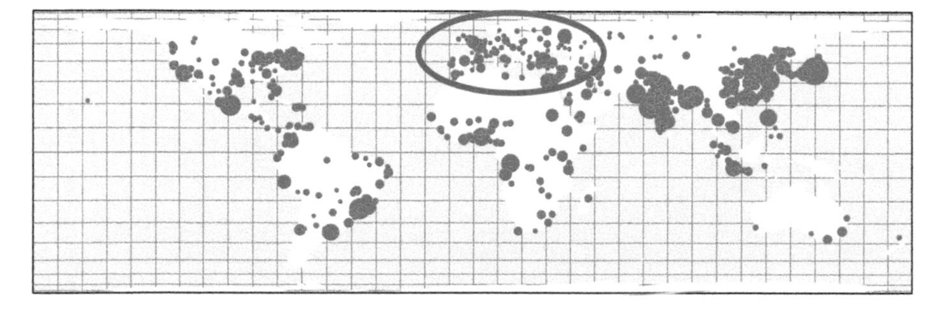

FIGURE 9.7
Population estimates for the large urban areas shown on the Lambert cylindrical equal area projection.

same length as the equator. As a result, this compression and stretching gives the appearance that the circles overlap and coalesce more than they do in reality, giving a false impression that the data in this region are clustered. Figure 9.8 shows the overall distortion pattern that includes a combination of scale and angular distortion.

Conformal Projections

Since conformal projections preserve angular relationships, this property is not necessarily desirous for proportional symbols and the associated visual tasks with these maps. Visually speaking, conformal projections contain considerable levels of distortion, which can greatly impact the appearance of landmasses. Like equal area projections, conformal projections tend to distort the overall shape of landmasses (especially those lying near the map periphery). Conformal projections usually exaggerate landmasses in the upper latitudes. This exaggeration can be seen in the Mercator conformal cylindrical projection in Figure 9.9. On this projection, distortion increases away from the standard line (the equator) toward the poles. Unlike the compression that was apparent to the landmasses in the upper latitudes in Figure 9.7, the Mercator projection actually exaggerates the sizes of these landmasses (Figure 9.10). The red ellipse shown in Figure 9.9 outlines the same region that was highlighted in Figure 9.7. Due to the exaggeration at this latitude, the map reader may actually be *better able* to distinguish the spatial distribution of the individual proportional circles throughout the red highlighted area than they would using the Lambert cylindrical equal area projection in Figure 9.7. However, we offer a cautionary note—while the exaggeration may enable the individual circles in the upper latitudes to be visually distinct, this separation may lead the map reader to the erroneous conclusion that the data are more dispersed than they are in reality. Despite this advantage of potentially being able to better distinguish individual symbols, the exaggeration also leads to excessive visual dominance of Greenland, Russia, and Antarctica, which may distract from the proportional circles.

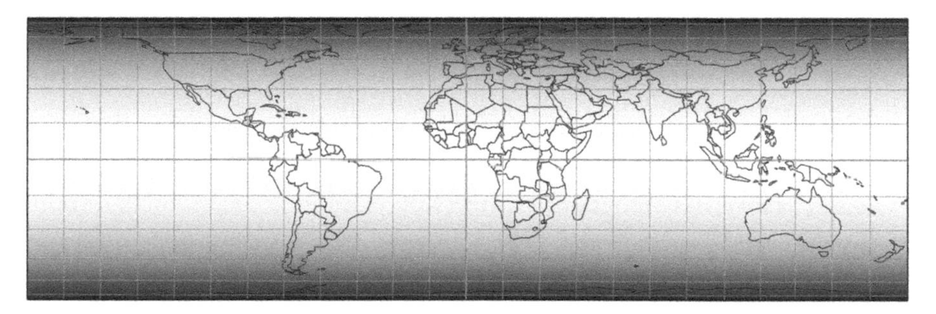

FIGURE 9.8
Overall distortion on the Lambert cylindrical equal area projection. Since this projection preserves areas, only angular and scale distortion is shown.

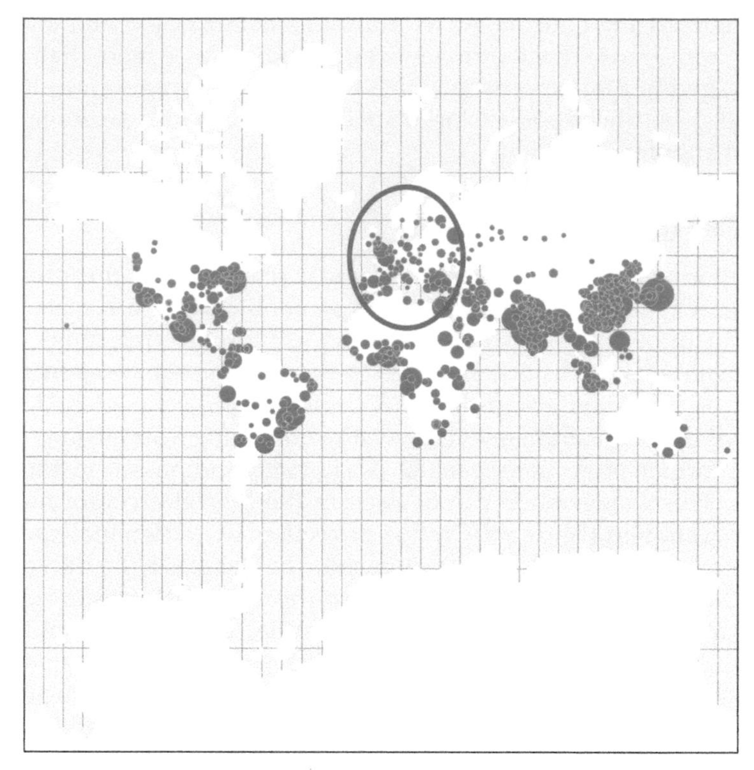

FIGURE 9.9
Population estimates for the large urban areas shown on the Mercator conformal cylindrical projection.

Compromise Projections

Compromise projections do not preserve any specific projection property and instead attempt to balance the exaggeration and compression to landmasses that is found on conformal and equal area projections, respectively. This balancing leads to a better approximation of the visual appearance of landmasses on globes than provided by either equal area or conformal projections. For example, Figure 9.11 shows the Gall stereographic compromise cylindrical projection. With this projection, the extreme compression and exaggeration of landmasses present with the Lambert cylindrical equal area (Figure 9.7) and Mercator cylindrical conformal (Figure 9.9) projections is absent. The Gall stereographic projection exhibits a slight north—south exaggeration in the landmasses, which increases the distance between the true point locations but provides more visual room for the proportional circles to be distinguishable. Note also the considerable east–west stretching in the upper latitudes that is characteristic of cylindrical projections. Again, this stretching allows for more visual room to display the individual circles in an otherwise crowded geographic space (particularly in Europe).

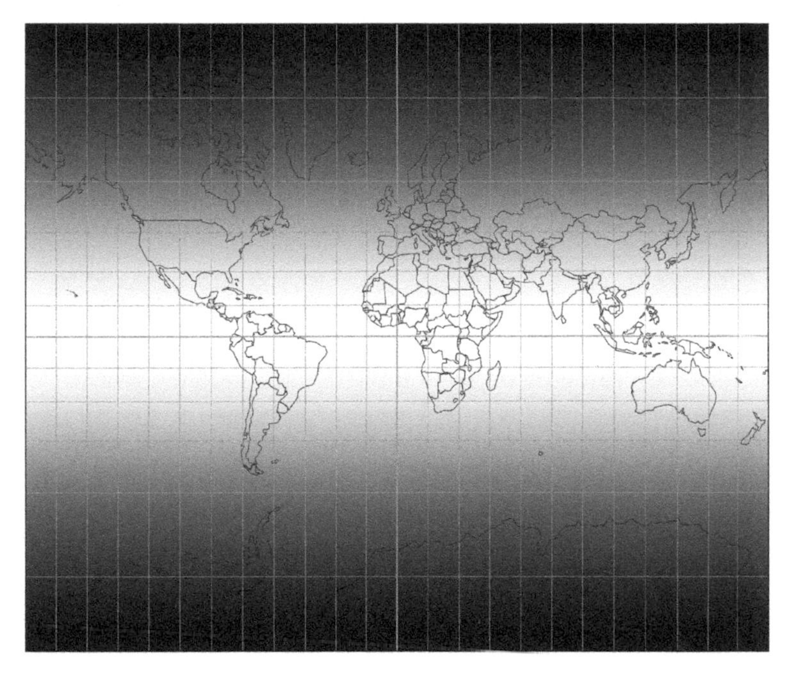

FIGURE 9.10
Overall distortion on the Mercator cylindrical conformal projection.

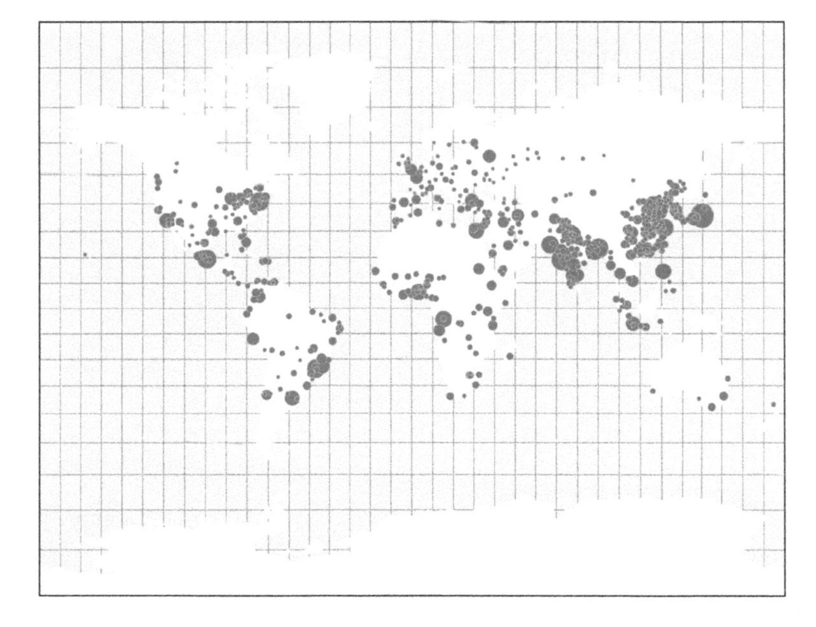

FIGURE 9.11
Population estimates for the large urban areas shown on the Gall stereographic cylindrical compromise projection.

The overall distortion pattern on the Gall stereographic projection is shown in Figure 9.12. On this projection, distortion increases as one moves away from the two standard lines which are located at 45° north and south. Since compromise projections generally balance areal and angular distortion, the overall green color values representing the blended areal and angular distortion in Figure 9.12 is lighter than in Figure 9.8 and Figure 9.10.

Equidistant Projections

Equidistant projections provide a notable benefit when the purpose of the map is to preserve distances across the map's surface. Unfortunately, this distance preservation can only occur in limited ways—along great circles from a single point or along meridians. Figure 9.13 shows the overall distortion patterns on common equidistant projections: plate carrée cylindrical, equidistant conic, and azimuthal equidistant. On the plate carrée and equidistant conic projections, distances are preserved along all meridians; this is fine if all distance measurements needed on a map are along exactly north–south lines. Distances on the azimuthal equidistant projection are preserved from one point (the projection's center) to all other points on the map. This projection affords a bit more flexibility in measuring distances on a map's surface as the projection's center can be set to any latitude and longitude value.

Equidistant projections can be successfully applied to proportional symbols. For example, imagine a cartographer wanting to examine the toxicity level of

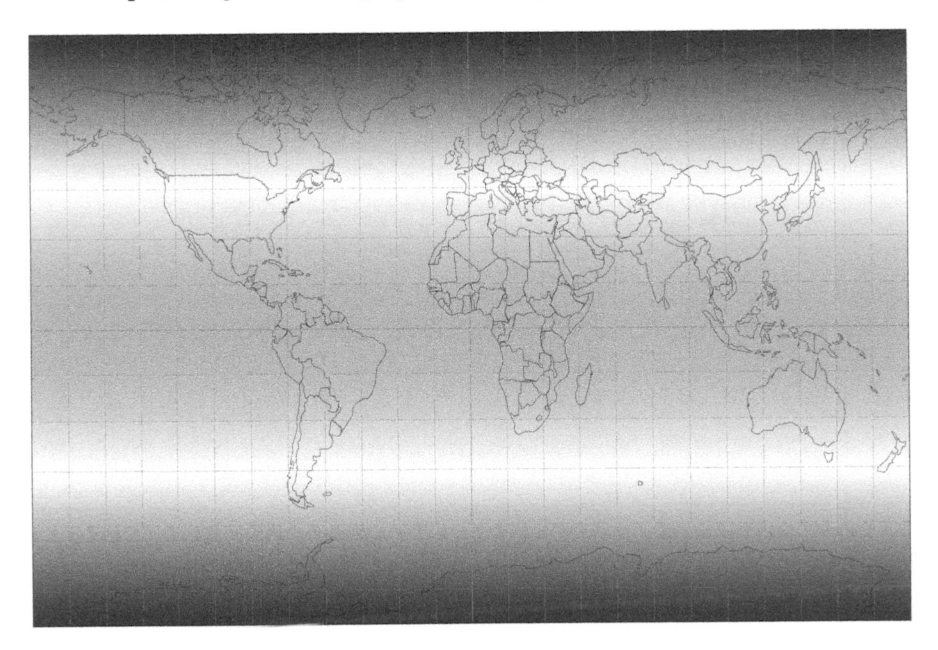

FIGURE 9.12
Overall distortion on the Gall stereographic cylindrical compromise projection.

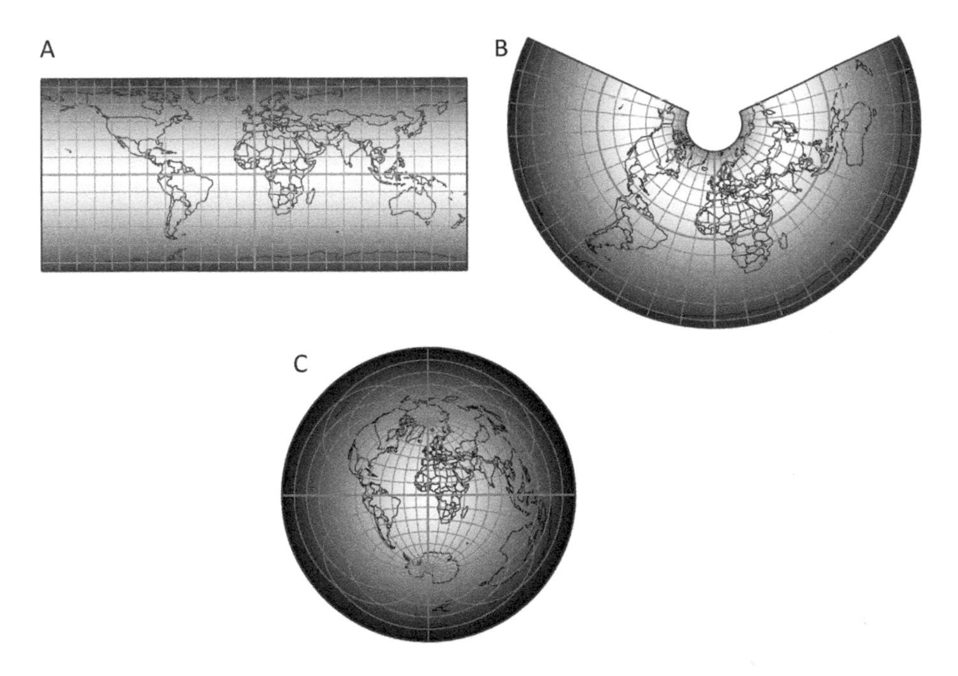

FIGURE 9.13
Overall distortion on the plate carrée (A), equidistant conic (B), and azimuthal equidistant (C) projections.

a pollutant at some distances from an origin. Centering the map on the origin point in question using an azimuthal equidistant projection would allow the map reader to measure accurate distances from that origin point. This distance measuring task could be helpful if the map goal was to get a sense of the spatial distribution of other sampled locations according to a specific distance.

Despite the utility of equidistant projections, measuring distances is not a common task that is needed when using proportional symbol maps, though it is important when we consider proportional flow maps. Figure 9.14 shows flow lines from Amsterdam to different destinations on the plate carrée equidistant, Gall stereographic compromise, and Lambert cylindrical equal area projections. Since the plate carrée projection is equidistant, the lines of latitude are equally spaced along any meridian. This equal spacing mimics what is found on the globe. To some extent, this equal spacing helps to display landmasses closer to their appearance on the globe than is seen on equal area or conformal projections where the spacing is not equal. Preserving the overall shape of the landmasses may be beneficial to some audiences, for example, in being able to recognize a country of interest on the map. Figure 9.14B shows the flow lines appearing on the Gall stereographic projection. On this projection, the landmasses are stretched out in a north–south manner more than is seen in Figure 9.14A or Figure 9.14C. This stretching also effects the appearance of the flow lines as the flow lines

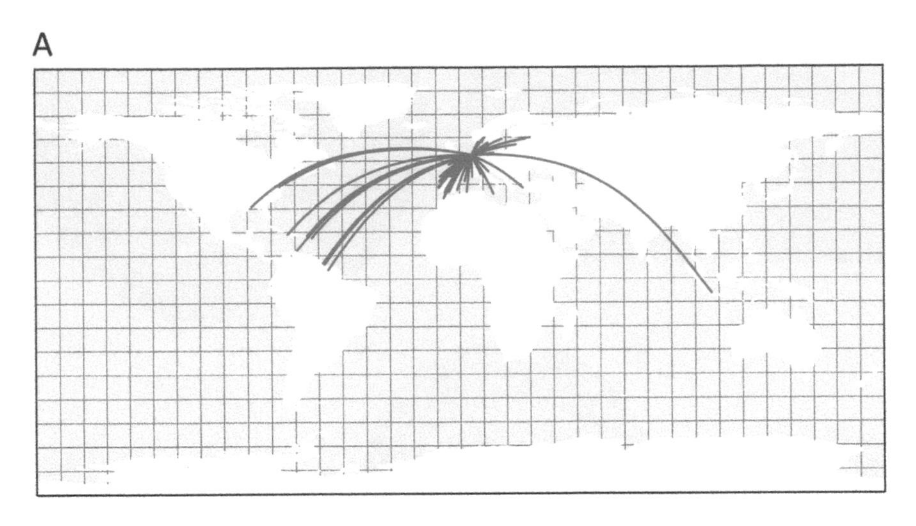

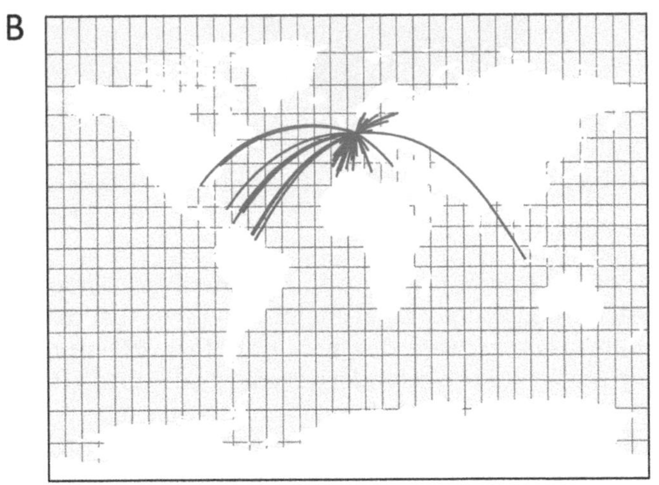

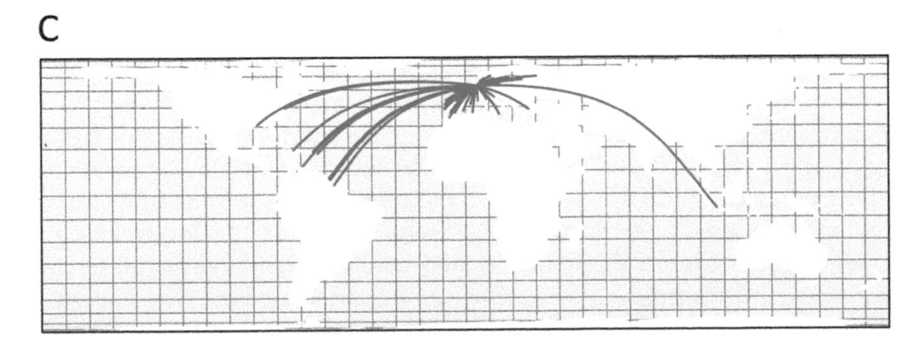

FIGURE 9.14
The flow lines mapped on the plate carrée equidistant (A), Gall stereographic compromise (B), and Lambert cylindrical equal area (C) projections.

appear to be more elongated in a north–south fashion than in Figure 9.14A and Figure 9.14C. The north–south compression of landmasses that appears in the Lambert cylindrical equal area projection (Figure 9.14C) results in a loss of distinction in some of the flow lines especially near the origin point of Amsterdam, making their recognition difficult.

Figure 9.15 shows the flow lines appearing on the azimuthal equidistant projection. On this projection, the map's center is aligned with the coordinate values of Amsterdam from where the flow lines originate, creating an oblique aspect. Since this projection is centered over the flow lines' origin, this projection presents the accurate distance of each flow line from its origin to destination.

Map Design Considerations on Proportional Symbol Maps

This section discusses four important design considerations when choosing projections for proportional symbol maps (we will discuss design considerations specific to flow maps in the next section). First, consideration should

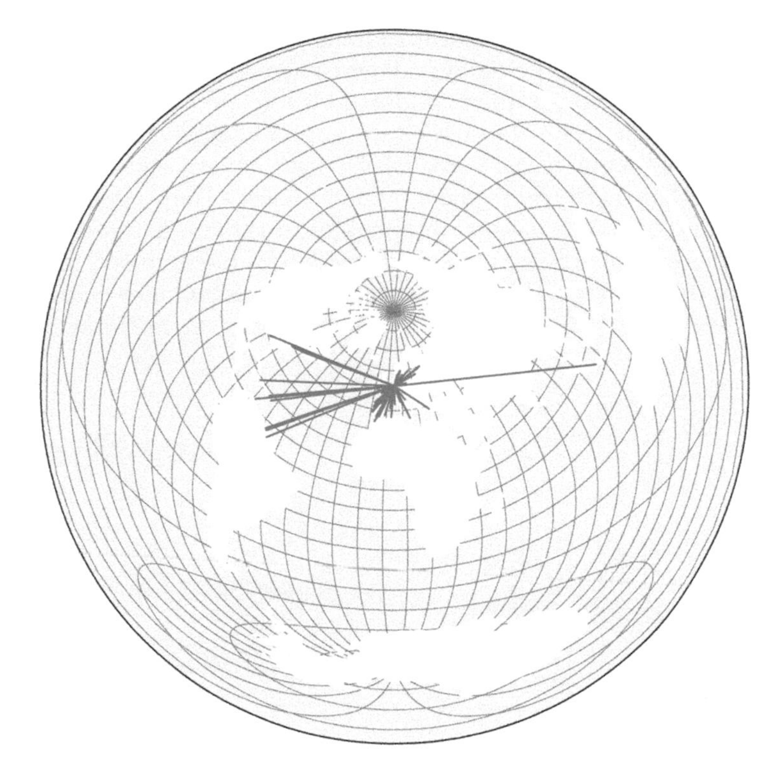

FIGURE 9.15
The flow lines mapped on the azimuthal equidistant projection.

be given to the projection class. While cylindrical, conic, and planar are commonly used classes, other classes afford interesting design options. Second, the projection aspect determines where the geographic center of the map is and can influence the arrangement of the proportional symbols around that center. Third, projections display the poles as points (as they appear on a globe) or pole lines of various lengths, which can be an aesthetic concern. Representing poles as points or lines also controls the curvature of the meridians and how they intersect the poles, which impacts the overall shapes of landmasses especially along a map's periphery. Fourth, projections can show interruptions that divide the oceans or landmasses into lobes. A separate discussion will cover specific projection applications when measuring distances and directions with flow lines.

Projection Class

The projection class can have a significant effect on the overall appearance of the proportional symbol map. We have already seen the proportional symbols appearing on the cylindrical class (Figure 9.14). Figure 9.16 shows the proportional symbols mapped using the Lambert conformal conic projection (Figure 9.16A) and the Lambert azimuthal equal area projection (Figure 9.16B) where the standard lines and central point, respectively, are selected to emphasize the Southern Hemisphere. The Lambert conformal conic projection creates an unusual global view (concave appearance about the South Pole) when the standard lines are both located in the Southern Hemisphere. In many cases, conic projections have their standard lines located in the Northern Hemisphere, which creates a "fan" appearance concave about the North Pole. Generally speaking, due to the extreme distortion that appears on conic projections, they are not frequently used to map the world. Displaying the world on an azimuthal projection also causes visual problems with data lying near the map's periphery. In this case, the data

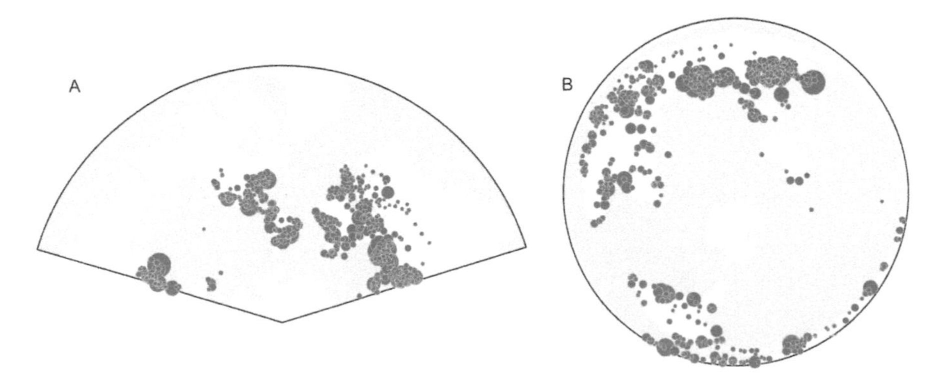

FIGURE 9.16
The Lambert conformal conic (A) and Lambert azimuthal equal area (B) projections.

associated with North America and Europe is compressed to the point of being unrecognizable.

To avoid some of these problems, other projection classes that are better suited to displaying the world can be considered for mapping proportional symbols. For example, pseudocylindrical and modified azimuthal classes are possible alternatives. The advantage of these classes is that the meridians are drawn using different mathematically defined curved lines, which better imitates the curvature of lines of longitude appearing on the globe. Three common pseudocylindrical projections include the sinusoidal, Mollweide, and Robinson projections. The sinusoidal and Mollweide projections are both equal area while the Robinson projection is compromise. The large urban area population estimates are shown on the Mollweide (Figure 9.17A), sinusoidal (Figure 9.17B), and Robinson (Figure 9.17C) projections. In Figure 9.17A–C, note that the data and landmasses are mostly distinguishable. Due to the way in which the poles are represented and the curvature of the meridians, the sinusoidal does cause some of the proportional circles to be visually compressed along the left-edge of the map, making the data pattern hard to distinguish.

While the parallels on the pseudocylindrical class all appear as straight lines, the modified azimuthal projections display the parallels as curves. Figure 9.18 shows three common modified azimuthal projections: the Briesemeister equal area (Figure 9.18A), Winkel tripel compromise (Figure 9.18B), and Wagner VII equal area (Figure 9.18C). The Briesemeister projection defaults to an oblique aspect centered at 10° E and 45° N. This aspect shifts the focus of the map so that the compressed landmasses found with other equatorial-centered projections is relieved. The Winkel tripel appears "taller" than other projections. This north–south stretching is actually beneficial in allowing more visual space for the individual proportional

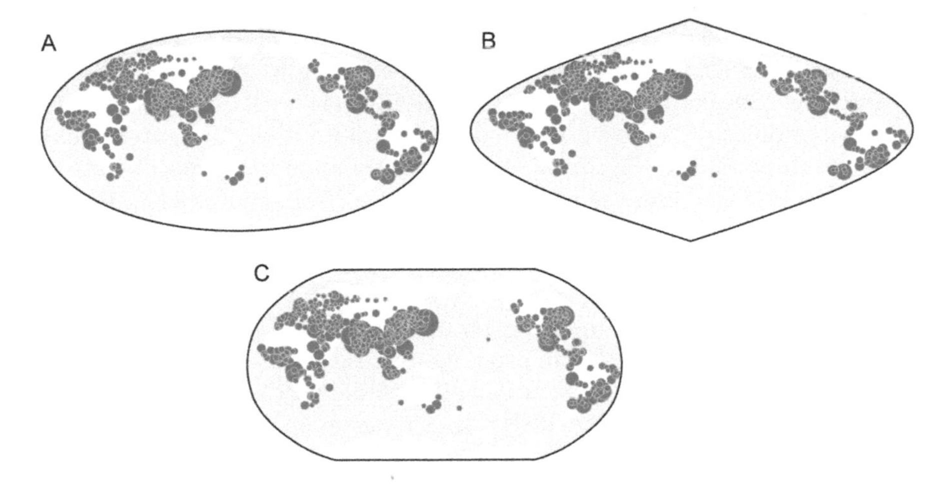

FIGURE 9.17
The sinusoidal equal area (A), Mollweide equal area (B), and Robinson compromise (C) pseudocylindrical projections.

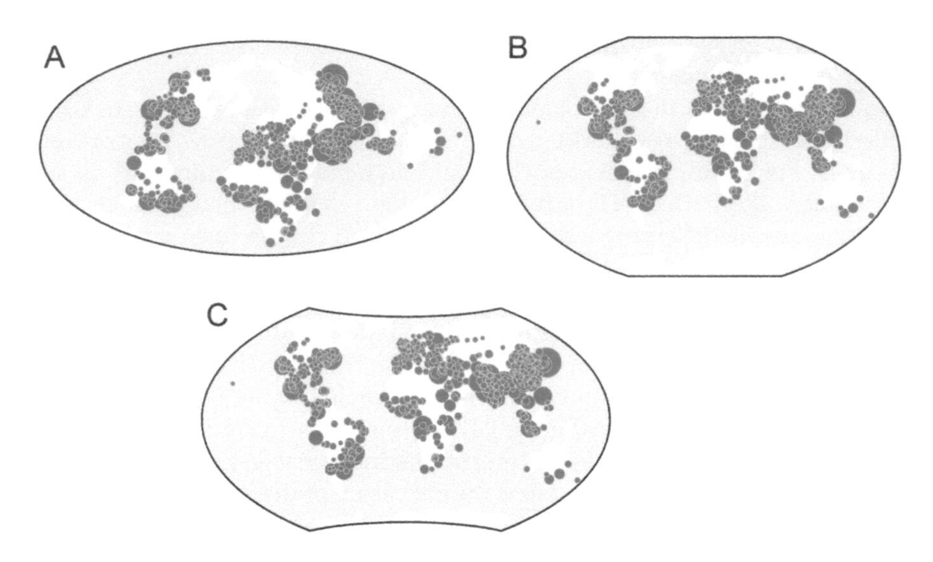

FIGURE 9.18
The Briesemeister equal area (A), Winkel tripel (B), and Wagner VII (C) projections.

circles to be displayed. The Wagner VII projection has a rather interesting appearance. The lines representing the poles are gentle curves and the length of the other lines of latitude are longer than they would appear on the globe, allowing more visual space for the proportional circles to appear.

Projection Aspect

The projection aspect can be thought of as centering the projection over a particular location. Any projection can be *centered*, which aligns a particular geographic area of interest to the map's center. As discussed in Chapter 2, the projection's aspect typically involves specifying a central latitude and central longitude coordinate value. For example, Figure 9.19 shows the large urban area dataset displayed on the Lambert azimuthal equal area projection with three aspects. Figure 9.19A displays an equatorial aspect where the central latitude is 0° and the central longitude is 15° E. Figure 9.19B displays a polar aspect where the central latitude is 90° S and the central longitude is 90° E. Figure 9.19C displays an oblique aspect where the central latitude is 20° N and the central longitude is 90° E. Changing the aspect brings different parts of the world into the center of the map and forces other parts to the periphery. This centering can produce striking visual impacts. For example, Figure 9.19A highlights the areas of Central Africa where large urban areas are not common; Figure 9.19B brings to the map reader's attention the stark difference in large urban area populations between the Southern and Northern Hemispheres; and Figure 9.19C focuses on the considerable dense urban areas that are present in the Indian subcontinent and nearby eastern Asia juxtaposed to the sparseness that is found in eastern Russia.

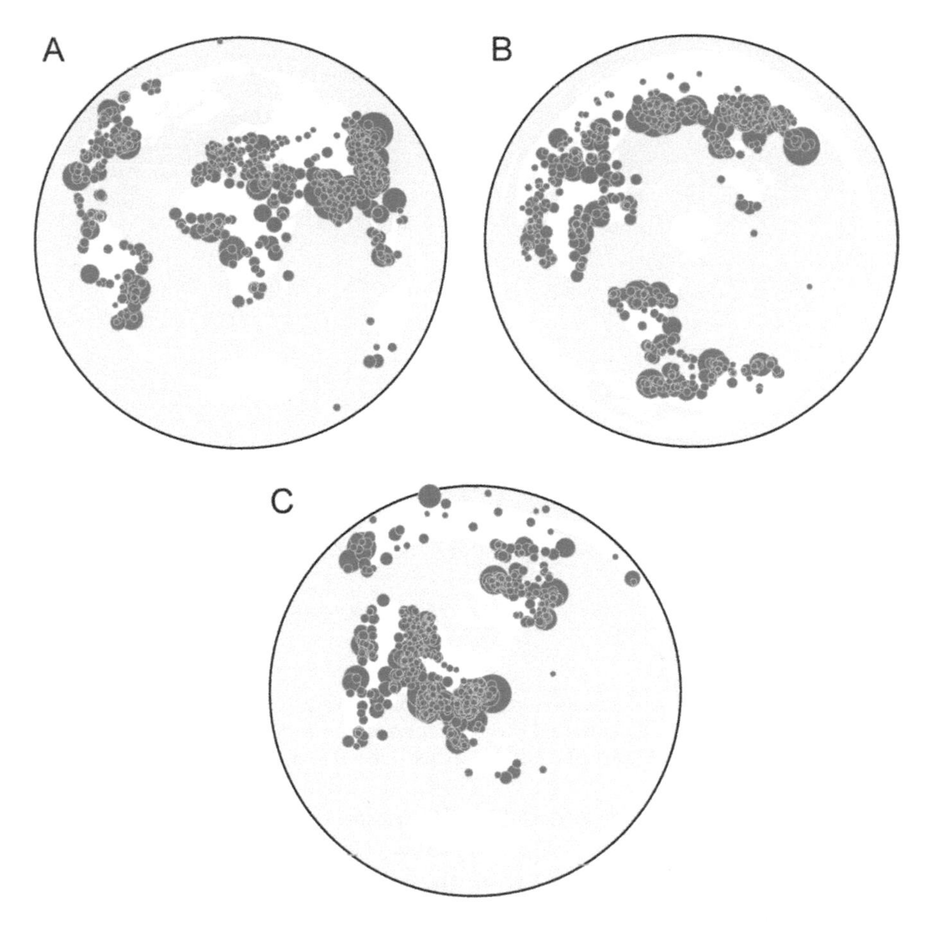

FIGURE 9.19
The large urban areas mapped on the Lambert azimuthal equal area projection showing equatorial (A), polar (B), and oblique (C) aspects.

Poles Represented as Points or Lines

On a globe, the North and South Poles are represented as points. Projections can preserve the poles as a point. Figure 9.20A–B show the Lambert azimuthal equal area and Aitoff modified azimuthal projections. On these projections, the poles are represented as points and the curvature of the lines of longitude create landmasses in the polar regions to be more similar to what is seen on a globe. As a result, the spatial distribution of the proportional circles looks close to how that distribution would appear if you were looking at the globe. Other projections represent the poles as lines of varying lengths, such as the Winkel I compromise pseudocylindrical (Figure 9.20C) and the Eckert III compromise pseudocylindrical (Figure 9.20D) projections. The Winkel I shows the lines of longitude as

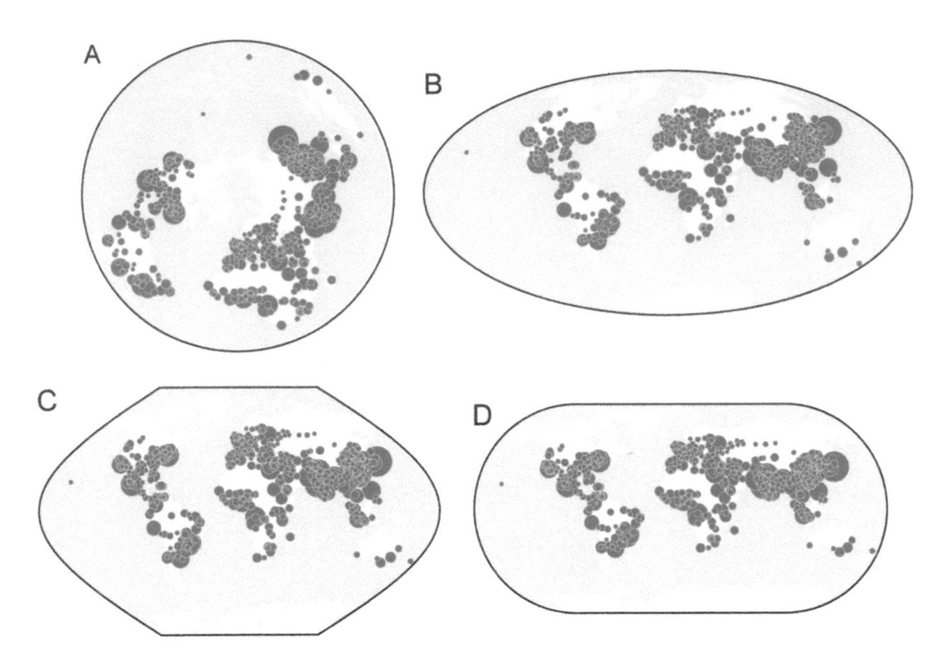

FIGURE 9.20
The North and South Poles represented as points on the Lambert equal area azimuthal (A) and Aitoff compromise modified azimuthal (B) projections and lines on the Winkel I compromise pseudocylindrical (C) and Eckert III compromise pseudocylindrical (D) projections.

terminating along lines representing the poles that are 0.61 the length of the equator. The Eckert III projection shows the poles as lines that are 0.5 the length of the equator. Aside from the general overall shape differences between the Winkel I and Eckert III projections, there is a visual difference in the appearance of landmasses in the polar regions and the apparent spatial distribution of the proportional circles. The gently curved lines of longitude (represented by sine curves) on the Winkel I near the polar regions creates a slight north–south stretching to the landmasses in these areas. The pole lines also create some east-to-west stretching to the landmasses in the upper latitude areas. The proportional circles near the North Pole on this projection also are impacted. The north–south stretching gives the appearance that the spatial distribution of proportional circles may be more expansive than reality would suggest. On the Eckert III, the lines of longitude are represented as semi-ellipses (smooth curves) that approximate the curvature of lines of longitude as on the globe. Since the poles on the Eckert III are represented by lines that are longer than found on the Winkel I, the landmasses in the upper latitudes exhibit greater stretching, which allows the proportional circles in these latitudes (e.g., Europe) to be more distinguishable on the Eckert III. The stretching of landmasses on this projection also helps the map reader better distinguish the individual proportional circles on the map.

Interruptions

Projections also can be interrupted. One of the well-known interrupted projections is the Goode homolosine equal area pseudocylindrical. On this projection, the interruptions occur along meridians and can be placed so that either landmasses or oceans are divided into lobes. Figure 9.21 shows the overall distortion on Goode homolosine projection with five interruptions. Figure 9.21A shows the overall distortion for interruptions over oceans while Figure 9.21B shows the interruptions over landmasses. Note that the white areas in each lobe in Figure 9.21A–B signify the areas that are least distorted.

While the Goode homolosine projection has seen considerable use in world thematic atlases, some argue that map readers are challenged by reconciling their mental map of the world to what is presented when looking at a projection. Chapter 4 discusses some of these concerns. Briefly, one concern relevant to proportional symbols is where to place the interruptions.

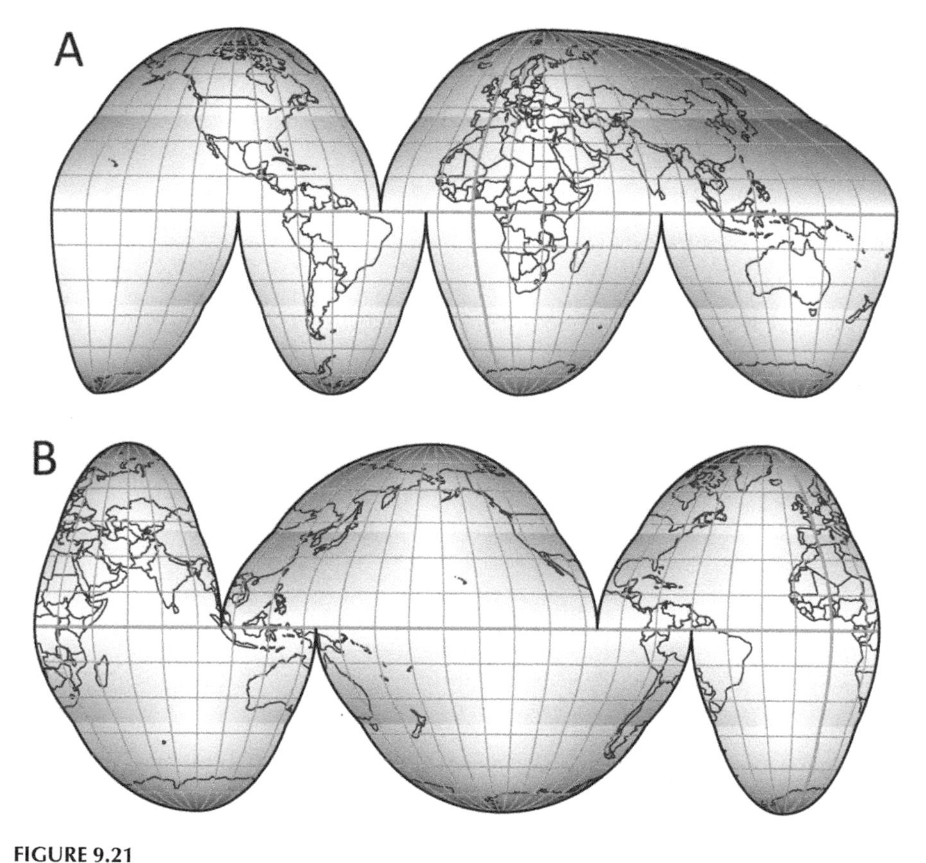

FIGURE 9.21
The overall distortion pattern on the Goode pseudocylindrical equal area homolosine projection for interruptions over ocean (A) and land (B).

Figure 9.22A shows a map of the proportional circles with the oceans interrupted. The lobes contain the landmasses as this is where the data and symbols are located. However, in Figure 9.22B, the divisions are not strategically placed and awkwardly divide some of the landmasses. In Figure 9.22B, the divisions are placed on the oceans, which can negatively impact the reader's ability to identify the symbols' sizes, especially those symbols juxtaposed next to the divisions. This same problem was discussed with respect to dot mapping in Chapter 8.

Using interrupted projections to map flow lines is not recommended. Figure 9.23 shows the flow lines appearing on the Goode homolosine projection. In this case, the flow lines are cut at the interruptions. This breaks the continuity of the flow lines. In turn, the interruptions make it difficult for map readers to mentally connect the flow lines from one side of the interruption to the other and therefore fully visualize the data pattern. Similarly, the

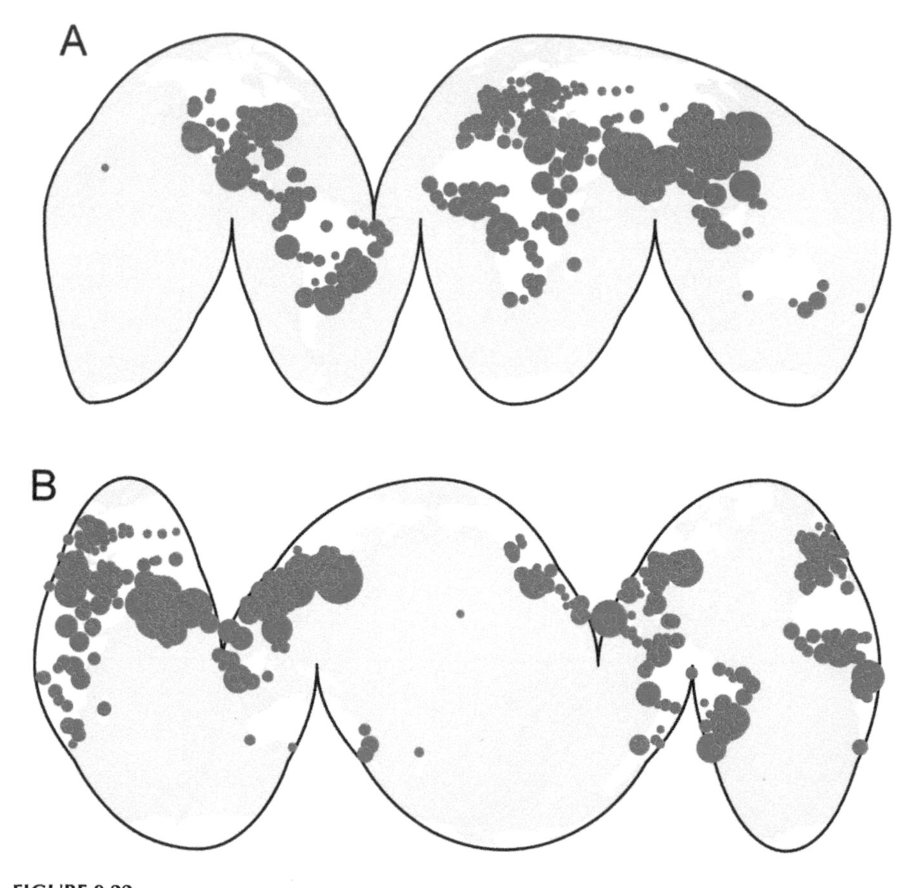

FIGURE 9.22
Proportional circle mapped on the Goode homolosine interrupted projection with the divisions placed over the oceans (A) and land (B).

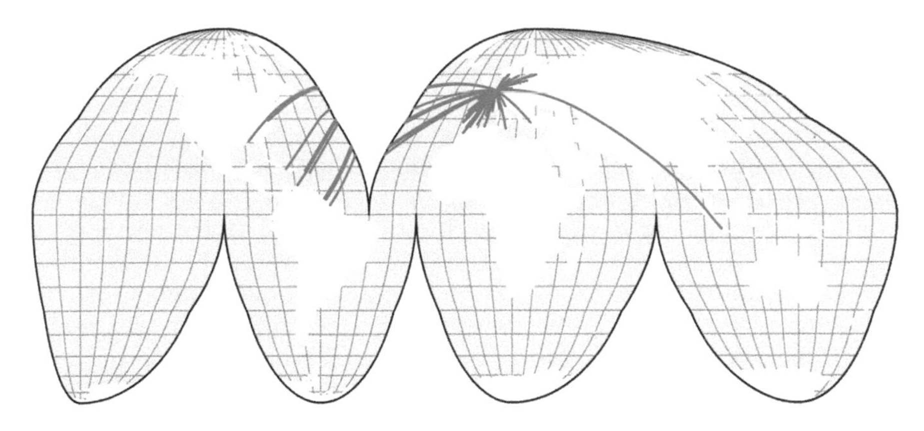

FIGURE 9.23
Flow lines mapped on the Goode homolosine interrupted projection with the divisions placed over the oceans (A) and land (B).

interruptions make it difficult for map readers to estimate relative distances of the flow lines. The interruptions may give the impression that there is "missing" distance when in fact the interruption line is actually the same line of longitude.

Map Design Considerations on Flow Maps

Choosing projections for flow maps can require additional consideration not found with other symbolization methods covered in this book. These are to facilitate the following:

- Calculating distance between the origin and destination of the flow lines
- Determining direction between the origin and destination of the flow lines
- Showing the path along which the data are "flowing"
- Enhance the appearance of the flow lines

Distance Calculation

In some instances, it may be necessary for map readers to not only know the quantity of data represented by the width of the flow lines, but the distance "traveled" by any given flow line. Measuring the distance of a flow line on a projection can be simple, provided that the flow *path* can be represented by a single smooth line such as a great circle. Accurately calculating distances on a map represented by complex flow lines is difficult. To measure distances from one point to another on a projection should begin by centering the azimuthal equidistant projection over the origin point of the flow line to be measured. By doing so, all lines from that single point are represented on the map as straight lines (great circles). These straight lines can easily be measured and accurately represent the distance of a given flow line (Figure 9.24A). If there are two origin points, as in the case of Figure 9.24B (Charleston, South Carolina, and Amsterdam), then the two-point azimuthal equidistant projection can be used. In this case, the two-point azimuthal equidistant projection allows accurate distances to be measured from two points to any

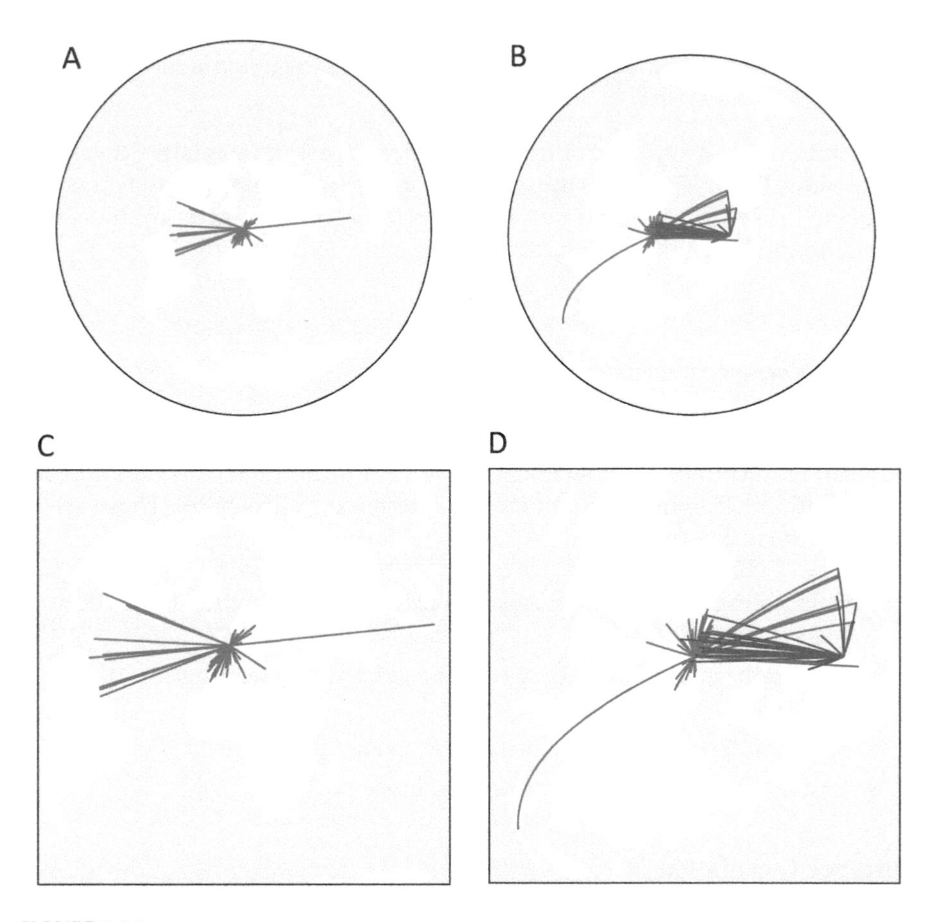

FIGURE 9.24
The azimuthal equidistant projection (A) and the two-point azimuthal equidistant projection (B). Zoomed in views of the flow lines are shown on azimuthal equidistant projection (C) and the two-point azimuthal equidistant projection (D).

other point on the map. It is important to remember that no projection allows for distances to be accurately measured from three or more distinct points.

Direction Calculation

Projections that preserve directions are called azimuthal and allow accurate directions to be measured from the projection's center to any other point on the map. The azimuthal equidistant, Lambert equal area, gnomonic, and orthographic are some of the more common azimuthal projections allowing accurate directions to be determined. To take advantage of the azimuthal property on any of these projections, the first step is to center the map over the flow lines' origin. In this case, the flow origin is Amsterdam. To determine a direction of a given flow line, connect an imaginary line from the North Pole to Amsterdam. Starting from this line, measure an angle clockwise until the flow line in question is intersected. The resulting angle (or azimuth) is the direction from Amsterdam to the flow destination.

Show the Flow Paths

In some cases, it may be necessary to show the actual paths of the flow lines. The gnomonic azimuthal projection is useful for this requirement as it shows all straight lines (drawn anywhere on the projection) as great circles. Figure 9.25 shows the flow lines from Amsterdam to other countries on the gnomonic projection. This special property is useful to visualize the path over which the shortest distance would travel assuming the travel path exactly follows a great circle. The disadvantage of this projection is that only less than a hemisphere can be shown (as evidenced by the flow line heading eastward toward Jakarta being cut off the map). Scale is also highly variable throughout this projection and because of this reason one cannot accurately measure distances of the flow lines. Another consequence of the considerable scale variation found on this projection is the exceptional exaggeration in the appearance of the landmasses. In some cases, the exaggeration is sufficiently extreme to prevent quick and accurate identification of a specific landmasses.

Enhanced Visual Appearance

While the previous discussion looked at how projections could fulfill specific measuring tasks, this paragraph focuses on characteristics of projections that can be used to enhance the visual appearance of the map. The orthographic azimuthal projection (Figure 9.26A) gives a view of Earth from out in space which can give an eye-catching appearance to the map.

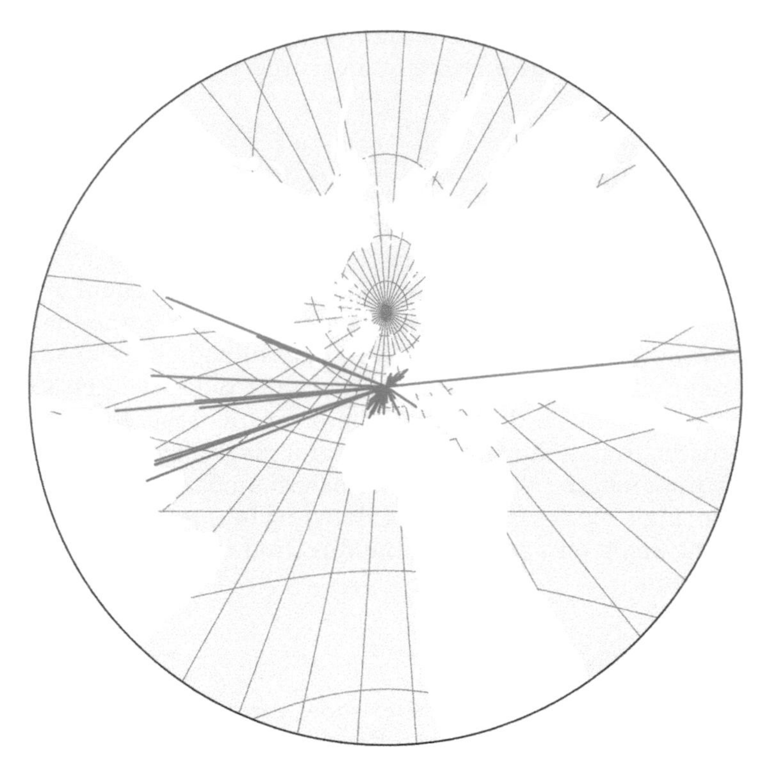

FIGURE 9.25
The gnomonic azimuthal projection.

The view shown in Figure 9.26A is centered over Amsterdam. Figure 9.26B uses the orthographic projection with a different center to give the perspective of the viewer looking from the flow destinations in the Western Hemisphere east toward Amsterdam. The general vertical perspective azimuthal projection appears in Figure 9.26C centered over Amsterdam. This projection has a height parameter which controls the extent of the geographic area to appear on the map. Heights that are more distant from Earth's surface take on the appearance of the orthographic (which is projected from a point located at an infinite distance from Earth's surface). Heights closer to Earth's surface give more of a "fish-eye" appearance to the map. Note that in Figure 9.26C, the height parameter of the general vertical perspective projection shows Europe at a larger scale than with the orthographic projection in Figure 9.26A. This larger scale allows more of the details to be seen and helps to highlight the flows ending in Europe. On either the orthographic or the general vertical perspective projection, scale is highly variable and should not be used in situations where measuring distances is required.

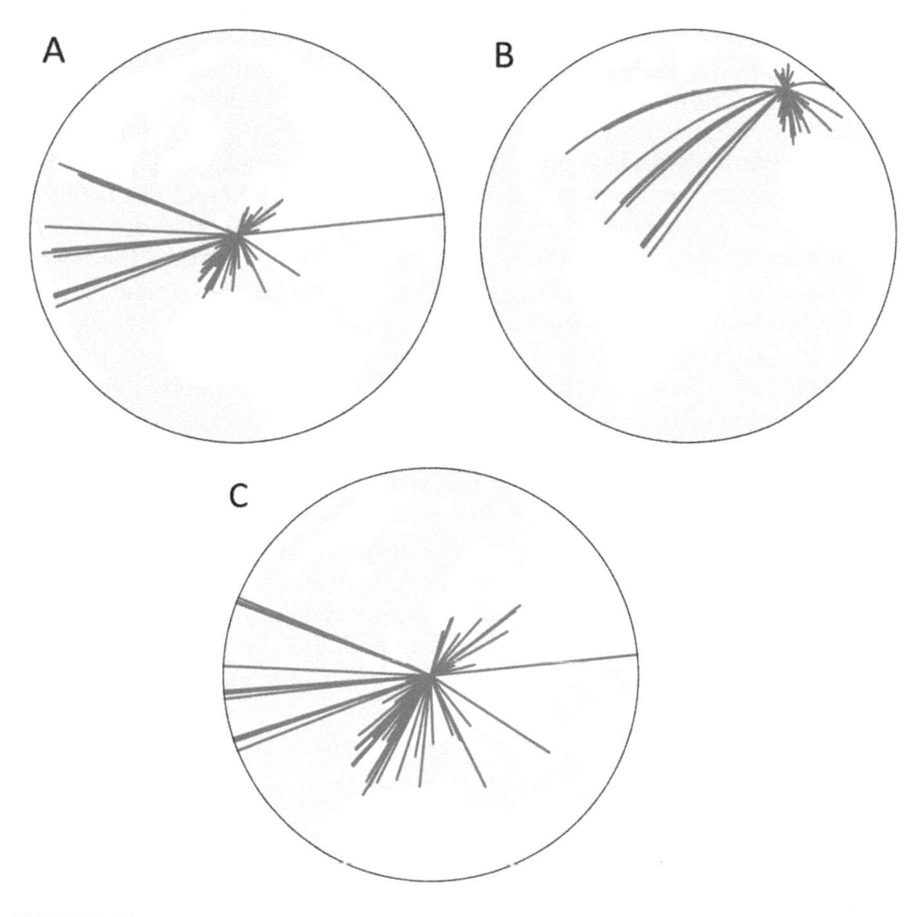

FIGURE 9.26
The orthographic azimuthal projection centered over Amsterdam (A) and in the central Atlantic Ocean (B), and the general vertical perspective azimuthal projection (C) centered over Amsterdam but focused on Europe.

Conclusion

In this chapter, we have examined the implication of projection on proportional point and flow maps (a map type for depicting discretely occurring and abruptly changing phenomena). These two map types tend to have different interpretation goals, so have different considerations for the projection properties that are most supportive of the common map reading tasks. For proportional point maps, map readers are typically focused on identifying patterns based on distance between points (e.g., when the symbols used start to coalesce) and on identifying values for specific locations. The challenge of accurate interpretation of patterns across the points is largely a distance-based task and is similar to the projection suggestions and needs discussed with dot maps. Since the symbols used in proportional point maps are not dependent upon the enumeration unit's size, preserving area is not critical, however, we still would list both equal area and compromise map projection as good choices. Since there are many different equal area and compromise projections to choose from, we recommend considering several options for each to identify the one that serves the distribution of your data the best—for instance, minimizes stretching and compression that may impact a map reader's ability to interpret the polygonal regions on the base map underneath your proportional point symbols. Map readers often evaluate the size of symbols and attach meaning to the symbol based on location on the map, and if the region is not recognizable due to stretching, shearing, or compression on the projection then it will limit the effectiveness of the map.

For flow maps, the key interpretation tasks are based on identifying value for the flow line (width of the line, which is not impacted by projection) and the distance and direction between locations. In general, we recommend using an azimuthal equidistant for distance measurements, and an azimuthal equidistant, Lambert azimuthal equal area, gnomonic, and orthographic when directions need to be determined. See Chapter 10 for deeper discussion of projections for measuring distances and directions between locations.

10

Special Maps

Earlier chapters in this section focused on the most common general use map types (isarithmic, choropleth, dot, and proportional symbol maps). While these map types are likely the most common, they are far from the only map types that are used in communicating spatial patterns. In this chapter, we turn our attention to some of the more specialized map types. It would be impossible to cover all potential map types that might be made, so we focus on ones that we think you would be most likely to encounter, including spatial bins, heatmaps, and path or navigational maps used for visualizing and/or measuring shortest routes or determining directions. Examples of these map types are shown in Figure 10.1. Through this chapter we will discuss each in more detail to provide background and guidance on projection selection.

In the earlier chapters, we provided more extensive detail and discussion behind projection choices. Since we are focusing on several specific map types or characteristics in this chapter, and they each have some level of overlap with other common map types, we will discuss each in slightly less depth. However, we will still include the relevant decision-making criteria for assessing appropriateness of different projections for each map type.

As a refresher, the basic guidelines for assessing appropriateness of a projection (Chapter 5) for any map type are based on:

- Considering the tasks that readers are likely to undertake with your map and identifying projection properties of importance.
- Evaluating geographic location, size, and shape of the region of interest.
- Considering broader design and analytic considerations such as the projection shape, graticule, and interpretability.

Spatial Bin Maps

Spatial binning is a mapping technique that is most often used when working with large, dense sets of spatial point data. In these instances, the point locations are sufficiently dense that the coalescence of point marks (due

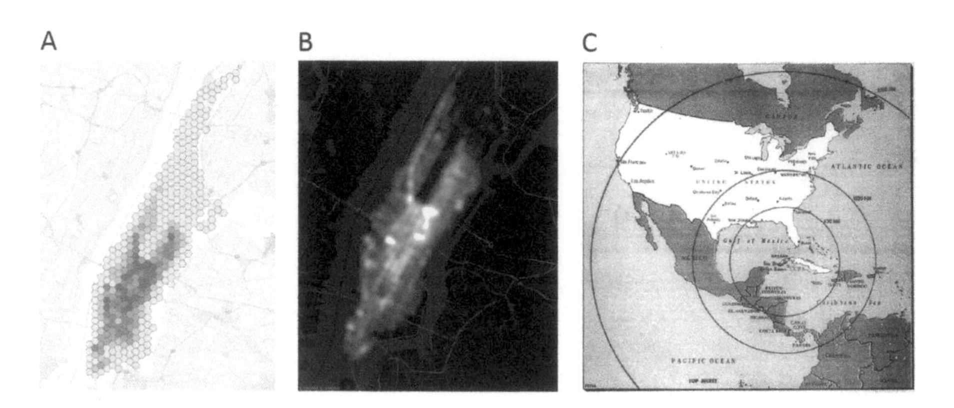

FIGURE 10.1
A spatial bin map of taxicab pickup locations (A). A heatmap showing density of taxicab pickup locations in Manhattan (B). A map showing range of Soviet missiles launched from Cuba to the United States (C). (Image retrieved December 28, 2018, from: https://govbooktalk.gpo.gov/2012/01/27/cia-world-factbook/.)

to a combination of number of points and size of symbols) masks the true patterns in the data, for instance, when considering a dataset with approximately 1 million points located in Manhattan. The raw point locations for this dataset (Figure 10.2A) are sufficiently congested that it is difficult to discern a pattern. The binned map (Figure 10.2B), simplifies the pattern so that we just see counts in individual, non-overlapping polygonal bins across the area of interest.

Spatial binning is simply a method for aggregating individual point locations into polygonal regions. These regions can be a regular tessellation of polygons (e.g., squares or hexagons) or irregular boundaries (e.g., political boundaries such as states or countries). Point data that have been aggregated into irregular boundaries are often simply referred to as choropleth maps (see Chapter 6). While, technically, hexagonal or square bin maps can also be considered choropleth maps, they differ in projection-related considerations for their creation and production. This section will detail these differences and discuss selection of appropriate projections for maps using regular spatial bins (we will just refer to these as "spatial bin maps").

Much like with isarithmic maps, consideration of appropriate projection is important in multiple parts of the spatial binning process—the projection in which the bins are created and applied, and the projection in which the binned results are displayed.

Tasks for Maps Using Spatial Bins

The general interpretation tasks for spatial bin maps are similar to those of choropleth maps: identify values at specific single locations of interest, compare values across multiple locations, and identify broad regional patterns

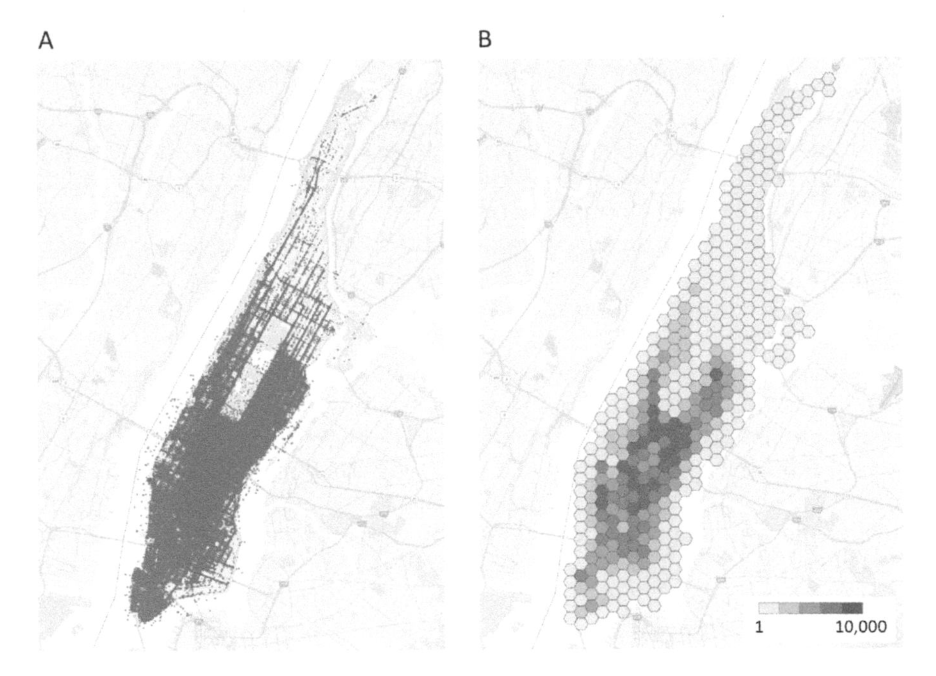

FIGURE 10.2
Taxicab pickup locations in Manhattan, N.Y., as raw point locations (A) and binned into a hexagonal grid (B).

through visual aggregation of groups of neighboring values. Since spatial bin maps rely on regularly sized and shaped polygons instead of the irregular polygons common in choropleth maps, the visible patterns of change across the map can also be considered somewhat in line with the *smoothly* changing phenomena shown in isarithmic maps. The regular grid of data allows the map reader to more easily focus on a smooth, general pattern in the data as it changes across the area of interest.

One of the benefits of spatial bin maps is that the enumeration units used present a continuous, regular grid (typically hexagonal or square polygons). Because the bins all visually appear the same shape and size on the projected map, it is easy to assume that the values represented by the bins illustrate the same relative count of points *or* density within each bin (since the area denominator for each bin appears the same). However, projection distortion creates a problem with the accuracy of this assumption. We will discuss this in limited detail in the next section on projection considerations for spatial bin maps; for a deeper consideration than we will cover here, we recommend "Shapes on a Plane: Evaluating the Impact of Projection Distortion on Spatial Binning" (Battersby, Strebe, & Finn, 2016).

Projection Considerations for Spatial Bin Maps

Much like with isarithmic maps, it is important to consider the projection implications when *creating* the spatial bin map/performing analytic calculations and when *visualizing* results of the binning process. Ideally, you would use the same equal area projection for both parts of the process (creating the bins and visualizing the binned data) so that the bins all represent the same geographic area *and* provide a regular grid to make visual analysis easier and more intuitive; however, we recognize that that isn't always possible. To explain our recommendation of an equal area projection, let's dig in a little more...

The first projection-related problem to consider when working with spatial bin maps is what projection to use when creating the spatial bins. It turns out that while both squares and hexagons, the two most common shapes to use for bins, tessellate perfectly on a plane (e.g., the flat map), they do not on the sphere. So, it is *impossible* to create a set of spatial bins that truly represent the same area and shape region on Earth's surface (see Chapter 2 for a refresher on the inability to preserve both areal and angular measurements on projected maps). To facilitate the primary task of helping map readers understand either the relative count or density of aggregated points on the map, we recommend using equal area projections for generating spatial bins. If it is not possible to use an equal area projection for generating spatial bins, you should examine the areal distortion across your region of interest in your projection of choice. There are mathematical ways of doing this (see Battersby et al., 2014), or you may opt to use a program like Geocart to visualize the distortion (we have used Geocart to make most of the distortion visualizations throughout this book).

Figures 10.3 to Figure 10.5 help explain our recommendation for using an equal area projection for binning your data. In this example, we begin by generating a set of 100,000 randomly distributed points and displaying them with two different projections: an equal area cylindrical projection (Figure 10.3A) and web Mercator projection (nearly conformal) (Figure 10.3B). On each of the two projections, we can create a set of regular hexagonal bins and overlay those bins on the map (Figure 10.4) to use in aggregating the points shown in Figure 10.3. As we mentioned earlier, while the hexagons appear to be the same size and shape in both projections shown in Figure 10.4, only the hexagons on the equal area cylindrical projection represent the same area on Earth.

To demonstrate the difference that projection makes with spatial binning, we binned the points shown in Figure 10.3 using the hexagonal bins shown in Figure 10.4 and symbolized them into quintiles (Figure 10.5). A quintile classification assigns the same number of points to each class, so the darkest blue shade on the map will be the bins representing counts in the top 20% of the distribution for each map, and the lightest blue will represent the bins with counts in the lowest 20% of the distribution. On the cylindrical equal

A

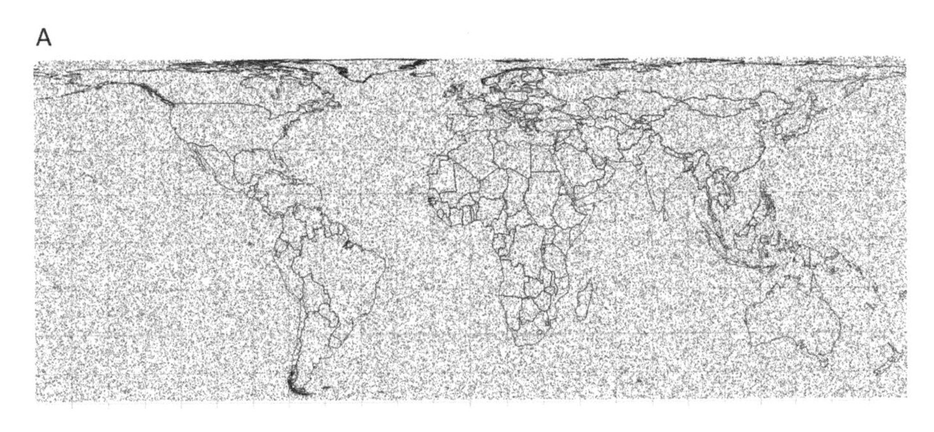

B

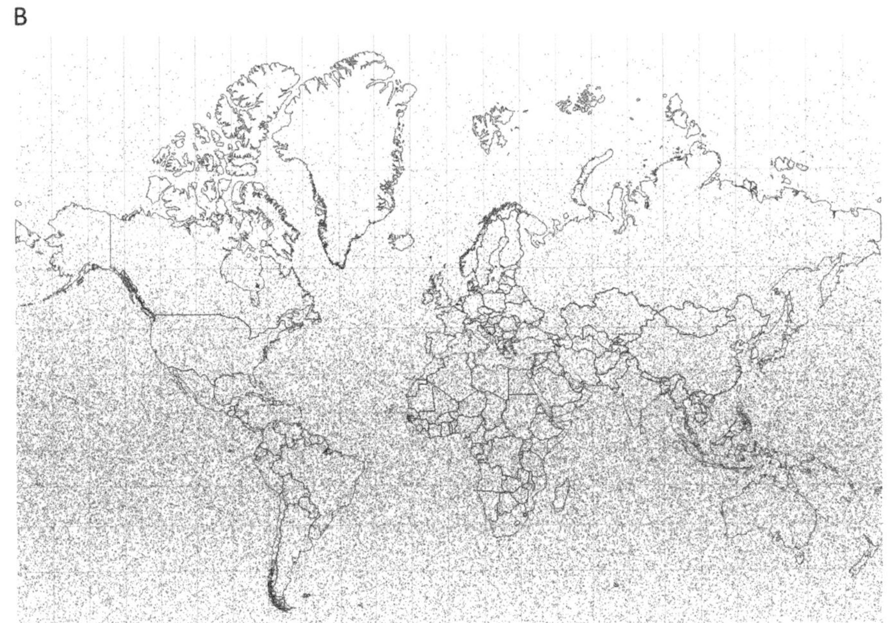

FIGURE 10.3
One hundred thousand randomly distributed points as seen on an equal area cylindrical projection (A) and on web Mercator (B). The distortion on web Mercator, with areal distortion increasing closer to the polar regions, leads to a more dispersed appearance to the points in the northern polar region on the map. Note that the southern polar region has been removed from the map.

area projection (Figure 10.5A), the distribution appears fairly random (as would be expected for binning a randomly generated point dataset), while on web Mercator (Figure 10.5B) the pattern clearly matches the pattern of areal distortion for the projection—the areal distortion increases approaching the poles, so while the hexagonal bins appear to be the same size, they actually cover a smaller and smaller ground area in reality; smaller bins tend

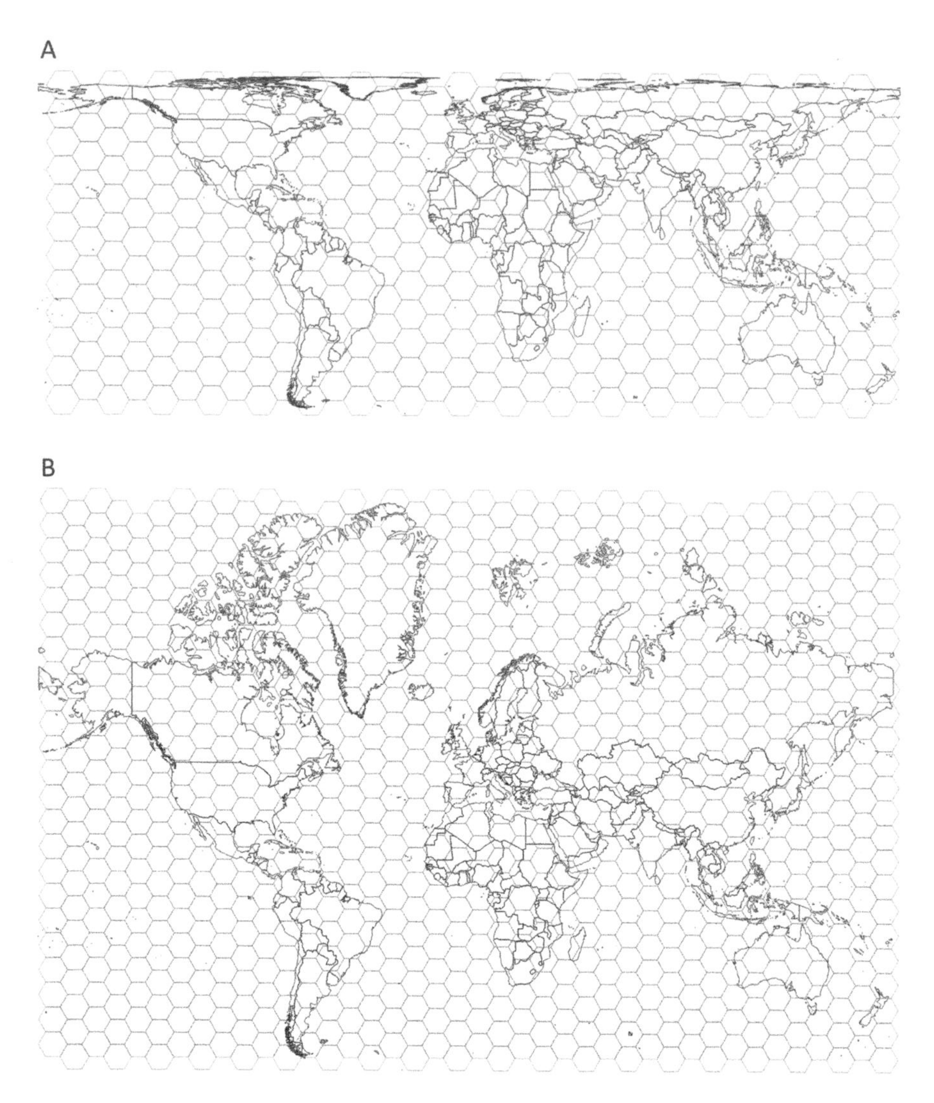

FIGURE 10.4
Regular hexagonal grids generated on an equal area cylindrical projection (A) and on web Mercator (B). While the hexagons in the grid appear to be the same size and shape on the projection, note that their shape and size on the sphere will differ due to distortion in the projection.

to contain fewer points in this spatial distribution. Even though these two maps are attempting to represent the spatial distribution of exactly the same random point dataset, they present very different patterns for the data distribution due to the distortion of area in the web Mercator projection.

When discussing general tasks for spatial bin maps, we noted that map readers may be interested in either the count of points *or* the density of points within each bin. Figure 10.5 demonstrates that the raw count of points will

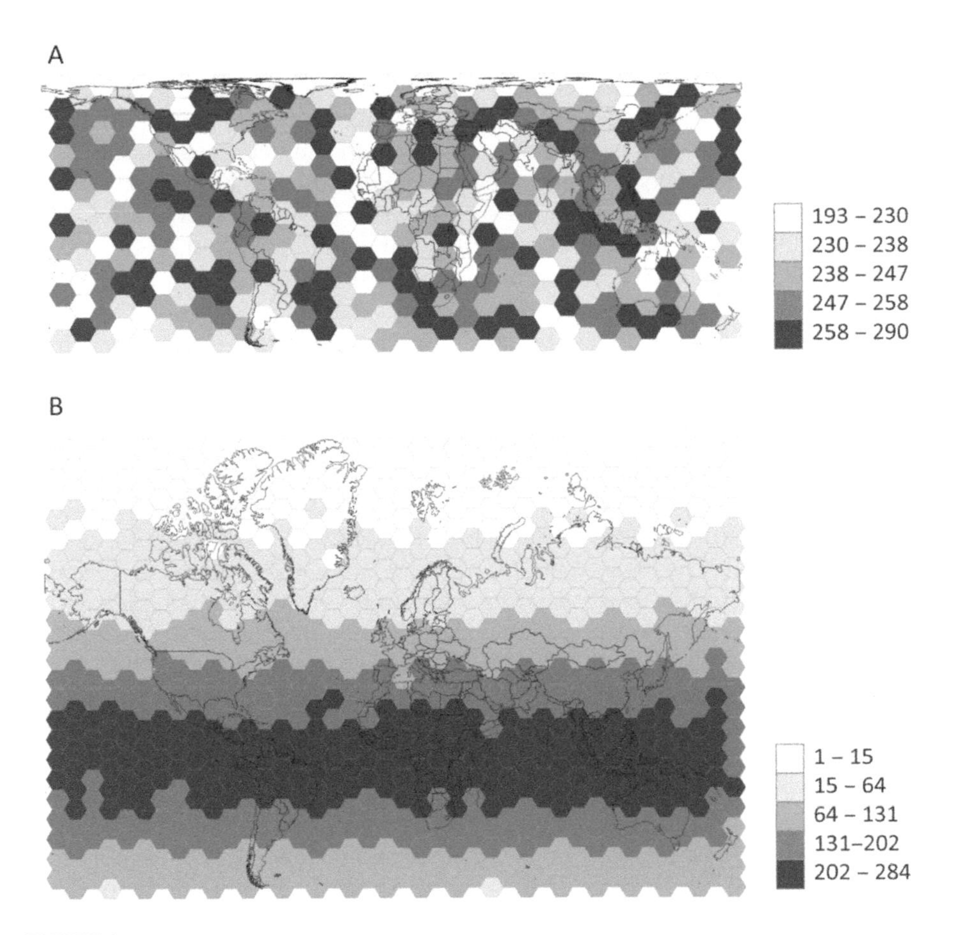

FIGURE 10.5
Count of points in each of the regular hexagons generated for spatial binning. Map A shows the result on a cylindrical equal area projection, while map B shows the result on web Mercator. Both maps are symbolized using quintiles, so that each class on the map shows 20% of the data distribution (e.g., top 20%, bottom 20%, etc.)

vary due to projection (e.g., spatial bins in areas with greater exaggeration of area, for instance the polar regions in web Mercator, will tend to have lower counts of points). A map reader's ability to estimate density will also be impacted by the projection distortion. Even though the hexagonal spatial bins shown in Figure 10.4 all look the same, the areas represented by each bin are quite different in the non-equal area projection (web Mercator). This means that the map reader's ability to *visually* estimate density will be completely incorrect. In web Mercator, for example, the spatial bins nearest the polar regions look the same size as the bins near the equator, but they represent a much smaller ground area. A map reader would not be able to visually estimate a density correctly unless they could also mentally calculate the appropriate area *accommodating for the projection distortion*. Even map projection experts can struggle with this task, so you should not expect your map readers to be able to successfully calculate density using non-equal area bins. Of course, if you are using an equal area projection to generate your spatial bins, they should represent the same ground area and the density measure *can be* visually estimated—because all of the bins really are the same size!

While we recommend equal area projections for creating spatial bins, other projection classes can be used successfully when working with small geographic areas. As we have seen with other mapping examples in this book, by carefully selecting your projection parameters you can establish a projection that minimizes, and virtually eliminates, both areal and angular distortion across a small area of interest—even if the projection is not equal area at the global-scale. This careful selection of projection parameters will allow you to reliably generate spatial bins on a non-equal area projection and feel confident that they are still providing an appropriate and fair unit for aggregating point data.

Regardless of the projection you select for creating and binning your dataset, you also need to be cautious of impacts of reprojecting the dataset for display (e.g., if you create the bins on an appropriate equal area projection, but then display the data online using web Mercator). In the case where you plan to display your data with a projection other than the one used to create the bins, be wary of the distortion that will be introduced. For example, Figure 10.6 shows the bins from the previously worked example, with the bins that were created in an appropriate equal area projection (cylindrical equal area) reprojected for display in web Mercator (near conformal) and sinusoidal (equal area) projections. The example using web Mercator (Figure 10.6B) greatly distorts the size and shape of the bins, particularly in the upper latitudes. While this map still presents accurate counts for points within each bin, any attempt to estimate a *density* would be grossly inaccurate as the bins near the polar regions appear to be much larger than those nearer the equator, but, in reality, represent a ground area that is much smaller; this would lead to huge underestimates in density in these areas. The example using the sinusoidal projection (Figure 10.6C) presents a different challenge. While the sinusoidal projection is equal area and all the bins represent the same "area," their shapes are distorted in the upper latitudes since the meridians converge to a point representing the pole, potentially leading map readers to *assume* that the bins are not equivalent (as they clearly are in the original cylindrical equal area projection).

The point here is that even if you choose an equal area projection to *create* the bins, remember that not all equal area projections *display* the bins equally. In other words, be careful as adjusting to a different projection to display your results will alter and distort your nice, regular spatial bins. Since one of the primary benefits of a spatial bin map is that the polygon bins are all the same shape and size, it is preferable to use the same projection for creation and display of the binned data—so, we recommend picking an equal area projection that you like and using that for all aspects of spatial bin mapping. If you cannot use an equal area projection, select a non-equal area projection but adjust the map parameters and limit the mapped area so that the areal distortion is minimized.

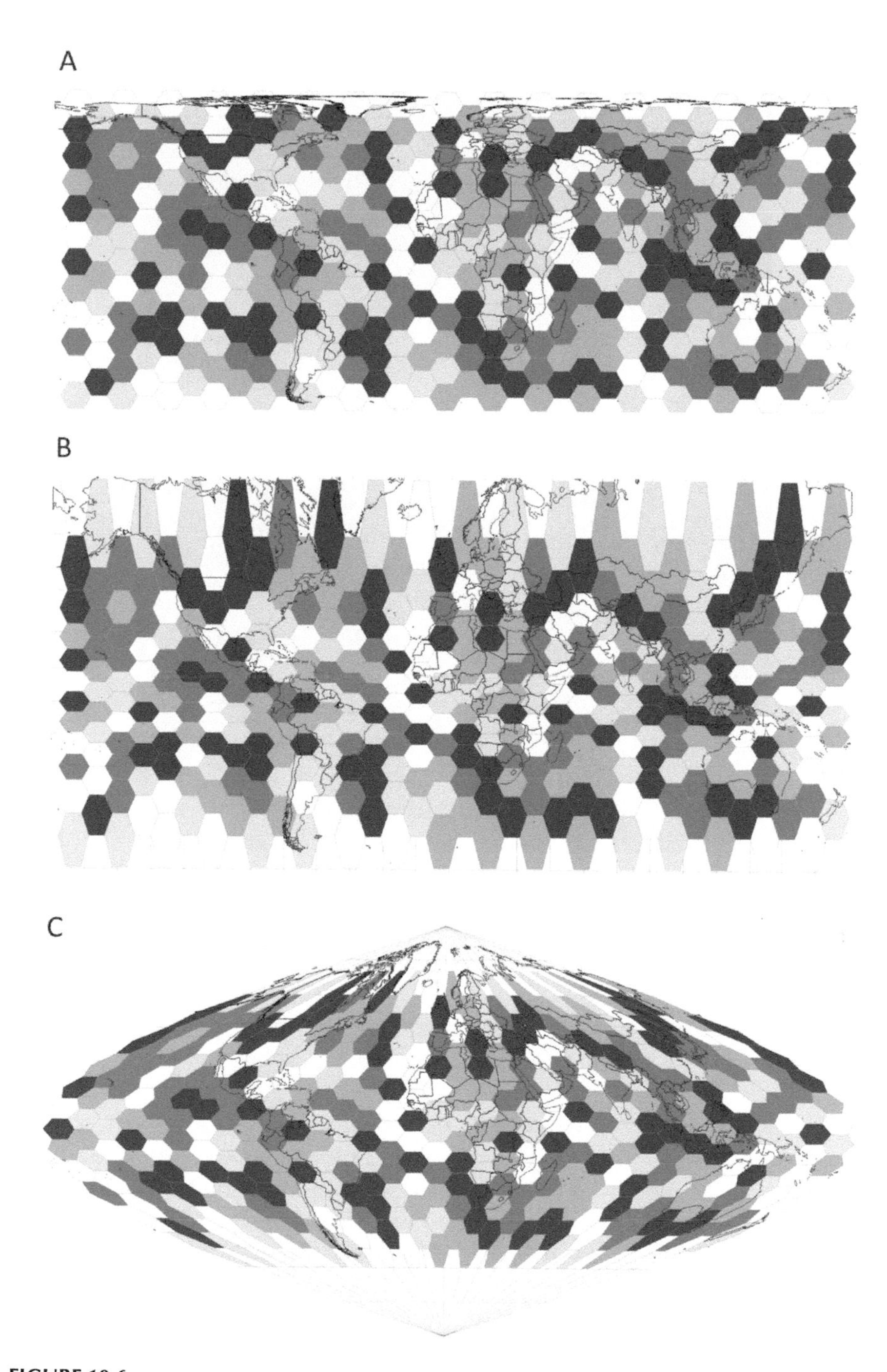

FIGURE 10.6
Spatial bins created in a cylindrical equal area projection (A), and reprojected for display in web Mercator (B) and sinusoidal (C).

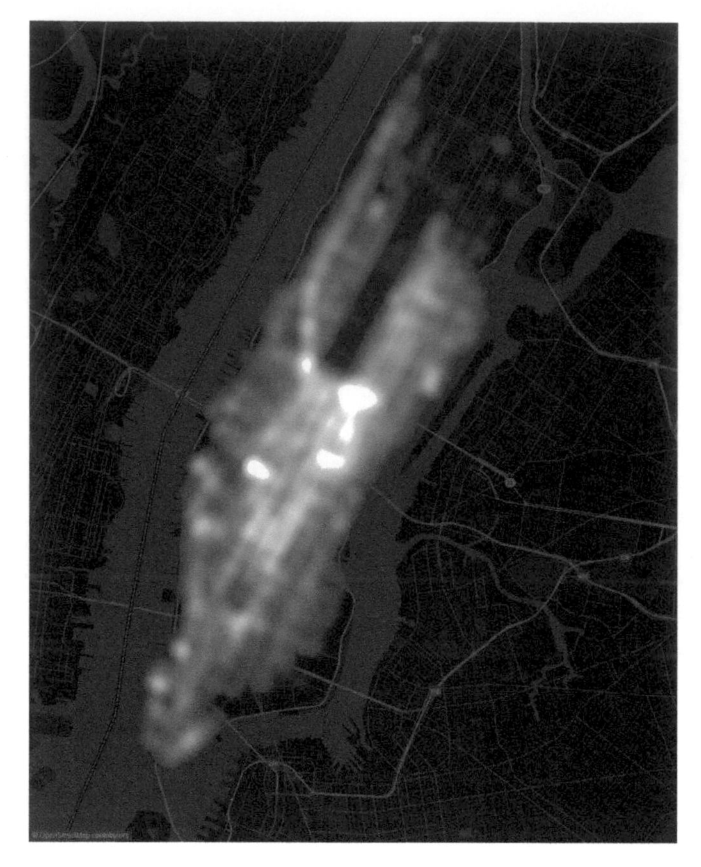

FIGURE 10.7
Heatmap showing density of taxicab pickup locations in Manhattan.

Heatmaps

A heatmap (Figure 10.7) is a mapping technique to present a smooth, continuous surface representing the density of point locations (note that the heatmap term may also be used to represent other visualization types outside of mapping, for instance, a cluster heatmap used to color encode 2-dimensional (2-D) matrices; see Wilkinson & Friendly, 2009). Often, a heatmap is based solely on a count of the number of points, but the weight of each point as it is used to calculate "heat" can be adjusted based on an attribute (e.g., a single point is counted more or less than other points based on an attribute at that location). Much like spatial bin maps, heatmaps can be a good choice when working with large, dense sets of spatial point data where an aggregated measure of density is needed for interpretation.

Where spatial bin maps are generated from aggregating an absolute count of points within a polygon region, heatmaps are based on a *kernel density estimation* using a "kernel" to tally the number of points in any given location. The kernel is used for every location on the map, and the value calculated for the kernel is assigned to the center of the kernel (pixel). This creates a smooth surface of density, instead of a density measure as represented in individual polygons with abrupt transitions in value between each polygon.

There are many factors that impact the resulting count of points for a location, including the size of the kernel (how big is the area being searched), the shape of the kernel (typically circular), and how points that fall within the kernel are weighted (e.g., every point has the same value, points are weighted based on distance from center of kernel, or points are weighted based on a non-distance based attribute at each location).

While the result of a heatmap is quantitative—every location on the map has a value returned from the aggregation of points within the kernel—the result is given at an ordinal level. The result is often normalized across the map (e.g., highest density = 1; lowest density = 0, and the relative "hotness" at every location is scaled within this range); however, quantitative values are rarely given for exact locations. Because of this, the results of a heatmap can only be interpreted at an ordinal level (e.g., higher density right here; lower density over there), as opposed to trying to identify an exact value, as a map reader would be able to do with an isarithmic map.

Tasks for Heatmaps

Heatmaps have many similarities with isarithmic maps—the map reading goal is typically to assess a dataset's overall spatial trend and to compare relative values between individual locations. Heatmaps specifically emphasize how the *density* of a phenomenon changes smoothly and continuously across an area of interest. This differs from isarithmic maps in that the isarithmic map uses isolines and/or filled contours to show interpolated values, allowing a map reader to create reasonable quantitative estimates of value at any location. A heatmap, on the other hand, provides a qualitative, ordinal-level approximation of value, which simply allows a map reader to estimate relative values of "more or less" between locations.

Projection Considerations for Heatmaps

Because heatmap values are based on a kernel density, there is a strong projection influence on the calculation of the "hotness" of any location. We have yet to see a software package or API that performs a kernel density function where the kernel is adjusted based on projection distortion (though, perhaps this does exist), so you should assume the kernel used will be a geometric shape *on the projected plane*. For this reason, we highly recommend generating heatmaps using equal area projections. When working with non-equal area projections, particularly for large regions where distortion cannot be minimized, the areal distortion can lead to grossly misleading visual patterns of density. Figure 10.8 emphasizes this by showing relative "hotness" at nine different locations on the globe, as displayed on a web Mercator projection. At each location on the map, the point values used in the calculation are equally spaced *on the sphere*; however, as areal distortion increases away from the equator on this projection, but the kernel size remains the same *on the plane*, the relative hotness decreases with the change in latitude. This presents a misleading pattern of the density of the data.

While we strongly recommend selecting equal area projections when working with heatmaps, there are exceptions to this suggestion. If the geographic area is relatively small, and the parameters for the projection (e.g., central meridian, standard line(s), etc.) have been appropriately selected to minimize areal distortion, then a non-equal area projection can be just as effective. The important factor here is to select appropriate projection parameters to minimize areal distortion in each area of interest. When working with multiple areas of interest and making comparisons, you will want to select different parameters relevant to each individual area of interest, regardless of how many different regions are mapped. If you can successfully do this with your projection, the calculations for your heatmap will be more accurate. Note, of course, that the *shape* of the kernel used will only be true on the projected plane where the heatmap was calculated, and not true on the sphere—but we see that challenge as both unavoidable and limited in importance, for a map where the visual representation offers, at best, an ordinal-scale comparison between locations. The equations provided by Battersby et al. (2016) for determining whether distortion will impact spatial bins on maps should be sufficient for evaluating the largest geographic area that is appropriate for generating a heatmap on web Mercator (and the equations can be adapted to other projections as needed).

Projections for Other Special Mapping Purposes

Hopefully, by reading through earlier chapters, it has become clear that there really isn't a single right projection for any mapping project. Each projection has been designed with specific properties in mind. We have discussed the "big" properties of whether or not a projection preserves area, angles, or distances, but within each of these categories, individual projections may have specialized properties that distinguish them—for instance, a projection might be designed to ensure that all rhumb lines (a path of constant compass bearing) appear as straight lines (e.g., Mercator projection),

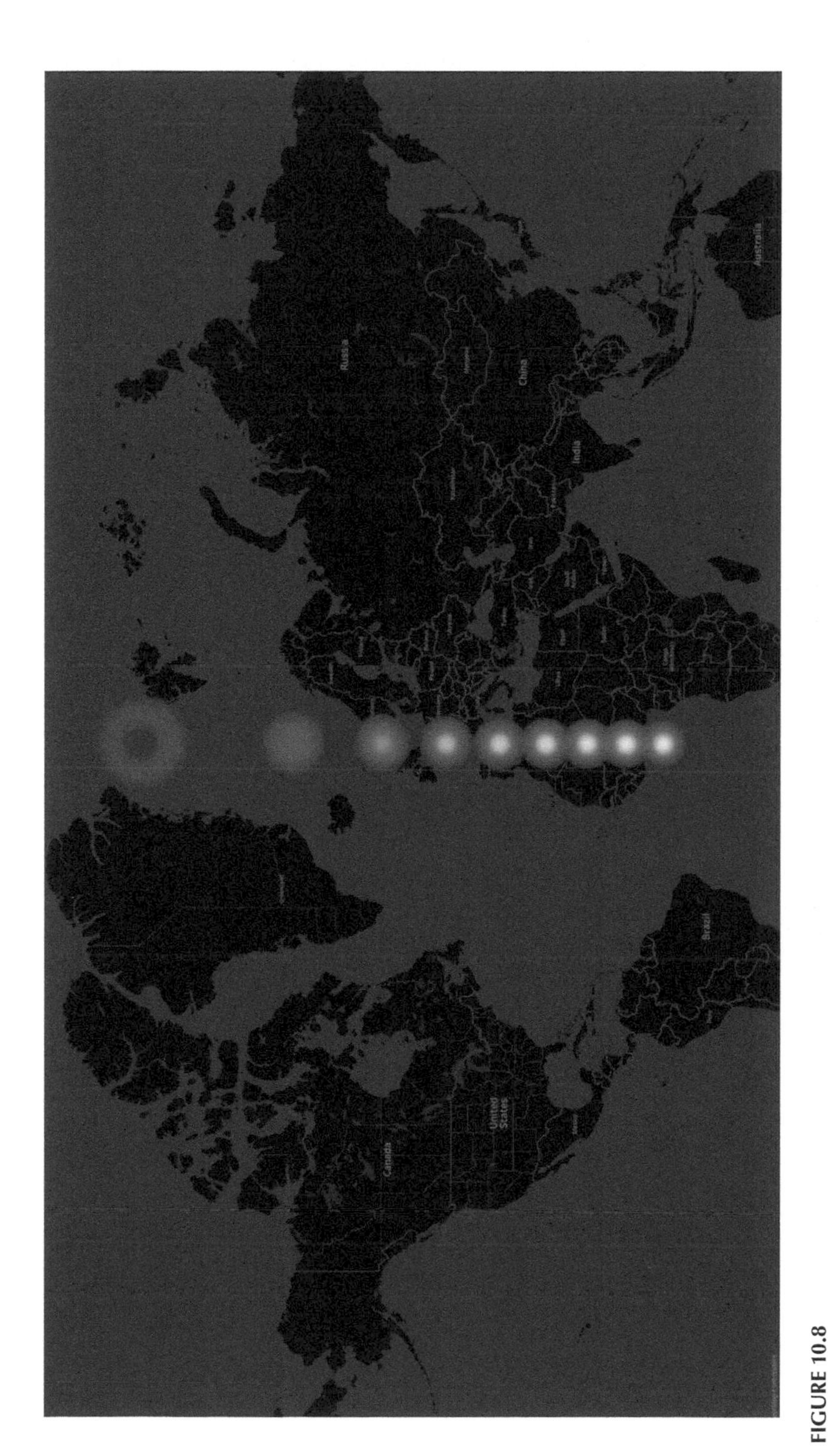

FIGURE 10.8
Heatmap showing spatial distribution of nine equally spaced points centered at varying latitudes on web Mercator (nearly conformal). Areal distortion increases toward the poles, leading to a corresponding decrease in visual "hotness" with higher latitudes on the map. On a globe, all of these hotspots should be exactly the same size, shape, and hotness.

or so that all great circle routes appear as straight lines (e.g., gnomonic planar projection). In addition, other projections permit the measurement of accurate directions or azimuths between locations. In this section, we discuss some of these special properties that you might need for mapping and suggest projections that are relevant for meeting these properties. To provide context for the discussion, we will look at measurement and mapping activities related to air travel, and focus on a hypothetical trip between Washington, D.C. (–77° 02' 07" W, 38° 53' 22" N) and Berlin, Germany (13° 28' 26" E, 52° 30' 54" N) as a reference.

Assume we are interested in traveling between Washington, D.C., and Berlin, Germany. Prior to taking our trip, we may want to visualize the two cities, their direction with respect to geographic north, and their relative distance apart. To view these two cities and their spatial context, we map them (Figure 10.9A). This figure represents the two cities as if looking at them on a globe. In fact, to provide this perspective, the map uses the orthographic projection. This projection is azimuthal, which means that any azimuth measured using the projection's center (in this case, Washington, D.C.) to any other point shown on the map is true. Using this characteristic, the true azimuth from Washington, D.C., to Berlin is 44.4°. However, this projection is not equidistant and the distance value shown in Figure 10.9A, (5,542.4 km) measured on the orthographic is less than the true Earth distance of 6,733.8 km.

Mapping Rhumb Lines

A rhumb line is a path of constant compass bearing and is usually desired by ship or aircraft navigators who wish to chart their course; the straight rhumb lines mean a navigator can mark off the path between two waypoints on a voyage and identify the bearing to travel between these locations. Following a path of constant compass bearing is advantageous as a rhumb line crosses every meridian at the same angle thus facilitating the navigator's "steering" process of the ship or plane by reducing the number of turns the ship or plane must make along the journey. To show rhumb lines as straight, Mercator is the recommended projection for this task, as it is conformal (preserves angular relationships throughout the projection) and has a special property of showing all rhumb lines as straight on the projection's surface. Figure 10.9B shows the rhumb line as a dashed straight line.

While depicting rhumb lines as straight lines is a valuable characteristic, the straight rhumb lines should not be confused with the shortest great circle route between locations. Figure 10.9B shows the great circle route as a curved solid line on the Mercator projection. In order to preserve angular relations on this projection, the ability to measure accurate distances is relaxed. If you measure the rhumb line distance between the two points on the Mercator (the dashed line) you arrive at a value of 10,301.1 kilometers which is considerably larger than the true Earth distance along this great circle route

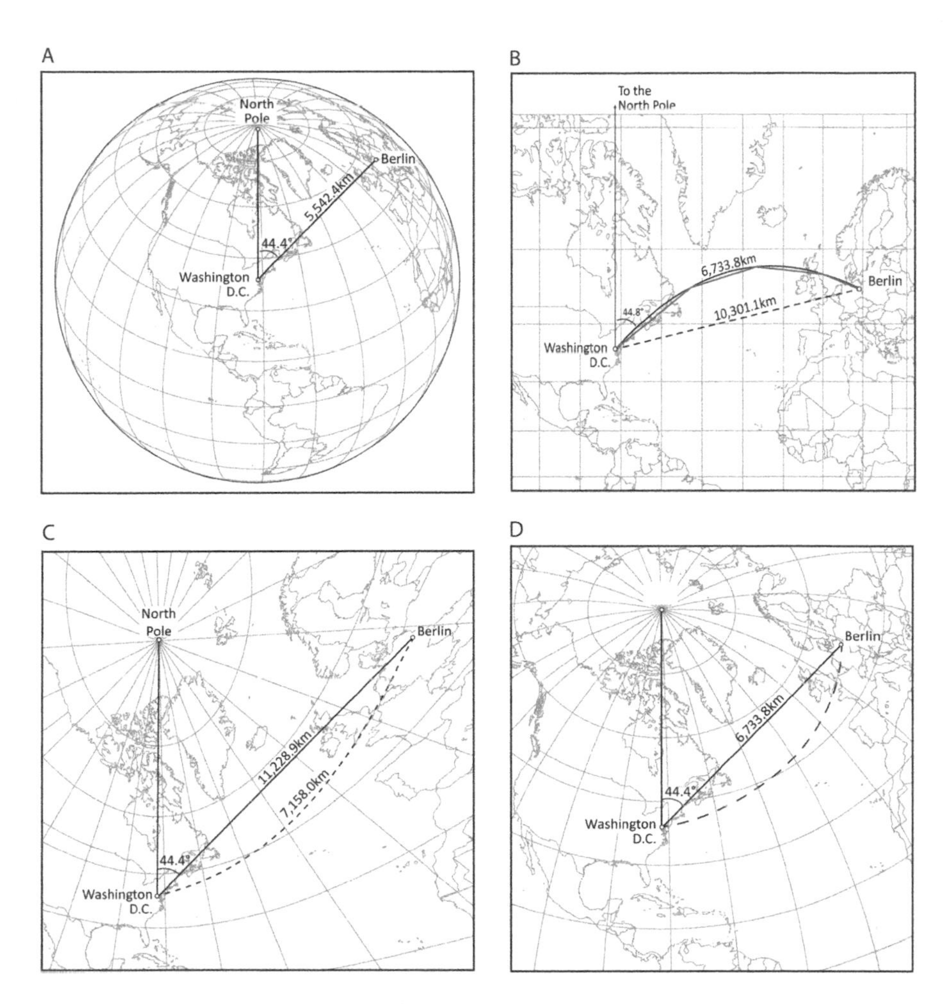

FIGURE 10.9
Four projections depict the path between Washington, D.C., and Berlin, Germany: orthographic azimuthal (A), Mercator conformal (B), gnomonic (C), and azimuthal equidistant (D). In (B–D) the great circle path and rhumb line are shown by solid and dashed lines, respectively.

(6,733.8 km). To accommodate for the distortion of distance in the Mercator and to use this projection for long-distance navigation, navigators will typically first plot their course on a gnomonic projection, which will show the great circle route as a straight line from the origin location to the destination. Once the great circle path has been plotted as a straight line on the gnomonic projection (appearing as a solid black line in Figure 10.9C), the route can then be approximated on the Mercator with a series of short, straight rhumb lines, which the navigator will use to travel. This approximation to the great circle path on the Mercator projection, shown as four short solid red line segments, is also shown in Figure 10.9B.

Mapping Great Circle Routes

A great circle is the shortest path between two locations on a curved spherical surface and is advantageous to navigation. Looking down at a great circle path drawn on a globe, that path appears as a straight line (Figure 10.9A). However, on most projections, due to the projection process of flattening the curved surface, the great circle route does not appear as a straight line. Representing the great circle route as a straight line is a special property that is preserved on two projections: gnomonic and azimuthal equidistant planar projections (Figure 10.9C and D, respectively). The gnomonic projection's use for navigation is advantageous in that *any* straight line drawn on the map's surface is a great circle. In other words, unlike other azimuthal projections whose center must be carefully selected, with the gnomonic, there is no need to center this projection over a specific point to be able to draw a straight great circle that connects two points. Compare this characteristic to the azimuthal equidistant projection where representing great circles is more restrictive. On this projection, a great circle is shown only as a straight line drawn from the projection's center to any point on the map.

Similar to our note in the previous section on rhumb lines stating that the line of constant bearing should not be confused with the shortest path, we must point out that the great circle route as a straight line on the gnomic projection should not be used to calculate the shortest distance! Figure 10.9C shows the measured distance along the straight line represented on the gnomonic projection as 11,228.9 km, while an azimuthal equidistant projection (Figure 10.9D), reports the great circle distance as 6,733.8 km. While this distance result may seem contradictory, the gnomonic projection greatly distorts distances—but by doing this, it facilitates the ability to map the shortest path (even if that path does not pass through the projection's center), which can then be measured on an appropriately centered azimuthal equidistant projection.

Measuring Distances between Locations

Preserving correct scale is one of the special properties needed for accurately measuring distances between locations. It is impossible to preserve scale at all locations on a projection (see Chapter 2), as scale is highly variable across most projections' surface. However, scale can be controlled in specific ways. For instance, equidistant projections preserve distances along all meridians or from a single point to all other points. This is particularly useful when measuring distance between locations. Even if the two locations you want to measure between are not aligned along a meridian, as long as they can be centered over a single point (e.g., Washington, D.C., in our flight mapping example), the distance can then be measured to any other point on the map (e.g., Berlin, Germany). For this task, we recommend using an azimuthal equidistant projection. Figure 10.9D shows the azimuthal equidistant projection with a thicker solid line connecting Washington, D.C., to Berlin, Germany. Carefully measuring the distance between these cities results in a measurement of 6,733.8 km. This corresponds to the true distance between these two cities. Note that the other projections shown in Figure 10.9 are not equidistant and the measured distances are either greater or less than the true distance resulting from the way in which scale distortion occurs across the projection's surface.

Measuring Directions between Locations

In addition to mapping the route and measuring the distance between our two points of travel, we are interested in knowing the direction of our journey. Recall from Chapter 2 that an azimuth is a direction measured as an angle clockwise from the North Pole to a point[1]. Any azimuthal projection can be used to measure directions from its center point to any other point on the map. To illustrate, in Figure 10.9A we see the solid line connecting Washington, D.C., and the North Pole; we can utilize this line to determine the azimuth by computing the angle between it and the vertex of the line connecting Washington, D.C., with Berlin. With the projection centered over Washington, D.C., we measure the azimuth to Berlin and arrive at an angle of 44.4°. This is the true Earth azimuth or direction the plane will fly between the two cities. Figure 10.9C and D show that the gnomonic and azimuthal equidistant projections both share the azimuthal property and report the true azimuth of 44.4°. Due to its conformal property, the Mercator projection in Figure 10.9B also reports the true azimuth of 44.4°.

While azimuthal projections preserve the true azimuth *from* the projection's center *to* any point on the map, *retro*azimuthal projections provide a different property of preserving directions. Retroazimuthal projections' special property provides the true azimuth *of* the map's center *from* any other point on the map. One particular retroazimuthal, developed by James Craig in 1909, is referred to as the "Mecca" projection as it sets Mecca as the map's center (21° 25′ 21.0″ N, 39° 49′ 34.2″ E). The Mecca projection allows followers of Islam to accurately determine the appropriate direction of Mecca to face when carrying out their prayers. This direction is known as the Qibla. Figure 10.10 illustrates the process involved in determining the Qibla. Assume we want to determine the direction to face Mecca from Berlin, Germany, and Singapore, Singapore. A vertical line parallel to the meridians is drawn through each location. Next, a straight line connecting each location to Mecca is drawn. The azimuth at each location to Mecca can then be determined. Individuals in Berlin and Singapore would direct themselves 136.6° and 292.9°, respectively, clockwise from north to face Mecca for their prayers.

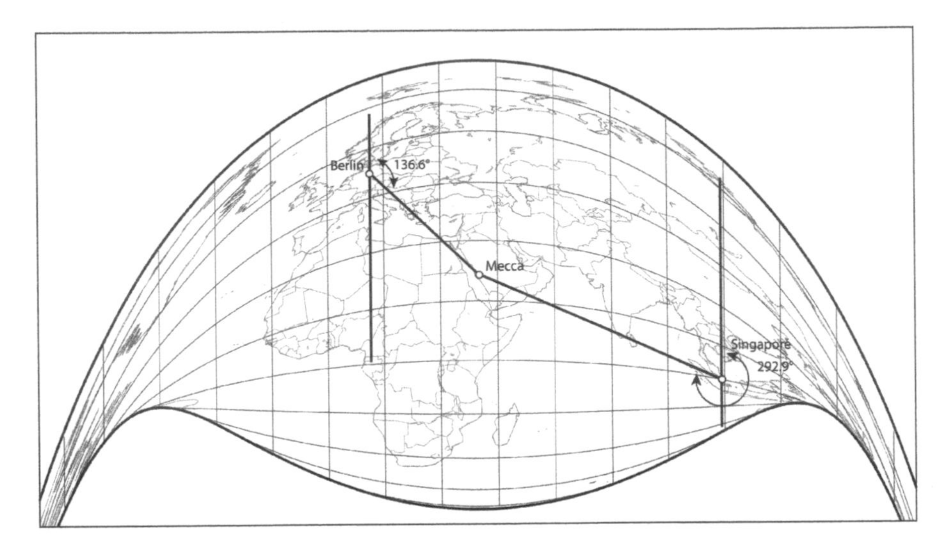

FIGURE 10.10
Mapping the Qibla on the Craig "Mecca" retroazimuthal projection showing accurate directions from Berlin and Singapore to Mecca.

Conclusion

In this chapter, we have examined the implication of projection on several different map types that you may encounter that are similar to maps discussed earlier in Part II of this book, but that have distinct characteristics or map reading tasks that require special consideration for selecting a map projection. The projection selection debates for spatial bin maps, heatmaps, and a few unique map/map reading tasks overlap many of the key points from earlier chapters in this section about more common map types (i.e., choropleth, isarithmic, dot, and proportional point and flow maps), however, adapts them to the tasks that are special to these particular map types. The most important part of the projection selection process is to identify the key tasks that a map reader might need to accomplish (e.g., measuring distances, comparing area covered by different data distributions, etc.), match that to the appropriate projection properties, and identify the specific parameters that might need to be adjusted to further minimize distortion for your particular map (e.g., shifting the central meridian or standard lines to tailor the projection). Additionally, while we have suggested specific projections for many different map types, keep in mind that there are likely numerous maps that will have the same properties (e.g., equal area or equidistant) and, ultimately, your final selection will be based on which projection with the property of interest looks the best for your application and for the space available for printing or digitally displaying your map (see Part I of this book for deeper discussion of the many factors that may influence selection of a projection to meet a particular "look" for your map).

Note

1. Measuring clockwise from north to determine an azimuth is not a universally accepted practice.

11

Web-Based Map Projection Resources

Introduction

Muehrcke et al. (2001, p. 586) stated that "[m]ore has been written about map projections than all other facets of mapping and map use combined, yet people still find the subject to be the most bewildering aspect of map appreciation." Part of this bewilderment is likely due to the often-complex mathematical underpinnings of projection formulae. Hundreds of books and articles have been written about projections. However, for the average cartographer or map reader, finding the right book to explain this mathematically focused subject in "plain English" is not easy (Figure 11.1).

The web has exponentially increased the amount of projection-related material available to the interested reader. A recent search produced more than 1,300,000 sites that contain the phrase "map projection," a number which certainly increases on a daily basis. Despite the plethora of immediately searchable resources, the reader interested in projections is still faced with the dilemma of finding one that explains the subject in understandable terms or contains the information they need to work with projections. We have sorted through the numerous online resources that we find particularly useful for working with and learning about projections. This chapter's contents are organized into four sections:

- Projection Tools and Applications
- Programming Languages, Libraries, and Tools
- Learning about Map Projections
- Map Projection Galleries

Projection Tools and Applications

Most cartography, and all GIS, is done with data. GIS software suites allow users to select from dozens of projections, specify a map center, and modify

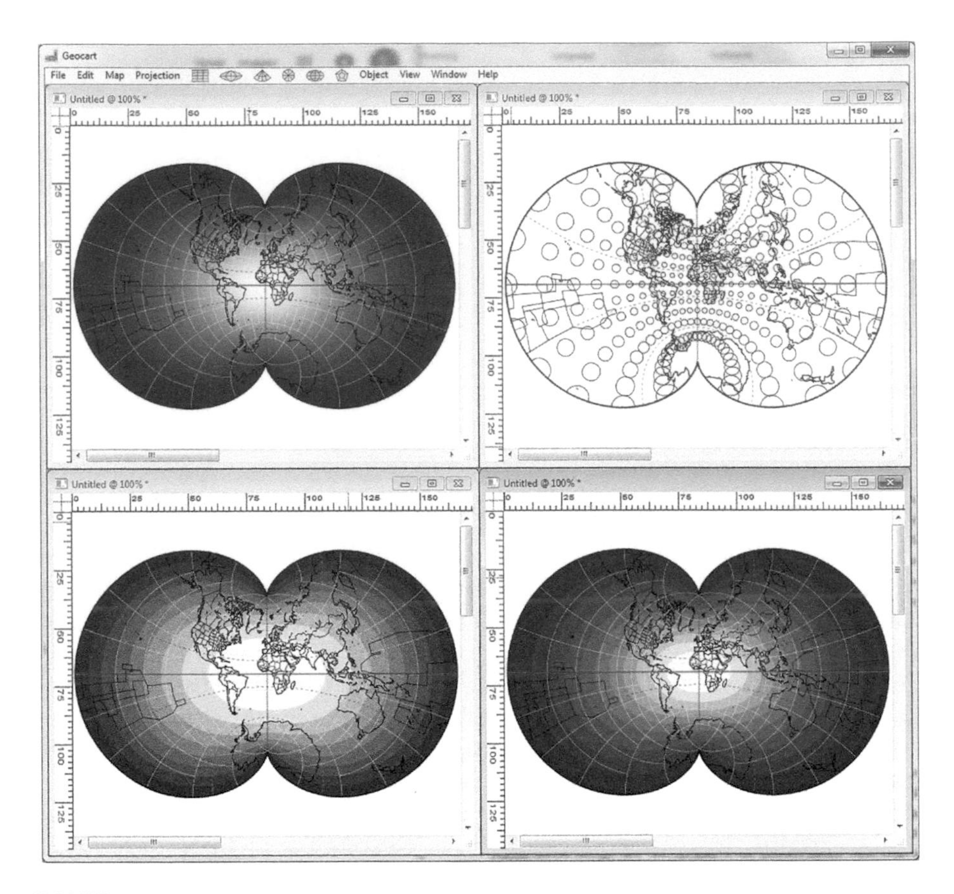

FIGURE 11.1
A screen capture from Geocart software, displaying the Eisenlohr epicycloidal conformal projection and its respective overall distortion (top left), Tissot's indicatricies (top right), scale distortion (bottom left), and areal distortion (bottom right).

a projection's parameters. Unfortunately, there is little guidance in the GIS environment that helps the end user work with or select suitable projections (we did find a plugin for QGIS to view projection distortion patterns*) These patterns can help users understand how distortion is distributed across the map's surface, helping to identify appropriate projections and how to adjust its parameters to suit the map's purpose. There are websites that provide specific projection tools and applications to provide this kind of information often lacking within GIS software.

This section presents a selection of websites where mapmakers can design their own projection, explore distortion patterns on projections, or receive guidance on selecting a projection. We have placed these websites into the following two categories: Visualizing and Selecting Projections, and

* Available at https://plugins.qgis.org/plugins/ProjFactors-gh-pages.

Customizing and Reprojecting Datasets. The first category presents websites where users examine different projections, view their distortion patterns, and choose a projection that is recommended based on the map purpose. The second category summarizes websites where users can specify projection parameters to see the outcome of either reprojecting an existing dataset or creating a new projection. Inside each category, we have organized the individual websites according to the knowledge level of projections the user is expected to have. The websites that require less knowledge about projections are presented first, followed by those that demand a greater understanding of projections.

Visualizing and Selecting Projections

Adaptive Composite Map Projections (ACMP)

Adaptive Composite Map Projections, developed by Bernhard Jenny (2012), demonstrates a method for selecting projections for web map services according to the latitude and geographic scale of the mapped area. By tailoring projection choices to these two conditions, instead of relying on a single projection for all latitudes and scales, the tool provides a method for lowering overall distortion. Tile-based web map services still rely on the web Mercator as *the* projection of choice, although there is ample criticism regarding its use (e.g., Battersby et al., 2014) due to the extreme areal distortion present in this projection. The ACMP tool shows that a better solution to delivering online maps through different projections is possible. It is freely accessible at www.cartography.oregonstate.edu/demos/AdaptiveCompositeMapProjections/.

Six projections are included in ACMP. Figure 11.2 presents how the adaptive map projections are applied to different latitudes and areas of interest,

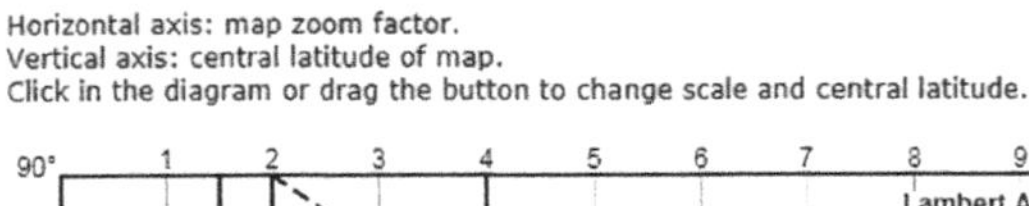

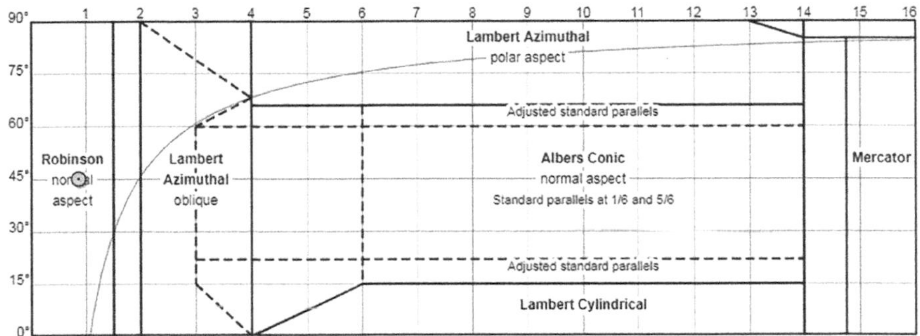

FIGURE 11.2
The arrangement of the different projections that are used to display Earth according to the latitude (shown on the left-hand axis) and geographic scale (shown on the top-most axis) of the map area.

using scale to stand in for the latter. The y-axis represents latitude, and the numbers along the x-axis correspond to different zoom levels. For a given intersection of latitude and zoom level, an appropriate map projection is selected. For example, the blue dot in Figure 11.2 indicates that the Robinson projection would be used for a central latitude of 45° and a zoom level of 1 (Figure 11.3). The ACMP tool is designed to transition seamlessly between the different projections. For instance, as the user zooms in at 45° latitude, the projection would transition from Robinson to Lambert azimuthal. As the user continues to zoom to increasingly larger scales, the projection transitions to Albers conic and then to Mercator cylindrical.

Projection Wizard

Projection Wizard was developed by Bojan Šavrič, Bernhard Jenny, and Helen Jenny (Šavrič et al., 2017) to provide a simple tool for selecting a map projection. Many individuals struggle when identifying an appropriate projection for their mapping needs. Unfortunately, there are few readily available resources that can provide clear guidance to a novice audience on how to select a projection. Projection Wizard (http://projectionwizard.org/) is a freely available interactive online tool where users enter basic information and a projection is recommended. Figure 11.4 shows the Projection Wizard interface.

Projection Wizard provides a simple approach to the complex problem of finding an appropriate projection. The Wizard requires two user inputs: the area to be mapped, and the desired projection property (conformal, equal area, equidistant, or compromise). First, the area of interest is defined by

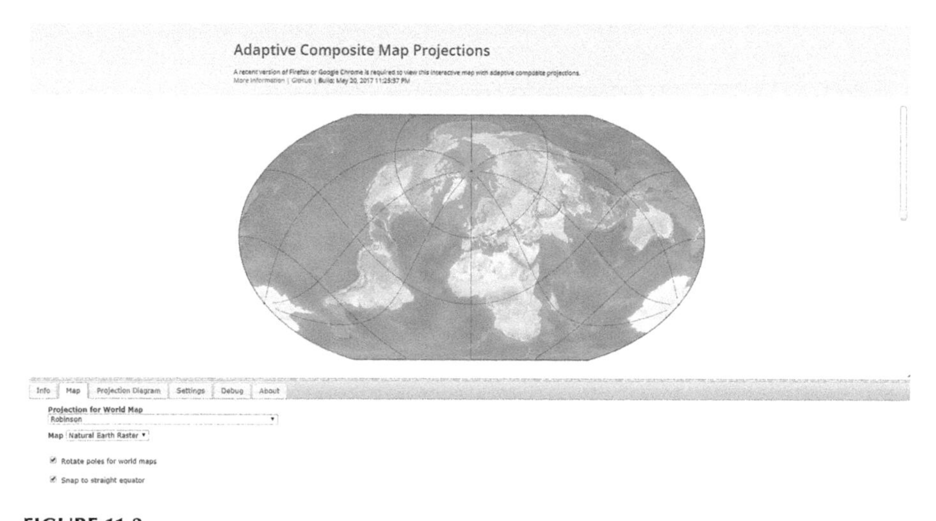

FIGURE 11.3
The interface of Adaptive Composite Map Projections. In this example, the Robinson compromise pseudocylindrical projection appears centered at 45° north and 0°.

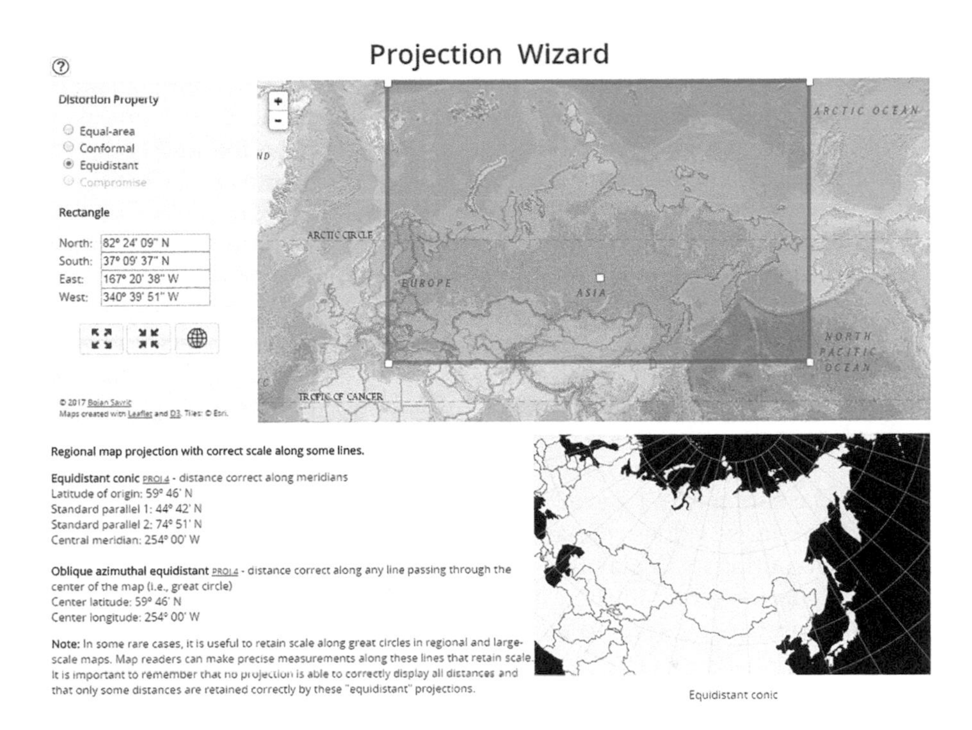

FIGURE 11.4
The recommended projections from the Projection Wizard for an equidistant map of Russia, outlined by the blue box (http://projectionwizard.org/).

resizing a window within the interface (Figure 11.4). Then the desired projection property is selected from a menu. A list of recommended projections is reported along with their parameters, and different maps of that area with those projections are shown.

With these options in an interactive environment, users with limited projection knowledge who need assistance when selecting a projection are able to experiment with different geographic extents and projection properties and view the results without having to repeatedly reproject the data in a GIS. Interactivity provides a level of information and understanding beyond guidelines presented in a book. We are unaware of any existing GIS software that provides assistance in selecting a projection. In addition, depending on the geographic area specified and the chosen property, Projection Wizard may recommend more than one projection. Thus, the user is still faced with selecting one projection, but from a limited set.

Geocart

Developed in 1992 by Daniel "daan" Strebe and Paul Messmer, Geocart (www.mapthematics.com/) is a commercially available software package, for both Mac OS and Windows, that provides map makers a way to visualize distortion patterns across a projection's surface. Geocart offers dozens of map projections from which to choose and provides a variety of symbolization methods to visualize projection distortion. We are particularly impressed with the intuitive color gradations. As an aside, we have found Geocart to be such an excellent resource for exploring map projections and displaying distortion metrics, that we used the software to create all of the colorized distortion figures in this book. Figure 11.5 shows the Distortion Visualization window. For each visualization option, the user can specify values that control the output.

The "Angular" option displays the range of angular distortion on a projection from 0° (no distortion) to 180°. Figure 11.6 shows angular distortion on the McBryde-Thomas flat polar pseudocylindrical equal area projection. In Figure 11.6, angular distortion is symbolized as a color gradation from white to darker magenta, where darker magenta equals more angular distortion. Graduated colors make it easier to see the range of angular distortion values, and how those values change across the projection's surface. Figure 11.5 includes the option to display the "Average scale factor (linear scale)" option. Here, scale factor refers to the ratio of the map scale measured at a given point on the map (referred to as local scale) divided by the nominal map

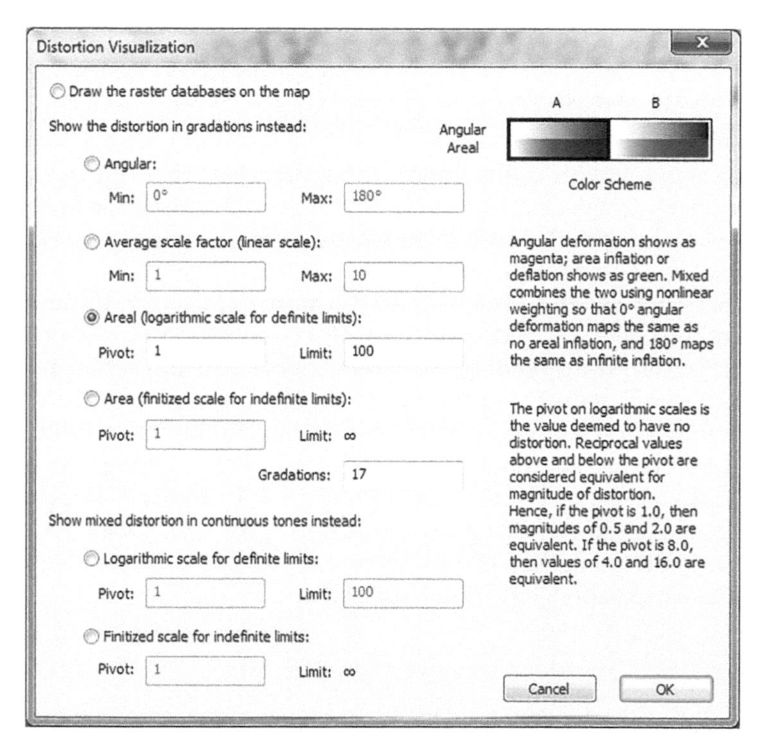

FIGURE 11.5
The Distortion Visualization window in Geocart.

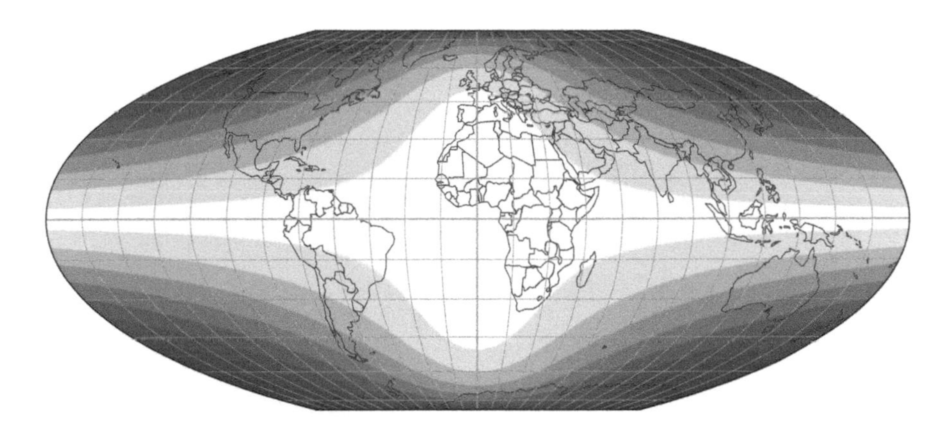

FIGURE 11.6
Angular distortion symbolized on the McBryde-Thomas flat polar pseudocylindrical equal area projection, shown in white-to-dark magenta color gradations.

scale (Equation 11.1). The nominal map scale is usually associated with the scale along a standard line or at a standard point. For example, a local scale may be measured on a map and expressed as 1:245,888. If the nominal map scale was 1:250,000, then the scale factor would be found by Equation 11.1.

$$\text{Scale factor} = \text{Local scale} / \text{Nominal map scale} \tag{11.1}$$

Substituting the scale values into Equation 11.1 we have the following scale factor:

$$1.0167 = (1/245{,}888)/(1:250{,}000).$$

The 1.0167 suggests that the scale at this point is exaggerated compared to the nominal map scale. A scale factor of 1.0 would suggest no scale distortion, while a scale factor less than 1.0 indicates a compressed scale compared to the nominal scale.

This option averages the scale factors at every point in all directions distributed evenly on the sphere, symbolized in grayscale. Darker shades indicate greater distortion. Figure 11.5 also offers two areal distortion options. The "Areal (logarithmic scale for definite limits)" option shows relative areal distortion according to a "pivot" value. In most cases, the pivot is set to 1, which corresponds to the nominal map scale, and suggests no areal distortion. Values less than 1 suggest deflation (areas on the map are displayed smaller than they are in reality), while values greater than 1 suggest inflation (areas on the map are displayed larger than they are in reality). The "Areal (finitized scale for indefinite limits)" option permits infinitely large area distortion to be displayed in a finite (or set) color range, where the number of gradations within the range can be changed according to the user's preference. Figure 11.7 shows the white-to-green color gradation on the August epicycloidal projection, where only one point is without areal distortion (the intersection of the equator and the central meridian). Of course, equal area projections would not show any areal distortion. There are two options to display the overall distortion on a projection's surface. The "Finitized scale for indefinite limits" option is shown in Figure 11.8. This option displays infinitely large areal and angular distortion using a

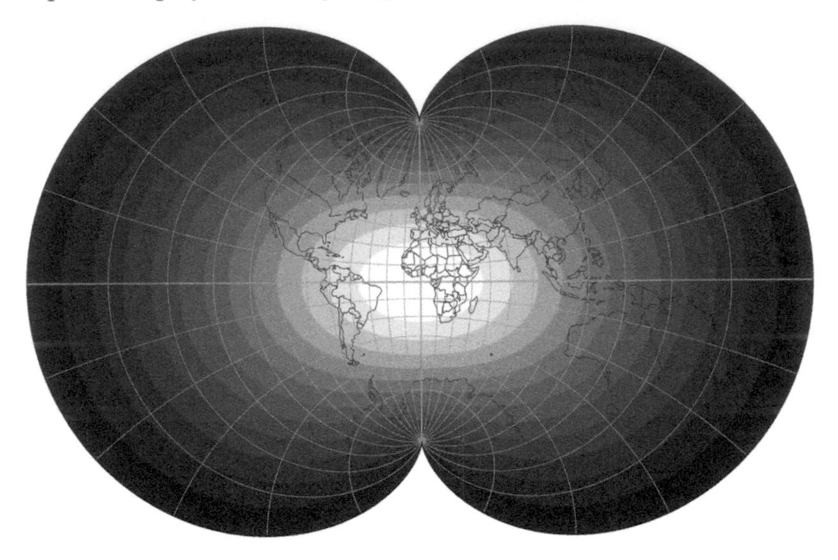

FIGURE 11.7
Areal distortion symbolized on the August epicycloidal projection, using white-to-dark green color gradations.

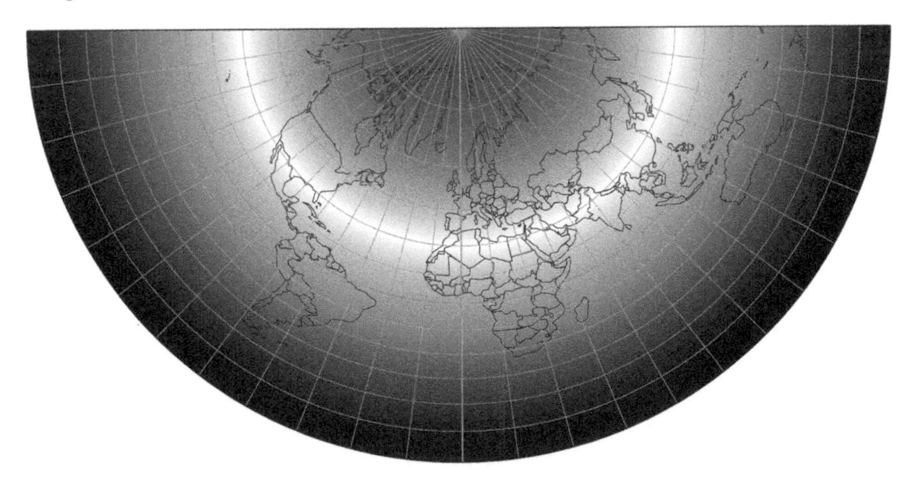

FIGURE 11.8
Overall distortion on the Braun stereographic conic projection symbolized with a green-to-magenta continuous tone.

finite color range. Figure 11.8 shows the overall mixture of areal and angular distortion pattern, mapped on the Braun stereographic conic projection. As with other visualizations in this software, lighter hues mean less distortion, while darker hues indicate more distortion. On this projection, angular distortion is more pronounced in the Southern Hemisphere, and areal distortion is more apparent in the Northern Hemisphere. Distortion is low around 30° north along the central meridian and the equator.

Geocart can also draw metric lines on the surface of the projection. Some of the metric lines include small circles, great circles, and rhumb lines. Each line type is defined in a user-generated XML file format. Figure 11.9 illustrates a portion of a rhumb (green), a portion of a great circle (red), and a small circle (blue). Displaying these metric lines can be useful for mapping routes or seeing how a projection distorts these lines compared to their appearance on the globe.

ICA's Map Projection Visualization Tool

Miljenko Lapaine and his colleagues (2014) created an online interactive tool that allows a user to explore a variety of cylindrical projections, their graticule arrangements, and their distortion patterns according to different properties. Their application can be found at http://ica-proj.

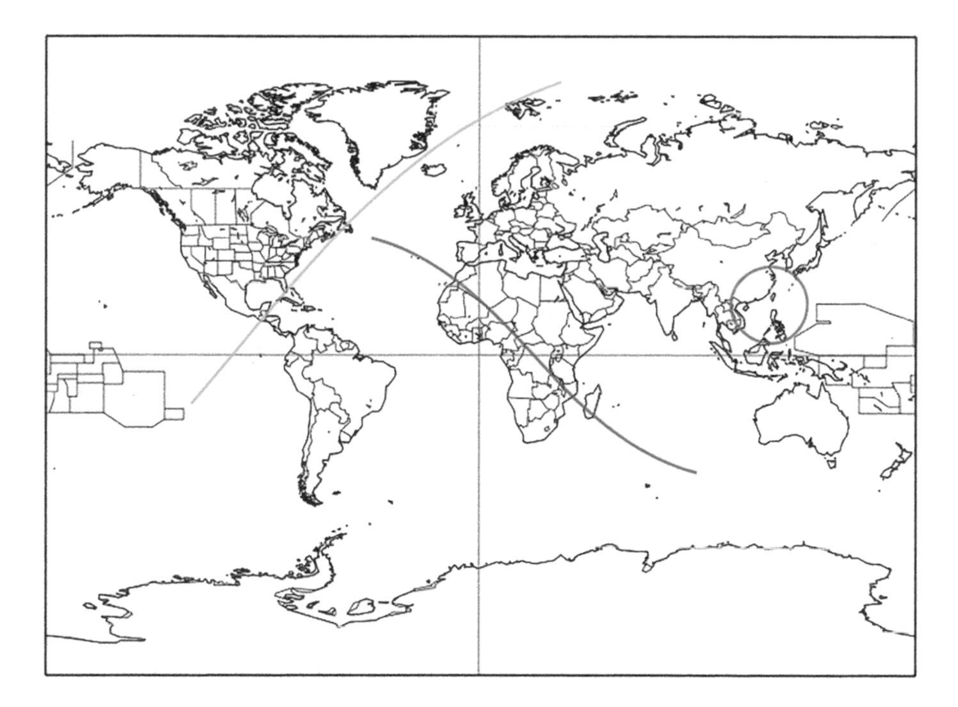

FIGURE 11.9
A rhumb (green), a portion of a great circle (red), and a small circle (blue) displayed on the Miller cylindrical compromise projection. Image produced in Geocart.

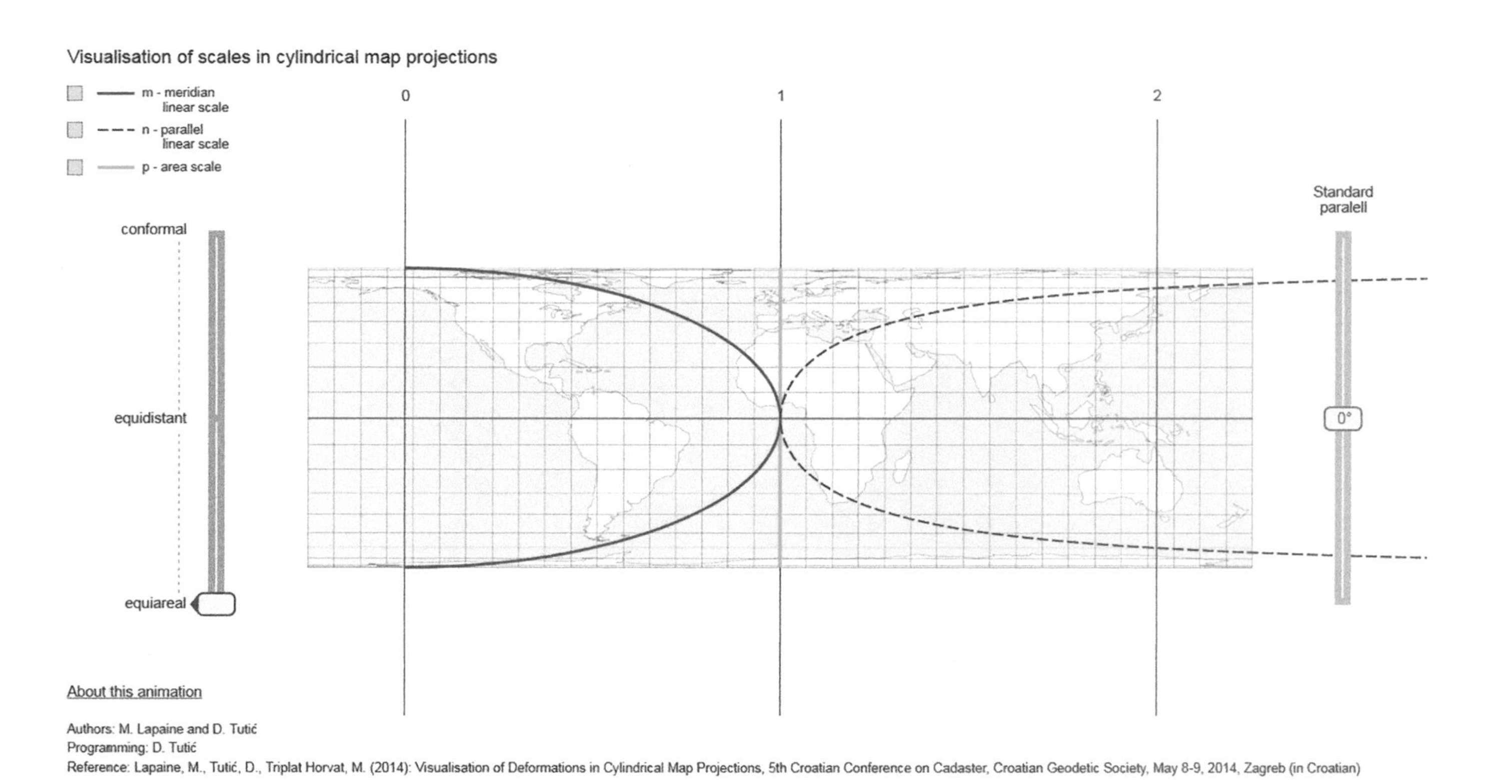

FIGURE 11.10
An interactive tool to explore map projection distortion on different cylindrical map projections and view the amount of areal and scale distortion.

kartografija.hr/tl_files/ICA%20Map%20Projections%20Commission/slike/cilindricne_final_en.xml.

Figure 11.10 shows the application interface. This useful tool visualizes different kinds of distortion patterns on cylindrical projections, and benefits map makers who are interested in choosing a projection other than web Mercator for their online web map service. One limitation of the application is that it only includes cylindrical projections. Also, users who are not familiar with the idea of linear scales along a meridian or parallel may have difficulties interpreting the significance of the isocols, which are lines of equal distortion.

The user first selects a projection property along a continuum between conformal and equal area, and then specifies the latitude of the standard line. After these inputs are defined, the maximum amount and types of distortion along a given line of latitude are shown as isocols with solid red, solid green, and black dashed lines. A distortion scale at the top of the window runs from 0 to 2, and the type of distortion (expansion or compression) along each isocol corresponds to the values. Values less than 1 mean that scale is being compressed at that point on the isocol, 1 represents no distortion, and values greater than 1 indicate scale expansion. The solid green line, representing areal distortion, has the distortion value 1, suggesting no areal distortion on this projection. The solid red line represents scale distortion along a meridian. This line is concave away from the Prime Meridian, with values less than 1, suggesting that scale is being compressed along a meridian. The red line coincides with 1 only at the equator. The black dashed line represents scale distortion along a parallel. This black line is also concave away from the Prime Meridian, but with values greater than 1, suggesting that scale is expanded. The black line intersects 1 only at the equator. On this projection, maximum and minimum scale distortion is at the poles and equator, respectively.

This tool can also be used to compare distortion patterns on different cylindrical projections. For example, Figure 11.11 shows two equal area projections with different standard parallels. The projection on the top has standard parallels at 15° north and south, while that on the bottom has standard parallels at 45° north and south. The standard lines on both projections appear as a solid and dashed blue line. On these equal area projections, the green line remains aligned with 1, indicating no areal distortion. The concavity of the red and black lines is similar to their appearance in Figure 11.10 (which was also of an equal area projection). Notice that for each projection, the latitudes where the red and black lines are both at 1 (representing no areal and no scale distortion) coincide to the locations of the standard lines (Figure 11.11).

Visualizing Map Distortion

Ian Johnson created an interactive online application that visualizes projection distortion on 29 projections. The application can be found at https://bl.ocks.org/enjalot/bd552e711b8325c64729. It is often challenging for map

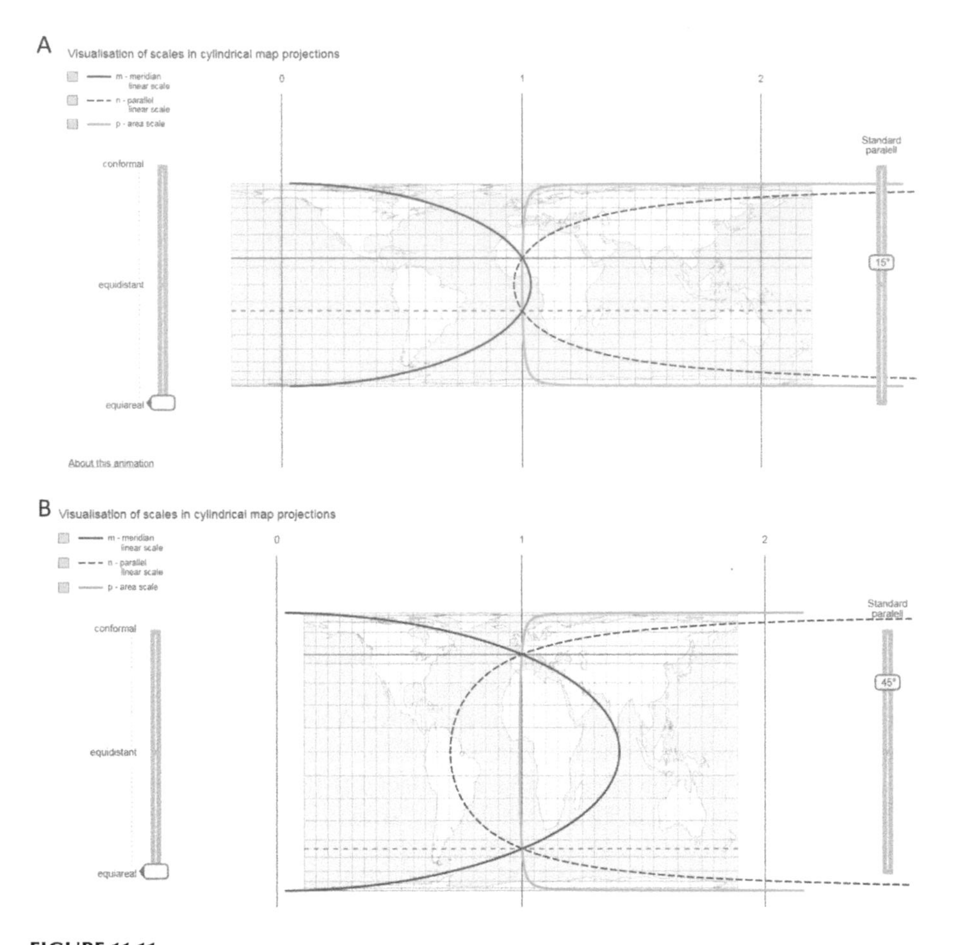

FIGURE 11.11
Two equal area cylindrical projections with standard parallels at 15° (A) and 45° (B) north and south.

readers to conceptualize how a projection distorts Earth's landmasses. However, Ian's approach is a little *backwards* in the sense that, while users usually want to see how a projection distorts an area of interest, the software provides a simple interface to visualize how a "square" area outlined on a given projection appears distorted on a globe.

Figure 11.12 shows the interface of Visualizing Map Distortion. The image on the left shows a map with the Eckert III projection, with a square area outlined by red dots. On the right, the same area is now shown on a globe (technically, an orthographic projection). As the user pans the map on the left, the shape remains square, but its shape on the globe adjusts according to the distortion on the projection's surface. Choosing a new projection will also change the overall shape of the red dots on the right map. Using this

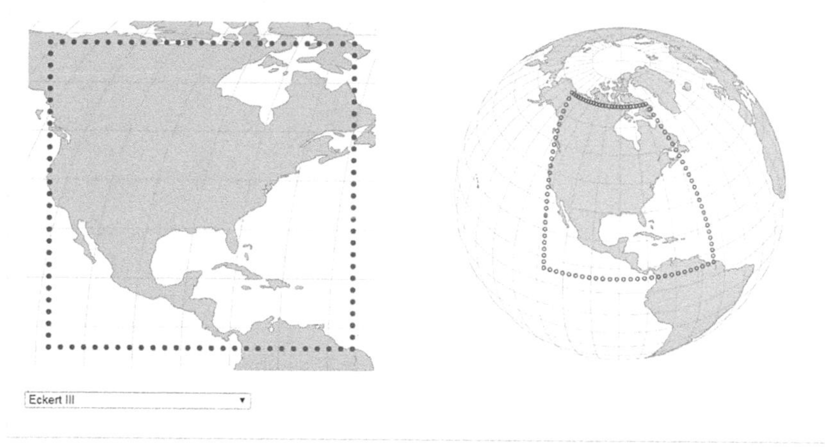

FIGURE 11.12
The Visualizing Map Distortion interface showing the Eckert III projection and its distortion impact on a globe.

software, users can learn about and explore the impacts that different projections have on geographic space.

Comparing Map Projections

Comparing Map Projections is a free, online interactive application that offers a way to quantitatively compare distortion across different projections, and is available at https://bl.ocks.org/syntagmatic/ba569633d51ebec6ec6e. This software is beneficial to users who need to explore projections' overall distortion metrics and select a projection that meets a minimum distortion threshold (e.g., a mean angular distortion value of 15°). However, users do need to understand how to interpret the distortion numbers.

Figure 11.13 shows the interface of Comparing Map Projections. A world map in a chosen projection is shown at the top of the screen. Below is a parallel coordinate plot with a list of several projections, and axes for different distortion metrics. When the mouse hovers over a projection, the map at top reprojects accordingly, and the plot highlights the line connecting the distortion metrics for that projection. It is a different kind of distortion visualization that permits fast comparison between projections based on the metrics of scale, areal, and angular distortion. One additional metric is the acceptance index, which is a number that summarizes the boundaries where angular distortion is limited to 40° and area distortion is limited to 150%.

Popular / About

Comparing Map Projections

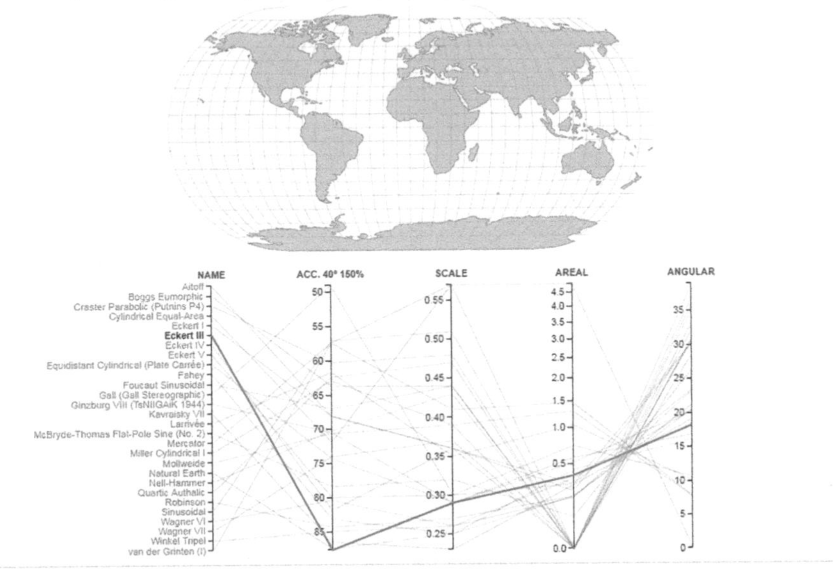

A mashup of Map Projection Distortions and transitions using the D3.js extended geographic projections plugin. Open

FIGURE 11.13
The Eckert III projection and its distortion impact on Earth's surface, compared to other projections in Comparing Map Projections.

Customizing and Reprojecting Datasets

G.Projector—Global Map Projector

G.Projector is a free standalone software package, developed by Robert Schmunk of NASA, for Mac, Windows, and Linux environments. It can be downloaded from www.giss.nasa.gov/tools/gprojector/.

It is a simple and inexpensive way to reproject a map, in image format, into one of 140 projections. Permitted input image formats include jpeg, gif, tiff, and bmp, among others. G.Projector only accepts inputs that are already in one of six expected projections, and the user must specify the image's bounding latitude and longitude values. A graticule and world coastlines can be added to the reprojected map, or they can be drawn independently and added to the map later.

For demonstration purposes, G.Projector includes sample image files. Figure 11.14 shows the software interface, with Earth's topography and bathymetry (imagery from National Geophysical Data Center's ETOPO2v2) in Clarke's twilight perspective projection. This far-side perspective projection

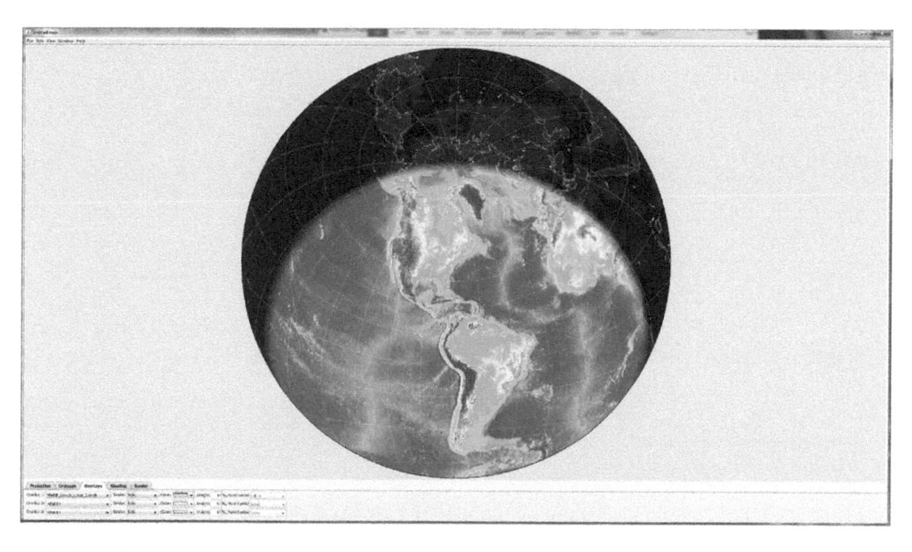

FIGURE 11.14
Clarke's twilight far-side perspective projection shown with Earth's dark side included.

is one of several that varies the angular range of how much "Earth" is shown to 108° (90° plus 18°). The 18° comes from the fact that when the sun dips 18° below the horizon, astronomical twilight is reached—hence the name "twilight" applied to this projection. The dark side of Earth's dark side is also shown.

G.Projector allows users options to customize the parameters that are associated with any projection. Users can import shapefiles to be displayed on top of the image, or add rivers, administrative boundaries, and coastlines that are provided by the software. The style, color, and weight of the shapefile features can be customized. The graticule spacing can be changed from the default of 15° to any desired interval.

Flex Projector

Flex Projector is free software developed by Bernhard Jenny, Tom Patterson, and Lorenz Hurni (Jenny et al., 2008). The various options and tools available with FlexProjector are extensive and can customize existing projections, visualize their distortion patterns, or create new projections that suit a particular mapping need. Figure 11.15 shows FlexProjector's interface displaying the Denoyer semi-elliptical modified pseudocylindrical projection. Below the projection are two plots that represent the distortion profiles along a meridian (Vertical Profile) and a parallel (Horizontal Profile). The right-hand panel includes options to customize an existing projection. The software can be downloaded from www.flexprojector.com/. Downloadable datasets such as coastlines, political borders, and hydrography can be viewed and projected in FlexProjector. Users can export a projection in one of several

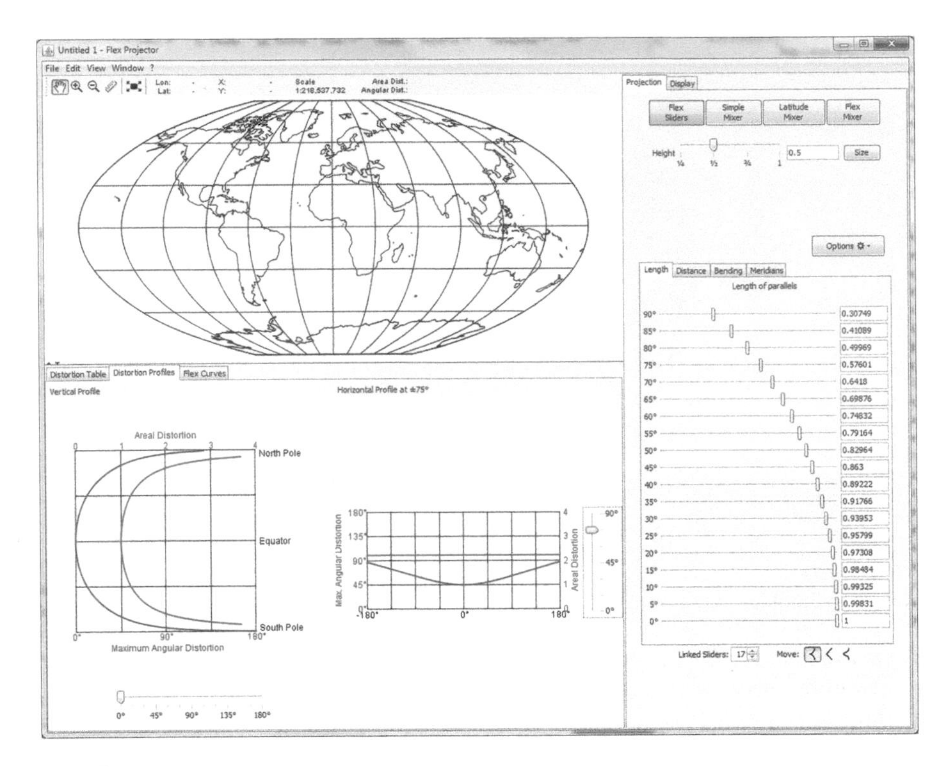

FIGURE 11.15
The interface of FlexProjector showing the Denoyer semi-elliptical modified pseudocylindrical projection.

formats (including shapefile) for use in other software. Experimenting with all of the options can be insightful and produce some very interesting projections. But, successful use of the software, and getting the projection to meet the needs of the map, does require the user to have a knowledge of projections in order to fully interpret all of the distortion metrics.

Users interested in customizing an existing projection (e.g., to redistribute distortion across the map) can start with one of the 30 included popular projections. By adjusting the length of one or more lines of latitude (shown every 5°) of any global projection, the user will alter the distortion pattern, which in turn changes the overall appearance of landmasses. The success of the customization is shown in a comparison of the before and after distortion profiles. In Figure 11.16, note that the pole line on the original Denoyer has been adjusted to a point. This changed the curvature of the meridians, which compressed the outlines of landmasses in the upper latitudes, making them harder to distinguish. This one adjustment also modified the distortion pattern. Compare the distortion curves reported by the Vertical and Horizontal Profile in Figure 11.15 and Figure 11.16. The curves on the Vertical Profile in Figure 11.16 are straighter than in Figure 11.15, suggesting that angular and areal distortion has been reduced along the central meridian (0°). On the

Horizontal Profile (along 75° latitude) in Figure 11.16, the angular distortion has greater variation than if the poles were represented as lines, but areal distortion is now reduced from about 220% to about 160%. Depending on the requirements of the final map, additional adjustments to the lengths of latitude on this projection could be made.

The interface also allows a comprehensive analysis of each projection's distortion pattern. Users can explore areal and maximum angular distortion across a projection's surface with different graphics. Figure 11.17 presents four distortion metrics symbolized by the software: Tissot's indicatrix (Figure 11.17A), isolines of areal distortion (Figure 11.17B), isolines of angular distortion (Figure 11.17C), and area of acceptable distortion (Figure 11.17D). The area of acceptable distortion is shown with blue shading, and is defined by a maximum angular distortion of 40° and a maximum areal distortion of 150% (these values are the software default, but can be adjusted).

The software provides capabilities to design new projections. For example, users can blend two projections into a new projection to achieve a better distribution of distortion. Figure 11.18 shows the Kavrayskiy VII pseudocylindrical compromise and sinusoidal pseudocylindrical equal area projections blended together as a new projection, which contains characteristics of

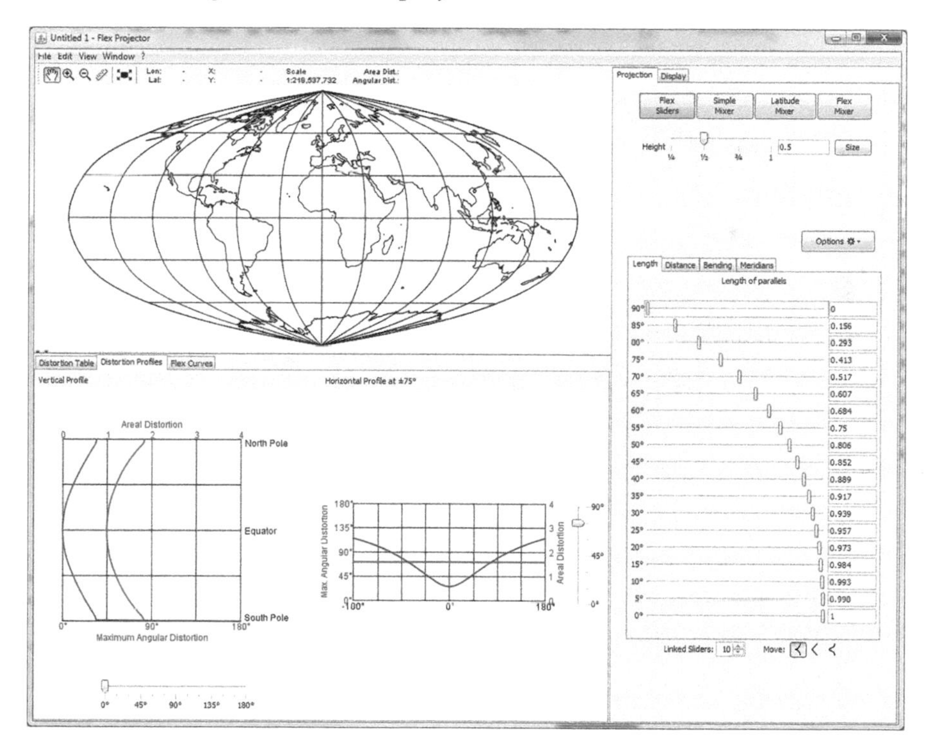

FIGURE 11.16
The Denoyer semi-elliptical modified pseudocylindrical projection with the lines representing each pole reduced to points.

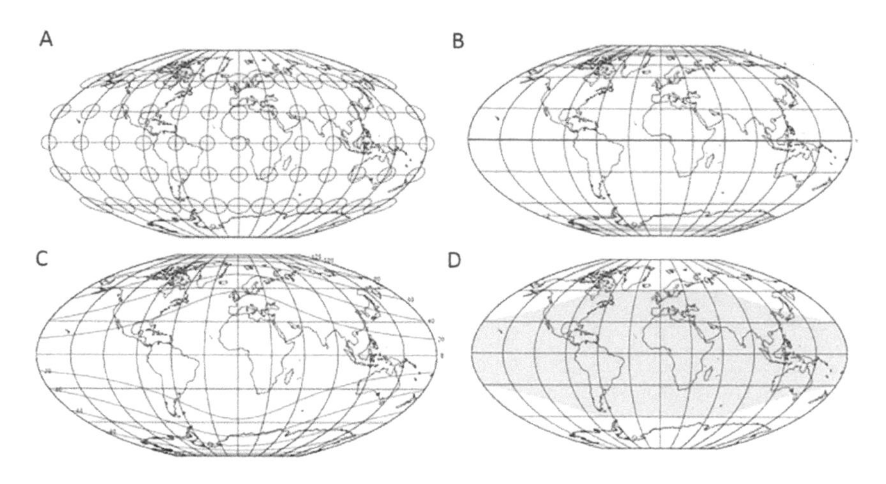

FIGURE 11.17
Symbolizations of distortion metrics available through FlexProjector: Tissot's indicatrix (A), blue isolines of angular distortion (B), magenta isolines of areal distortion (C), and blue-shaded region of maximum acceptable distortion (D).

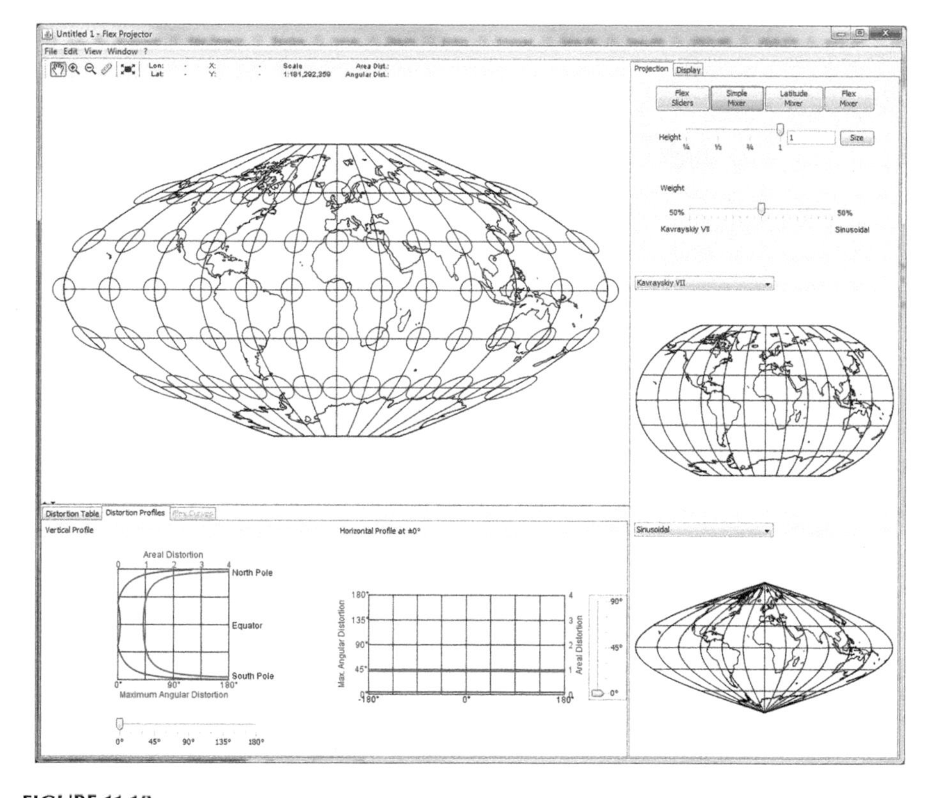

FIGURE 11.18
Blending the Kavrayskiy VII and sinusoidal projections together to create a new projection in FlexProjector.

both original projections. Blending projections may achieve a more favorable distribution of distortion than any single projection. Use the Vertical and Horizontal Profiles, or the other distortion visualization methods, to assess the success of the blending process.

Programming Languages, Libraries, and Tools for Projections

This section presents a brief overview of programming languages, libraries, and tools that are suitable for incorporating projections into an application. There are dozens of programming languages that can be used to code projection equations that project latitude and longitude coordinates into x and y Cartesian values. In addition, there are many code libraries that provide resources to someone interested in using projections in the web environment who does not want to convert complex projection formulae into code. There are also specialized tools for those who need to work with projections (e.g., projecting an image from one coordinate system into another). We do not attempt to discuss all of the possible languages, libraries, or tools. Rather, our focus is on those that we feel have already demonstrated sufficient staying power and utility.

Common Programming Languages

For the past several years, there has been a trend of developing spatial applications in Python, R, and JavaScript, instead of in traditional compiled languages like C/C++ or Java. Thus, we focus our discussion on these first three languages. Python (www.python.org) is a high-level open-source programming language that was created in the early 1990s. The language is flexible and supports both object-oriented and structured programming. Its code syntax is fairly simple, and it can be used in multiple platforms without recompiling the source code. The Python coding community, Python Package Index (PyPI), is a repository where coders freely distribute their software to other Python users (pypi.org).

Similar to Python, R (cran.r-project.org/) is a free, object-oriented language that also got its start in the early 1990s. Originally developed as a language for statistical computing, R has rapidly expanded into other fields, and has added functionality such as dynamic and interactive graphics. A strength of R rests with users being able to create and share packages. Once created, these packages are made available to other R users through the Comprehensive R Archive Network (CRAN).

JavaScript or JS (www.javascript.com/) is another high-level, object-oriented and platform-independent language. In addition to HTML and CSS, JS is one of the three core building-block languages of the web. With JS, web pages can be interactive and dynamic. Unlike compiled languages like C or C++, JS's syntax is considered to be easier to learn and implement.

Code Libraries

A code library, or simply library, is a collection of code developed by other programmers posted to a central location or repository. Once in the repository, that code is available to other programmers who then use it in their applications (and sometimes modify it). The primary benefit of a library is that it permits easy incorporation and extension of functionality without requiring the user to write what could be a complex algorithm.

For coders who use Python, there are two useful libraries that support working with projections. First, pyproj allows the map projection of a dataset to be defined and changed, and it can transform datums. There are two classes in pyproj that handle coordinate system operations, Proj and Geod. Proj class converts latitude and longitude values to map projection coordinates (forward method) or derives latitude and longitude from a native projection (inverse method) for reprojection. The Geod class handles various geodetic operations, such as determining forward and backward azimuths, and calculating geodesic or great circle distances.

Most major GIS software vendors identify coordinate systems with a spatial reference system code. Each coordinate system (with preset parameters) has a unique code, making it unambiguous. Use of the identifying code eliminates the possibility of making mistakes from typing in values and parameters one by one. Software vendors either develop their own codes (e.g., Esri's projected coordinate system codes; http://resources.esri.com/help/9.3/arcgisserver/apis/rest/pcs.html) or rely upon an authority. One authority of note is the European Petroleum Survey Group. This group developed and maintains a Coordinate System Registry (www.epsg.org/EPSGDataset/DownloadDataset.aspx). This freely available registry, delivered in a Microsoft Access database, is a collection of coordinate system definitions, datums, reference ellipsoids, and transformations. While the registry is comprehensive, using it to find a specific code can be challenging, as the interface is not necessarily intuitive. Python users can load the python-epsg library (https://pypi.org/project/python-epsg/), a GUI-based API that accesses coordinate system metadata from the registry. Alternatively, on Spatial Reference (http://spatialreference.org/) one can search for EPSG and other vendor-specific codes. Searching for EPSG 3857 (the code for the web Mercator projection) returns a list of definitions of this coordinate system by different vendors in various formats (e.g., Proj.4, JSON, GML, .prj file, GeoServer, and PostGIS). The user chooses the format they need, and then the code block defining that projection is provided.

The popular language R offers its users several packages that provide tools to work with coordinate systems. We will briefly review three here. The mapproj package projects latitude and longitude coordinates into over two dozen projections and includes various projection utilities. Documentation is available at https://cran.r-project.org/web/packages/mapproj/mapproj.pdf. The sp package (https://cran.r-project.org/web/packages/sp/sp.pdf) provides several functions that operate on spatial coordinate systems. For example, the spTransform function converts from one projection into another or transforms datums. Rgdal is a package that works in tandem with GDAL (discussed later).

One of the more prevalent projection code libraries, Proj.4 (proj4.org) offers a range of coordinate system transformations, allowing coders to project from one coordinate system to another, carry out a datum transformation, and conduct forward and inverse projection routines. For example, a coordinate transformation in Proj.4 is called via the proj-strings, which holds the parameters of a given coordinate transformation:

$$+\text{proj} = \text{cea} + \text{long\_0} = -96 + \text{lat\_ts} = 30.0 + \text{ellps} = \text{WGS84}$$

This syntax defines the cylindrical equal area projection (cea) whose central longitude is −96° (long_0=−96) and whose standard parallels are 30.0° north and south (lat_ts = 30.0), and which uses the WGS84 reference ellipsoid (ellps = WGS84). This particular type of cylindrical equal area projection is called the Behrmann projection.

Data-Driven Documents, also called d3 (d3js.org), contains JavaScript-based projection available to coders who wish to integrate projections into a web application. Projections in d3 are classified as standard abstract projections, standard projections, and raw projections. Standard abstract projections allow the user to modify projections by inverting, rotating, centering, or scaling. For instance, projection.invert(point) projects Cartesian coordinates, defined as pixels, into latitude and longitude coordinates. The following website illustrates these modifications: https://bl.ocks.org/d3indepth/f7ece0ab9a3df06a8cecd2c0e33e54ef. Standard projections are projections that have predefined functions such as d3.geo.albers() or d3.geo.azimuthalEquidistant() that define the Albers conic and azimuthal equidistant projections, respectively. Raw projections take coordinates in *raw* latitude and longitude values (in radians) and project them into Cartesian coordinates, also in radians. A brief example of how d3 projection modules can be implemented is the website www.jasondavies.com/maps/transition. This page shows an animation of the world seamlessly morphing from one projection to another. Users can click on the projection and drag to recenter the map. Unfortunately, d3 does not support datum transformations or other geodetic computations, such as forward or backward azimuths computations on an ellipsoid.

Similar to Python and R, Geospatial Data Abstraction Library (GDAL) is a free and open-source code library for processing and transforming raster and vector geospatial data (www.gdal.org/), with a complete set of functions. GDAL works with a wide variety of formats, such as GeoTIFF, MrSID, JPEG2000, .shp, KML, PostGIS, and GML. Two GDAL classes

(OGRSpatialReference and OGRCoordinateTransformation) handle projection and datum conversions and transformations. Coordinate system definitions and assignments are handled through the many functions available in the OGRSpatialReference (www.gdal.org/classOGRSpatialReference.html) class. GDAL uses well-known text (WKT) to define some coordinate systems like NAD27 or WGS84, but users can specify coordinate systems by their EPSG codes. Conversion between coordinate systems are done by functions available in the OGRCoordinateTransformation. For example, OGRCreateCoordinateTransformation() transforms data between two projected or two geographic coordinate systems. A gentle introduction to GDAL can be found at https://medium.com/planet-stories/a-gentle-introduction-to-gdal-part-1-a3253eb96082.

Coders working in Python can access GDAL through the GDAL/OGR project, while R users can access GDAL by adding the rgdal package. JavaScript users can point to the native GDAL binding from Node.js, an open-source and freely available server environment.

Tools

Generic Mapping Tools (GMT), originally developed by Paul Wessel and Walter Smith in 1988, contains approximately 80 commands that manipulate raster and vector data (https://github.com/GenericMappingTools/). The software has many available mapping functions (e.g., creating isarithmic maps, labeling, specifying color spaces), graphing functions (e.g., plotting histograms), and animation options (e.g., flying over topography) that produce high quality PostScript output files. GMT can project from latitude and longitude to one of thirty common projections. A Python/GMT interface is available at http://gmt.soest.hawaii.edu/projects/gmt-python-api.

The National Geodetic Survey has developed a suite of tools referred to as the Geodetic Toolkit (www.ngs.noaa.gov/TOOLS/). While many of the tools at this site are designed for geodetic applications (e.g., computing the geodetic azimuth and ellipsoidal distance between two points given their latitudes and longitudes), some are more tailored to working with projections, such as conversion between state plane coordinates, UTM coordinates, and X, Y, and Z to latitude and longitude values.

Learning about Map Projections

There are many websites that provide introductions to projections, although site content varies from brief overviews to detailed textbook-like discussions. Explanations of the different projection classes, aspects, cases, distortion patterns, and properties are common. Many websites include small illustrations of different projections, with a simple coastline and graticule. Additionally, some websites include links to other online resources, references to reading materials, and projection software.

Geographer's Craft

One of the earliest web-based educational approaches to learning about map projections was The Geographer's Craft. In 1995, Peter Data created the site to reach a broader audience, posting a series of lecture notes on various geography topics. One of those topics was projections. The lecture included a brief introduction to what a projection is and the properties it has. Illustrations of different projections were presented, grouped according to the cylindrical, pseudocylindrical, conic, azimuthal, and miscellaneous classes, and a short descriptive explanation of each was provided. As of this writing, the Geographer's Craft pages are no longer available.

Geokov: Map Projections

http://geokov.com/education/map-projection.aspx

Geokov discusses projection classes according to the developable surface concept, and their different cases and aspects. Geokov also contains sections on scale factors, standard lines, and distortion. A separate section illustrates several projections with Tissot's indicatrix. For each projection, a short discussion explains the factors that controls the appearance of the indicatrix.

Axis Maps

www.axismaps.com/guide/general/map-projections/

This website covers the basics of projections, with a focus on projections as a design variable. The discussion begins with an overview of projection properties, referencing area, distance, and direction. In an odd usage, the term "form" is also introduced as a property that is associated with the ability of a projection to show landmasses as if they were on a globe. Finally, directions are provided on how to choose a projection, and projection parameters are discussed.

Geocart

www.mapthematics.com/ProjectionsList.php

This companion website to the Geocart software presents a brief overview of each projection available in the software. Each projection is described in terms of its classification, scale characteristics, distortion, usage, and other features. Distortion diagrams for each projection are shown using color gradations as seen throughout this book to illustrate overall projection distortion.

USGS Map Projection Poster

https://store.usgs.gov/assets/mod/storefiles/PDF/16573.pdf

This poster was originally created in print form by the USGS. You can still order the print version from https://store.usgs.gov/product/16573. The poster was designed to summarize some of the projections in John Snyder's tome *Map Projections: A Working Manual*, and to assist the reader in choosing a projection. Each projection appearing on the poster has a short description of its class, history, intended uses, distortion description, and graticule spacing, and is accompanied by a small illustration. Three tables summarize recommendations for projections by property (e.g., conformal, equal area, equidistant), area to be mapped (e.g., world, hemisphere, continent), and the type of map (e.g., topographic, geologic, and thematic). A short dictionary of projection terminology is provided.

Radical Cartography

www.radicalcartography.net/index.html?projectionref

Created by Bill Rankin, this website categorizes map projections by map use and coverage: world maps for wall display, hemispheric and continental coverage, navigation of large areas, and regional and local maps. For each category, a table lists suitable projections and where they have been used. Other details about each projection are the author and date of development, a small image of the projection with coastline and graticule, alternate names of the projection, and its properties.

ICA's Commission on Map Projections

The International Cartographic Association (ICA) is an organization whose mission is to promote the practice of cartography and GIScience (icaci.org/). The ICA organizes commissions to consolidate and promote cartographic scholarship and technique. Each commission focuses on a single topic, and functions as a repository of knowledge and a means to disseminate that knowledge to the broader cartographic community. The Commission on Map Projections (CoMP) is a group of professionals and academics who take interest in the study and application of map projections. Many members of this commission publish articles and books related to projection research. The CoMP website has links to various educational resources on projections (http://ica-proj.kartografija.hr/map-projections.en.html).

Map Projection Galleries

Projection galleries are just what the phrase suggests—a web page containing illustrations of different map projections often accompanied by a short description of the projection's properties, class, and special characteristics.

There are quite a few of these online galleries. Below are some of the more useful ones.

A Gallery of Map Projections

www.csiss.org/map-projections/index.html

A website created by Paul B. Anderson that contains dozens of projections drawn in simple black and white illustrations with downloadable 8.5" by 11" PDFs.

Map Projections—Complete Directory of Map Network Designs

www.boehmwanderkarten.de/kartographie/is_netze_projection_register.html

A website developed by Dr. Rolf Böhm with material on 324 map projections. Each projection has a color illustration, and source code describing it (although the code is written for a specific software package).

Wolfram MathWorld: Map Projections

http://mathworld.wolfram.com/topics/MapProjections.html

MathWorld has the equations used in 35 popular projections. These equations generally represent the spherical form, rather than the ellipsoidal form. A small illustration of the projection's graticule arrangement also accompanies each equation.

MicroCAM

www.csiss.org/map-projections/microcam/(click on MicroCAM Graphics Gallery link)

Developed by Dr. Scott Loomer, MicroCAM (Cartographic Automated Mapping) is a freely available Windows-based software projection software package. Unfortunately, the software is no longer being supported. The most recent version was created for older 32-bit operating systems. However, the site provides small images of dozens of projections. Two distortion symbolization options are available, Gedymin profiles and Tissot's indicatrix (Figure 11.19). Gedymin profiles present a "projected" human head in profile. The contortions of the chin, nose, ears, and other parts suggests where distortion is more or less severe.

Institute of Discrete Mathematics and Geometry: Picture Gallery of Map Projections

www.geometrie.tuwien.ac.at/karto/

Developed by Hans Havlicek, the site provides a semi-interactive gallery of dozens of map projections as simple graphics, with a black graticule and

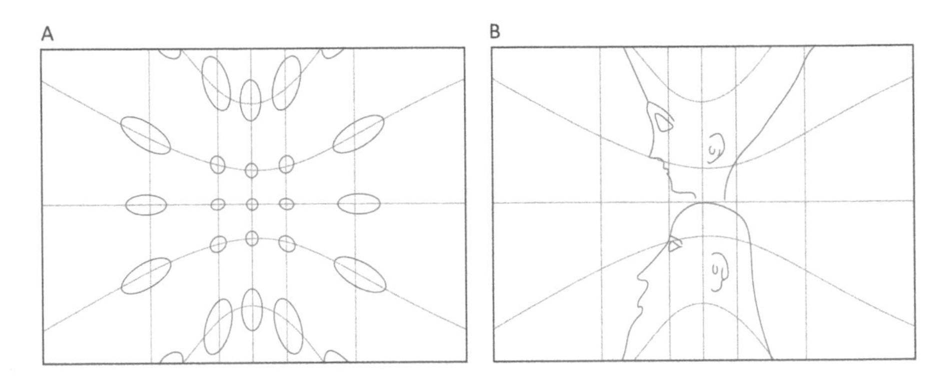

FIGURE 11.19
Gedymin profiles (A) and Tissot's indicatrix (B) displayed on the gnomonic azimuthal projection.

dark blue coastlines. The user can select different aspects, rotations, and display options, and the image changes with the selection.

Compare Map Projections

https://map-projections.net/

Created by Tobias Jung, Compare Map Projections allows a visual comparison of the similarities and differences between two projections. In Figure 11.20,

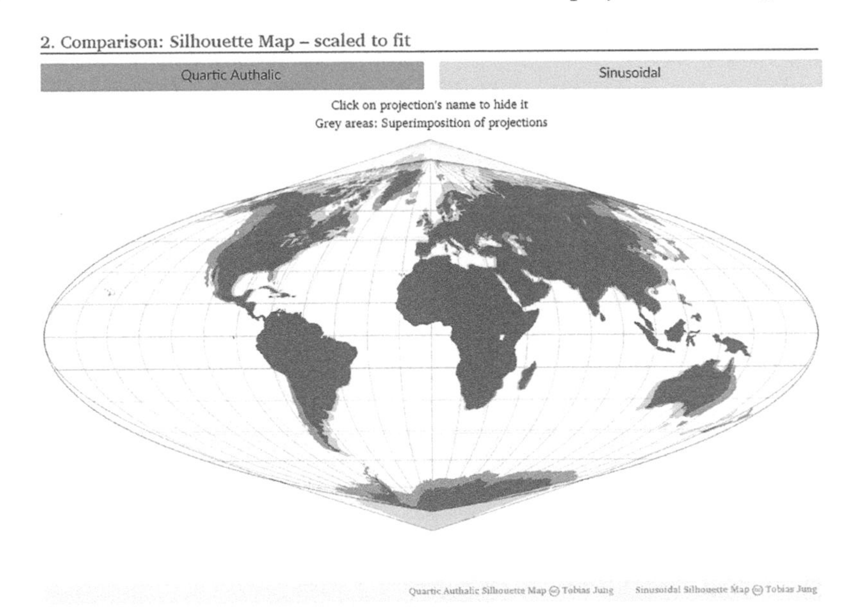

FIGURE 11.20
A comparison between the quartic authalic and sinusoidal projections on the Compare Map Projections website.

we show the comparison between quartic authalic and sinusoidal, both of which are equal area and pseudocylindrical. One projection is overlaid on the other, and the gray areas show where landmasses in the two projections overlap. The magenta areas are those visible only on quartic authalic, while the green areas are those unique to sinusoidal.

Appendix

This appendix presents a gallery of map projection names and their distortion patterns that were used as examples throughout this textbook. These projections were included in this book for two reasons. First, approximately 50 projections have seen frequent use in mapping throughout history. Second, several projections possess specific characteristics or parameters that we wished to highlight in a given chapter.

Each projection included in this appendix is listed alphabetically by its common name. Next, the projection property is specified. Then, the projection's class is stated. Maps of the world, illustrating each projection's pattern of areal, angular, scale, and overall distortion patterns, are presented. For each distortion pattern, green, magenta, and gray tones represent areal, angular, and scale distortion, respectively. Lighter shades of a specific color indicate lower distortion while darker shades of a given color represent greater distortion. Depending on a given projection's property, images illustrating overall distortion generally show a mixture of color hues and tones across a projection's surface. If a distortion image is not shown for a projection in the table, then that projection preserves that property and there is no distortion to show (e.g., an equal area projection does not possess any areal distortion).

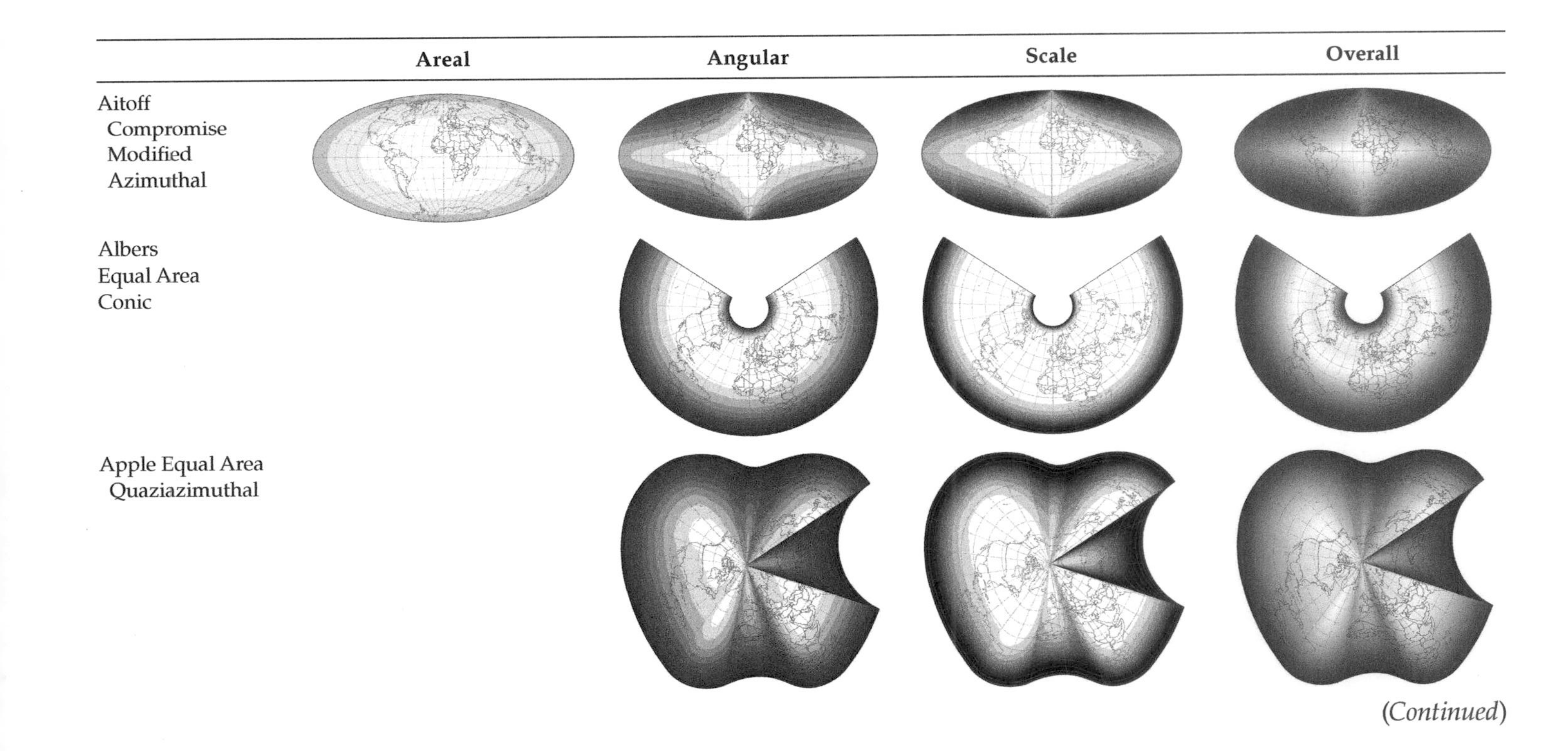
Areal
Angular
Scale
Overall
Aitoff Compromise Modified Azimuthal
Albers Equal Area Conic
Apple Equal Area Quaziazimuthal
(Continued)

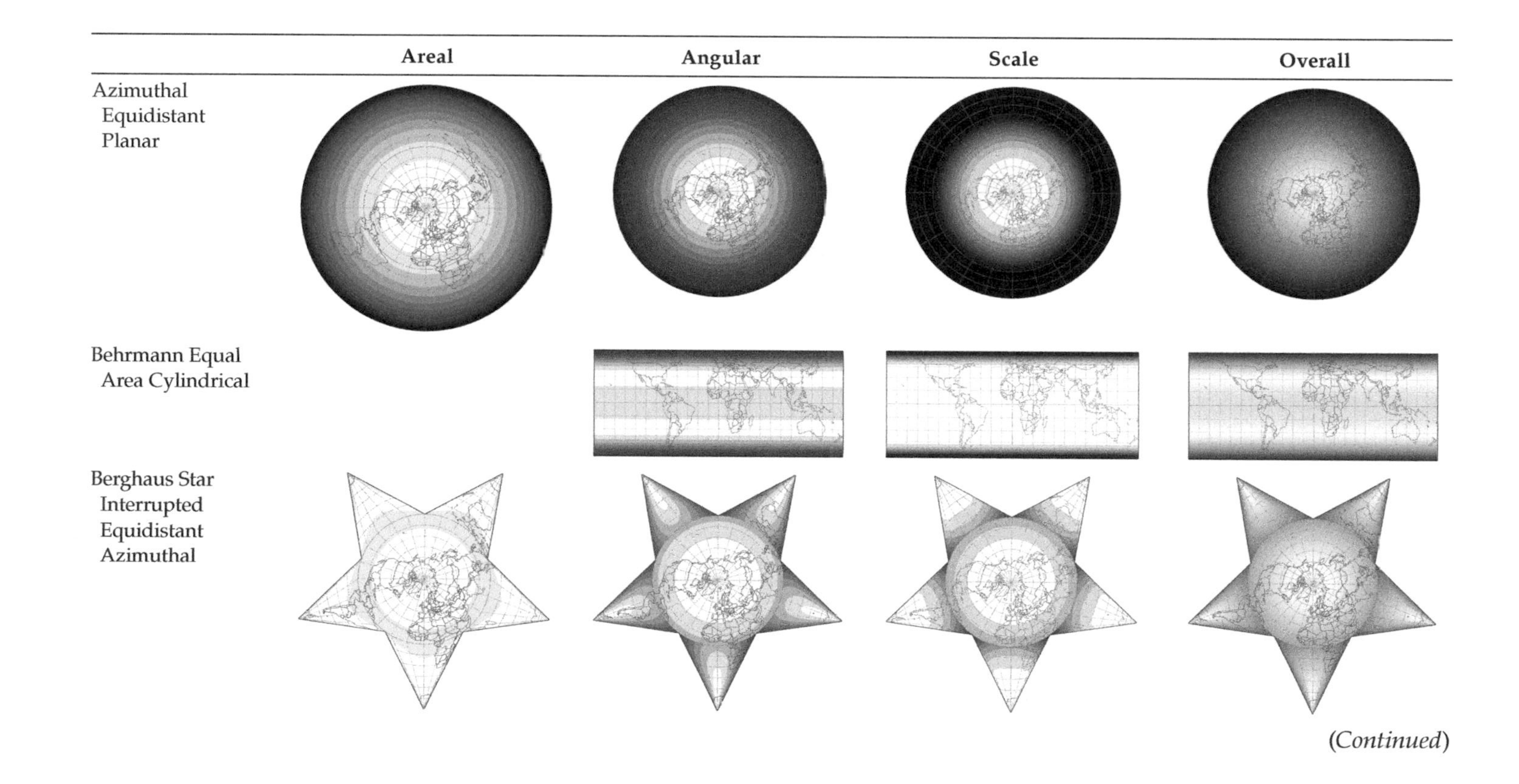
Areal
Angular
Scale
Overall
Azimuthal Equidistant Planar
Behrmann Equal Area Cylindrical
Berghaus Star Interrupted Equidistant Azimuthal

(Continued)

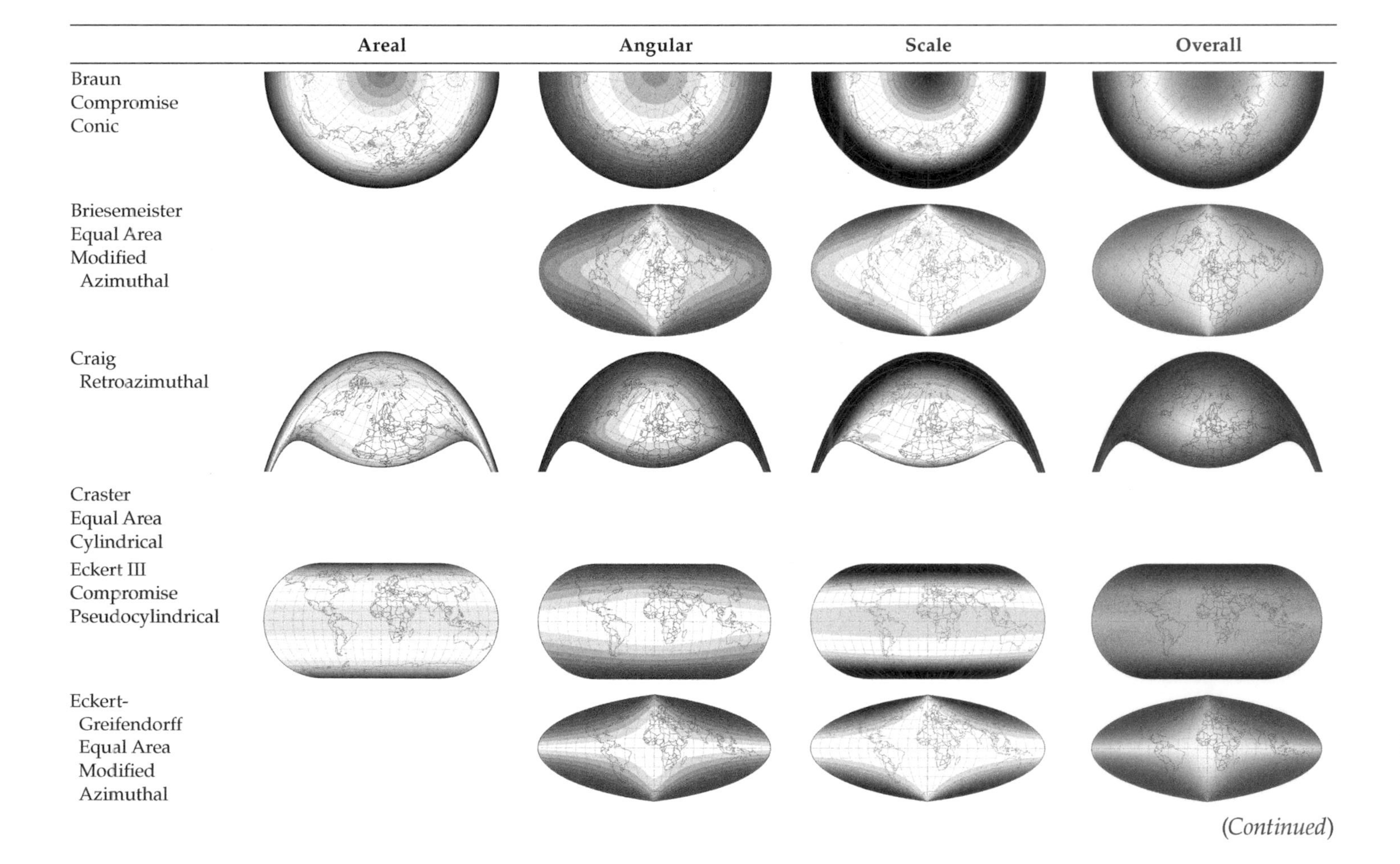
Areal
Angular
Scale
Overall
Braun
Compromise
Conic
Briesemeister
Equal Area
Modified
Azimuthal
Craig
Retroazimuthal
Craster
Equal Area
Cylindrical
Eckert III
Compromise
Pseudocylindrical
Eckert-
Greifendorff
Equal Area
Modified
Azimuthal

(Continued)

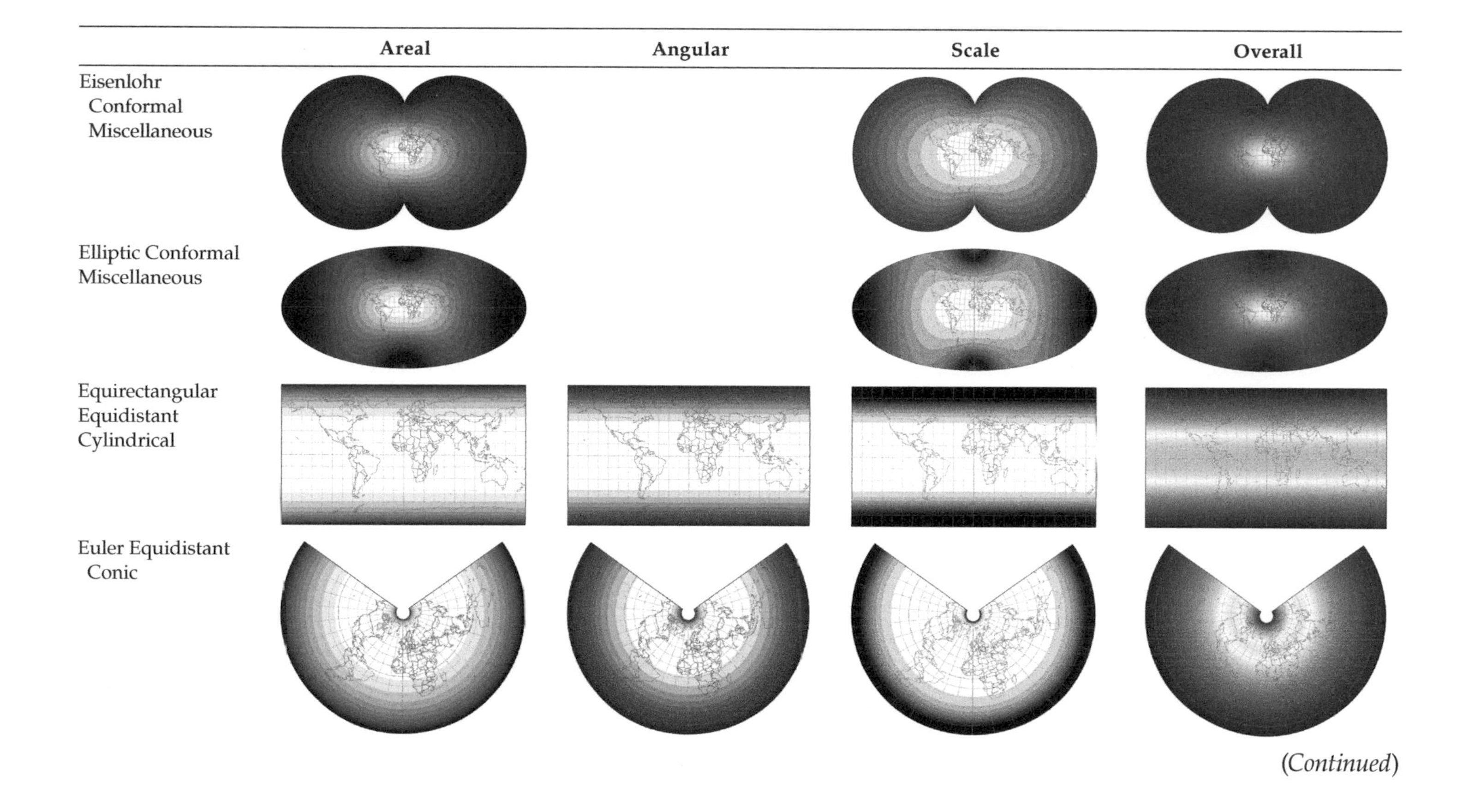
Areal
Angular
Scale
Overall
Eisenlohr
Conformal
Miscellaneous
Elliptic Conformal
Miscellaneous
Equirectangular
Equidistant
Cylindrical
Euler Equidistant
Conic

(Continued)

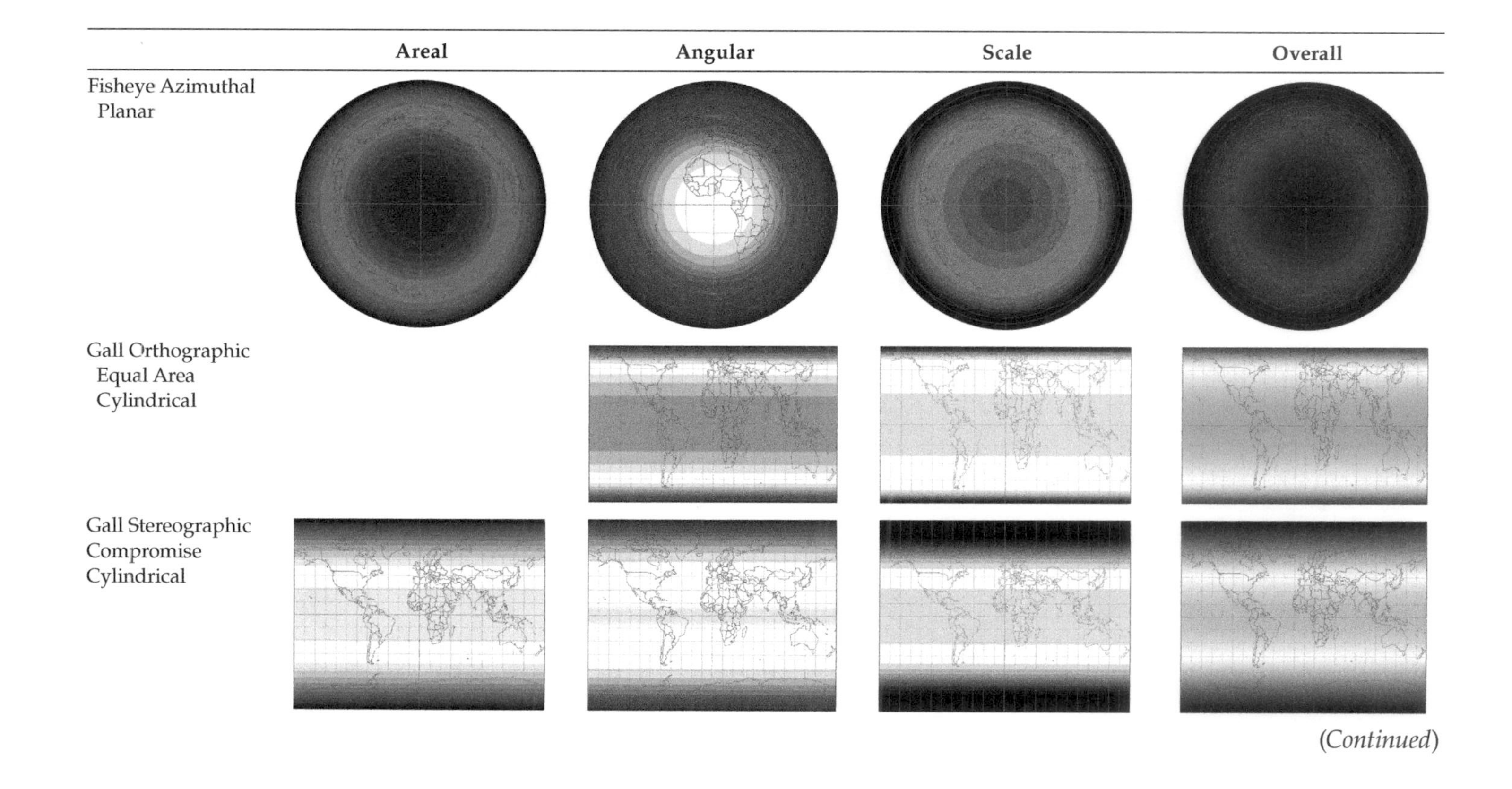
Areal
Angular
Scale
Overall
Fisheye Azimuthal Planar
Gall Orthographic Equal Area Cylindrical
Gall Stereographic Compromise Cylindrical

(Continued)

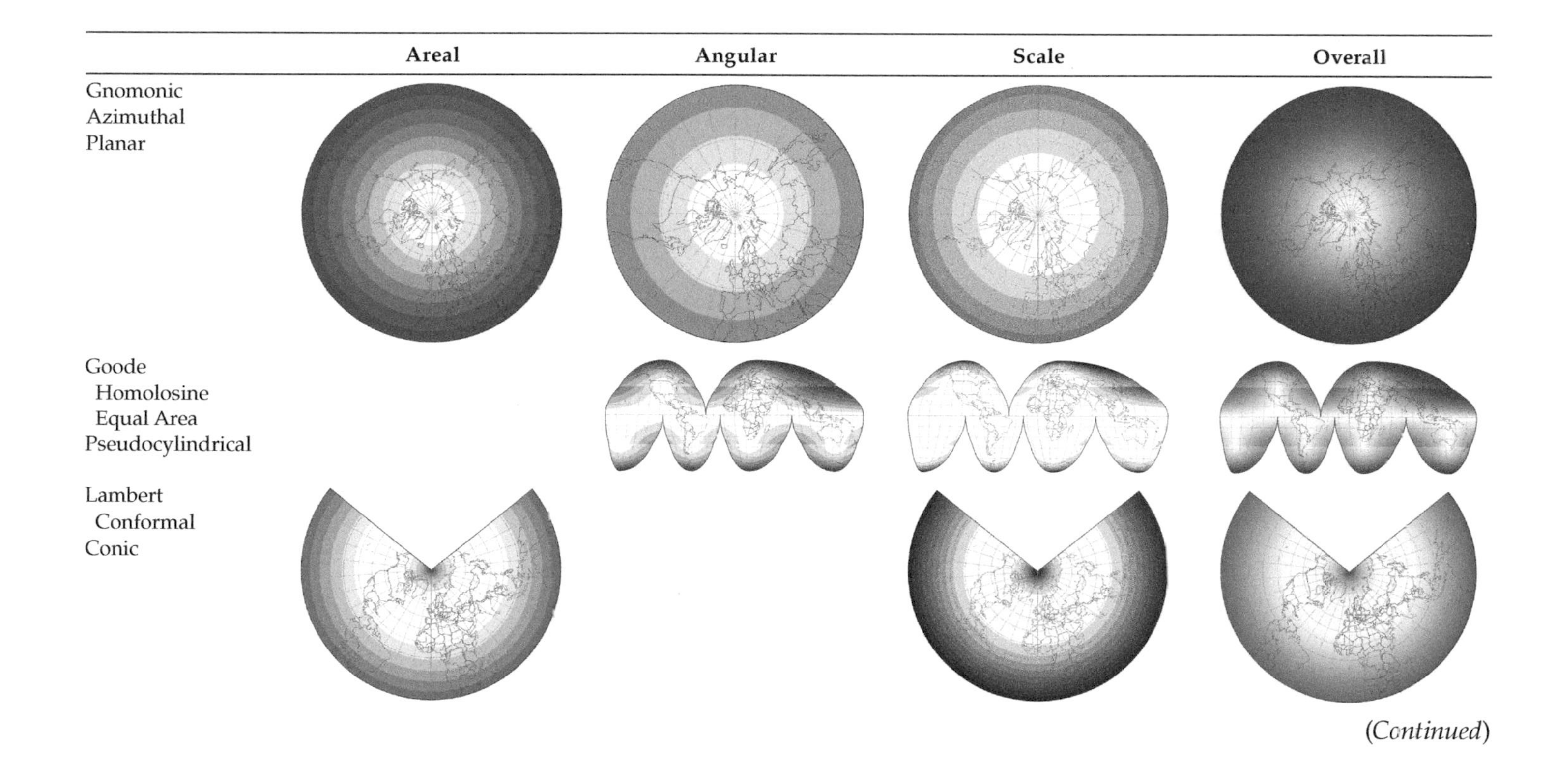

Areal
Angular
Scale
Overall
Gnomonic
Azimuthal
Planar
Goode
Homolosine
Equal Area
Pseudocylindrical
Lambert
Conformal
Conic

(Continued)

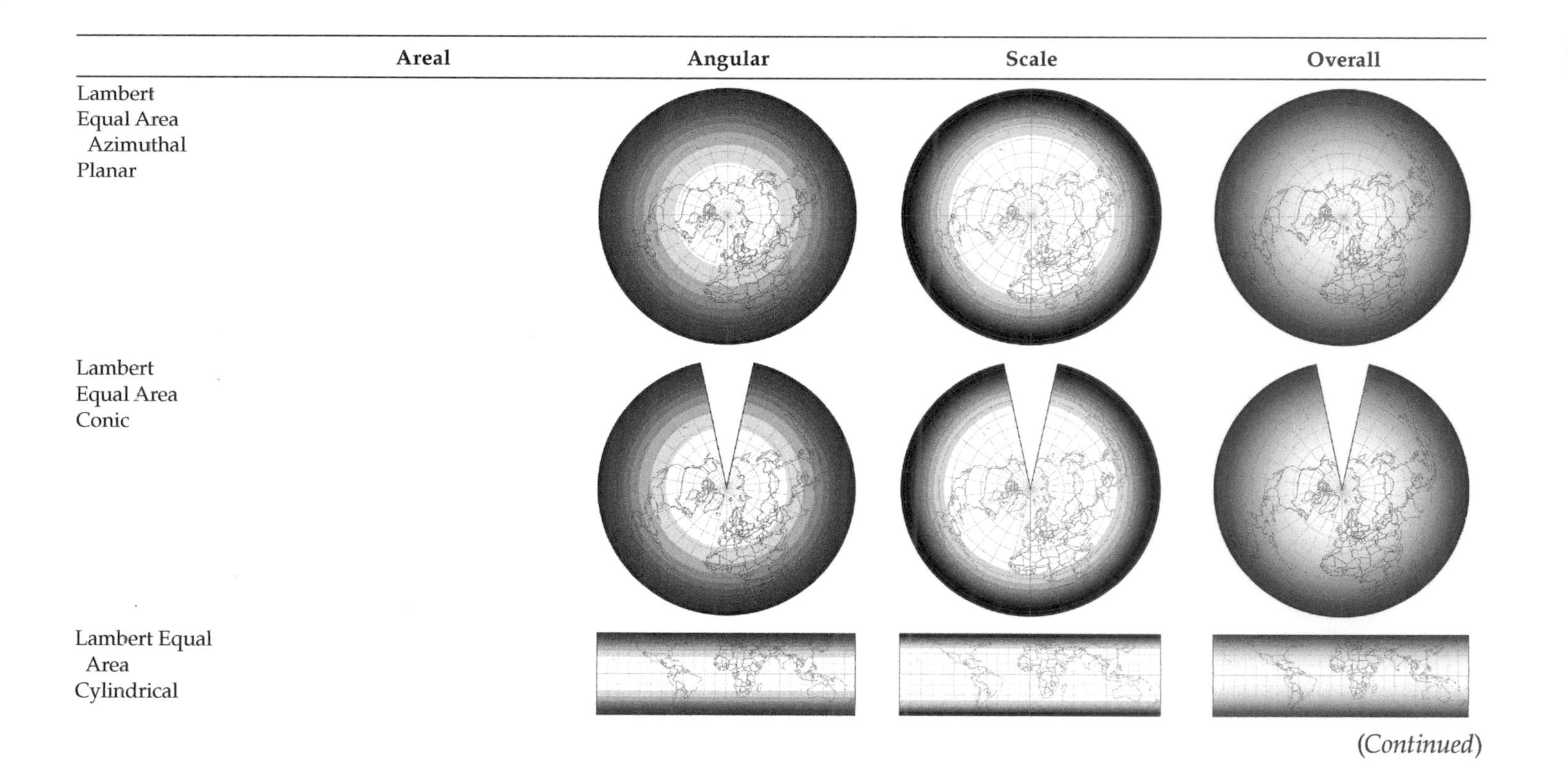

	Areal	**Angular**	**Scale**	**Overall**
Lambert Equal Area Azimuthal Planar				
Lambert Equal Area Conic				
Lambert Equal Area Cylindrical				

(Continued)

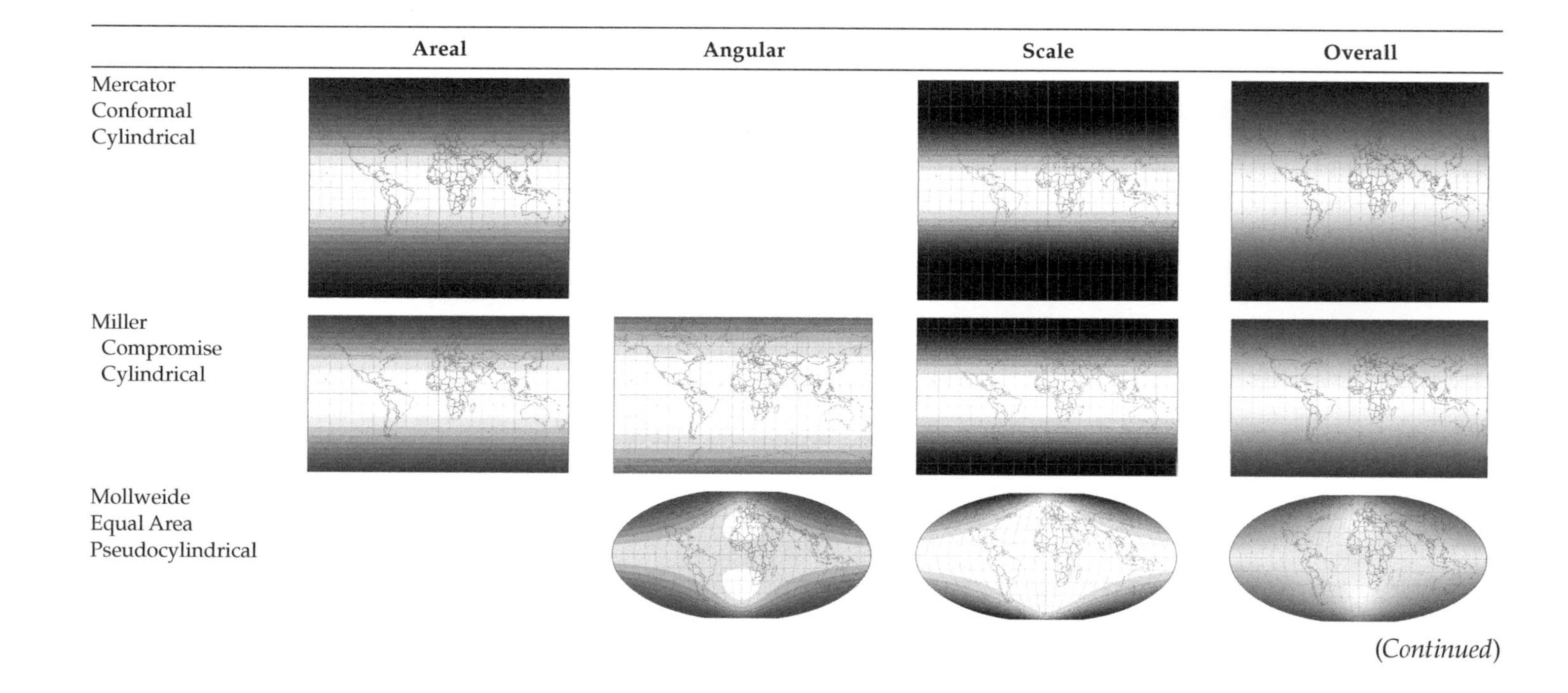
Areal
Angular
Scale
Overall
Mercator
Conformal
Cylindrical
Miller
Compromise
Cylindrical
Mollweide
Equal Area
Pseudocylindrical
(Continued)

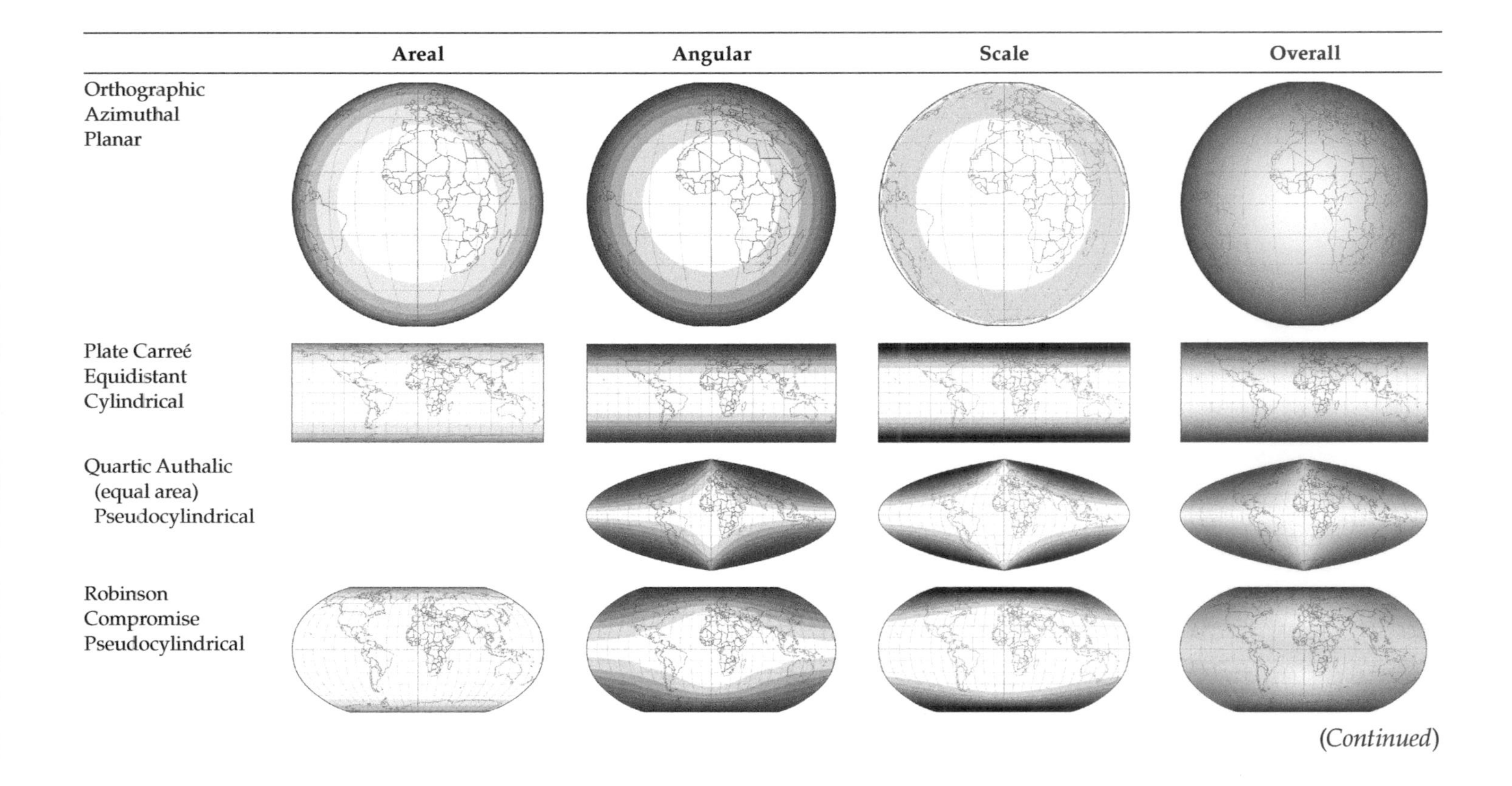

	Areal	Angular	Scale	Overall
Orthographic Azimuthal Planar				
Plate Carreé Equidistant Cylindrical				
Quartic Authalic (equal area) Pseudocylindrical				
Robinson Compromise Pseudocylindrical				

(Continued)

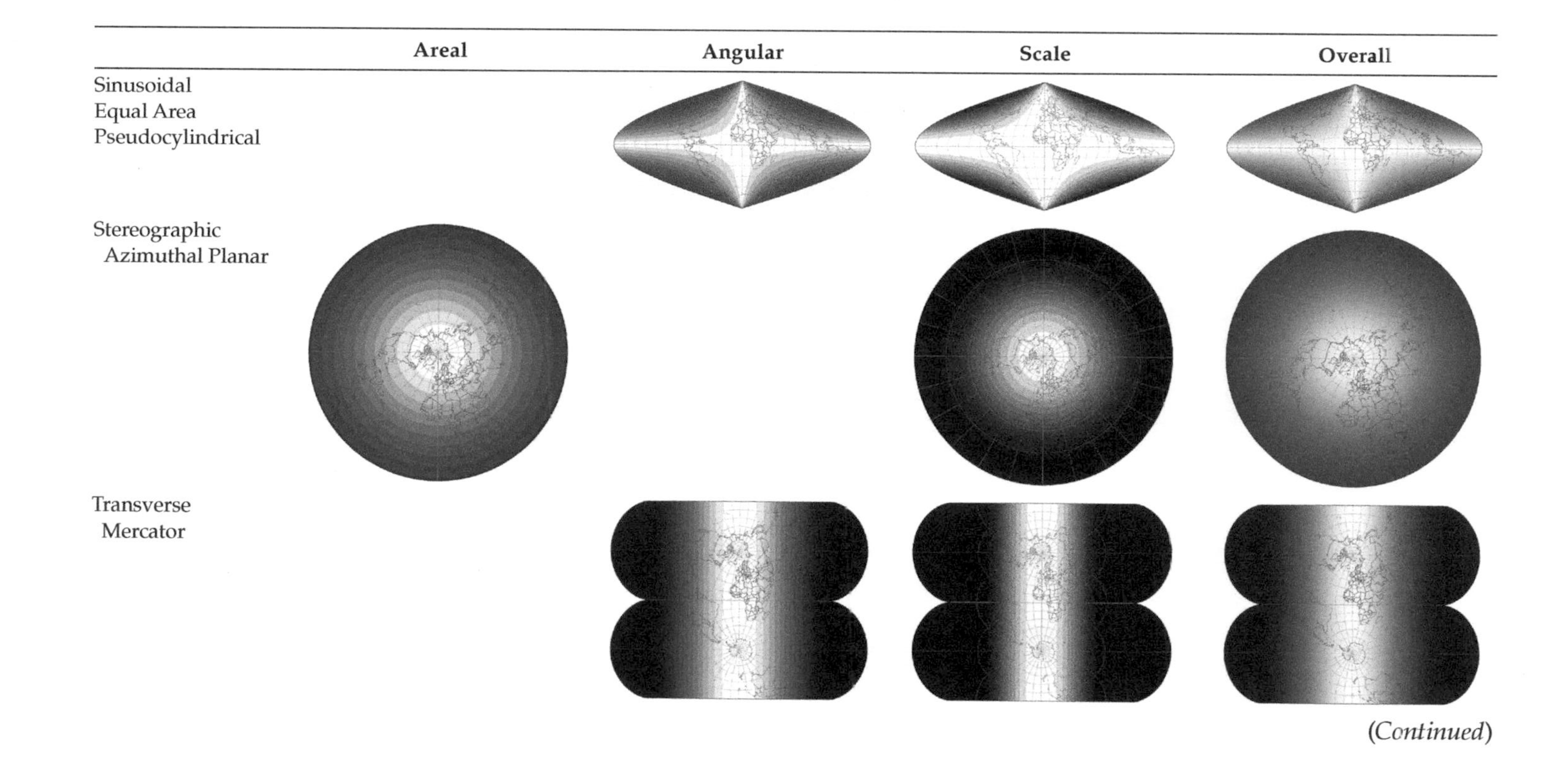
Areal
Angular
Scale
Overall
Sinusoidal
Equal Area
Pseudocylindrical
Stereographic
Azimuthal Planar
Transverse
Mercator
(Continued)

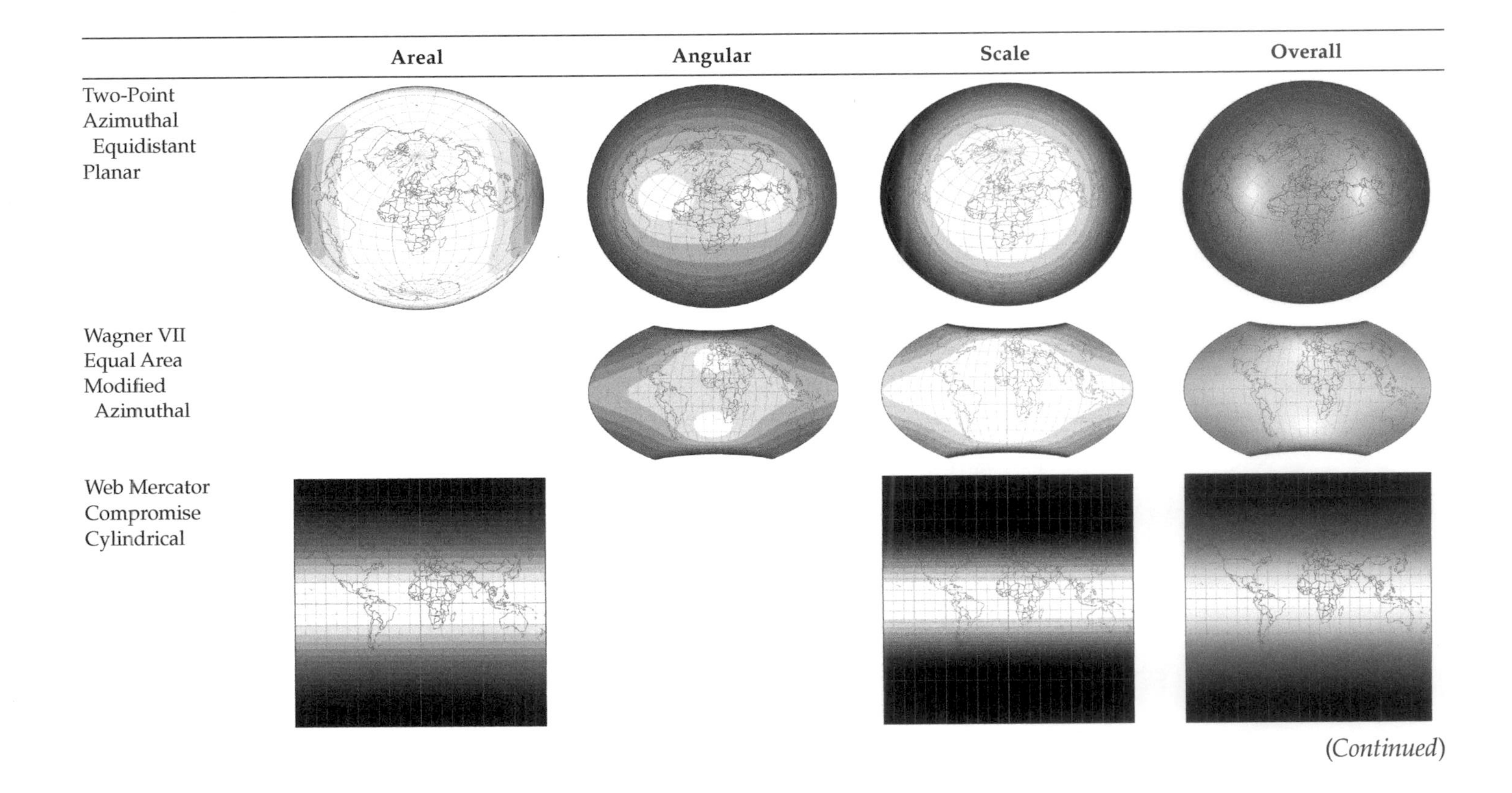
Areal
Angular
Scale
Overall
Two-Point Azimuthal Equidistant Planar
Wagner VII Equal Area Modified Azimuthal
Web Mercator Compromise Cylindrical

(Continued)

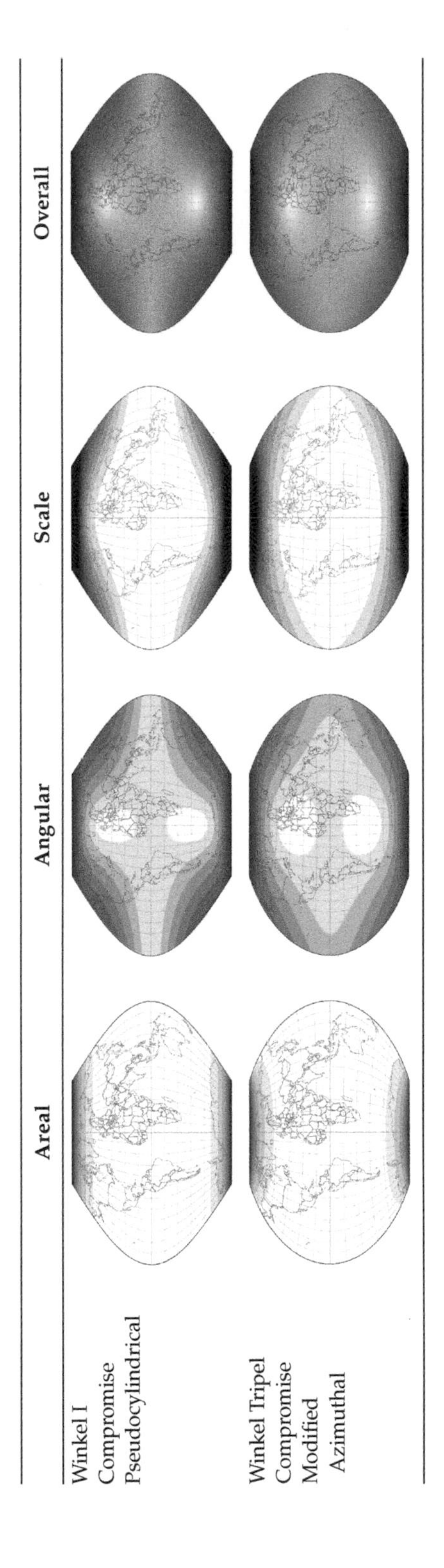
Areal
Angular
Scale
Overall
Winkel I
Compromise
Pseudocylindrical
Winkel Tripel
Compromise
Modified
Azimuthal

Bibliography

American Cartographic Association, American Geographical Society, Association of American Geographers, Canadian Cartographic Association, National Council for Geographic Education, National Geographic Society, and Special Libraries Association, Geography and Map Division Resolution. 1989. Resolution regarding the use of rectangular world maps. *American Cartographer* 16(3): 223.

American Society of Civil Engineers, American Congress on Surveying and Mapping, and American Society for Photogrammetry and Remote Sensing. 1994. *Glossary of the Mapping Sciences*. New York, New York: American Society of Civil Engineers.

Anderson, K., and G. Leinhardt. 2002. Maps as representations: Expert novice comparison of projection understanding. *Cognition and Instruction* 20(3): 283–321. DOI: 10.1207/S1532690XCI2003_1.

Battersby, S., and D. Montello. 2009. Area estimation of world regions and the projection of the global-scale cognitive map. *Annals of the Association of American Geographers* 99(2): 273–291. DOI: 10.1080/00045600802683734.

Battersby, S., D. Strebe, and M. Finn. 2016. Shapes on a plane: Evaluating the impact of projection distortion on spatial binning. *Cartography and Geographic Information Science* 44(5): 410–421. DOI: 10.1080/15230406.2016.1180263.

Battersby, S., and F. Kessler. 2012. Cues for interpreting distortion in map projections. *Journal of Geography* 111(3): 93–101. DOI: 10.1080/00221341.2011.609895.

Battersby, S., M. Finn, E. Usery, and K. Yamamoto. 2014. Implications of web Mercator and its use in online mapping. *Cartographica: The International Journal for Geographic Information and Geovisualization* 49(2): 85–101. DOI: 10.3138/carto.49.2.2313.

Battersby, S.E. 2009. The effect of global-scale map-projection knowledge on perceived land area. *Cartographica* 44(1): 33–44. DOI: 10.3138/carto.44.1.33.

Bertin, J. 1983. *Semiology of Graphics: Diagrams, Networks, Maps*. Madison, Wisconsin: University of Wisconsin Press.

Beşdok, E., A. Geymen, C. Özkan, and H. Palancıoğlu. 2012. Animation-based learning of map projections in geomatics engineering. *Computer Applications in Engineering Education* 20: 666–672. DOI: 10.1002/cae.20436.

Chiodo, J. 1997. Improving the cognitive development of students' mental maps of the world. *Journal of Geography* 96(3): 153–63. DOI: 10.1080/00221349708978777.

Deetz, C., and O. Adams. 1944. *Elements of Map Projections with Applications to Map and Chart Construction*. 5th edition. Washington, D.C.: U.S. Coast and Geodetic Survey. Special Publication #68.

Egenhofer, M., and D. Mark. 1995. *Naive Geography*. COSIT: Conference on Spatial Information Theory. Semmering, Austria: Springer-Verlag.

Gescheider, G. 2013. *Psychophysics: The Fundamentals*. 3rd edition. New York, New York: Psychology Press.

Gilmartin, P. 1983. Aesthetic preferences for the proportions and forms of graticules. *The Cartographic Journal* 20(2): 95–100. DOI: 10.1179/000870483787073125.

Hruby, F., M. Avelino, and R. Ayala. 2016. Journey to the end of the world map: How edges of world maps shape the spatial mind. *GI Forum* (1): 314–323. DOI: 10.1553/giscience2016_01_s314.

Hsu, M. 1972. The role of projections in modern map design. *Cartographica: The International Journal for Geographic Information and Geovisualization* 18(2): 151–186. DOI: 10.3138/9821-m648-7189-0088.

Jenny, B. 2012. Adaptive composite map projections. *IEEE Transactions on Visualization and Computer Graphics* 18(12): 2575–2582. DOI: 10.1109/tvcg.2012.192.

Jenny, B., B. Šavrič, N. Arnold, B. Marston, and C. Preppernau. 2017. A guide to selecting map projections for world and hemisphere maps. In *Choosing a Map Projection* (pp. 213–228). Cham, Switzerland, Springer International Publishing.

Jenny, B., T. Patterson, and L. Hurni. 2008. Flex Projector–interactive software for designing world map projections. *Cartographic Perspectives* 59: 12–27.

Kessler, F. 2018. Map projection education in general cartography textbooks: A content analysis. *Cartographic Perspectives* 90: 6–30. DOI: 10.14714/CP90.1449.

Krider, R., P. Raghubir, and A. Krishna. 2001. Pizzas: π or square? Psychophysical biases in area comparisons. *Marketing Science* 20(4): 405–425. DOI: 10.1287/mksc.20.4.405.9756.

Lee, L. 1944. The nomenclature and classification of map projections. *Empire Survey Review* 7(51): 190–200. DOI: 10.1179/sre.1944.7.51.190.

MacEachren, A. 1995. *How Maps Work*. New York, New York: Guilford Press.

MacEachren, A. M., and D. DiBiase. 1991. Animated maps of aggregate data: Conceptual and practical problems. *Cartography and Geographic Information Systems* 18(4): 221–229.

Maling. D. 1968. The terminology of map projections. *The International Yearbook of Cartography* 8: 11–65.

Maling, D. 1992. *Coordinate Systems and Map Projections*. Oxford, England: Pergamon Press.

Monmonier, M. 2004. *Rhumb Lines and Map Wars: A Social History of the Mercator Projection*. Chicago: University of Chicago Press.

Morrison, J. 1978. Towards a functional definition of the science of cartography with emphasis on map reading. *The American Cartographer* 5(2): 97–110. DOI: 10.1559/152304078784022845.

Muehrcke, P., and J. Muehrcke. 1978. *Map Use: Reading, Analysis, and Interpretation*. Madison, Wisconsin: JP Publications.

Muehrcke, P., A. J. Kimerling, and O. Juliana. 2001. *Muehrcke. Map Use: Reading, Analysis, and Interpretation*. Revised 4th Edition. Madison, WI: JP Publications.

Mulcahy, K., and K. Clarke. 2001. Symbolization of map projection distortion: A review. *Cartography and Geographic Information Science* 28(3): 167–181. DOI: 10.1559/152304001782153044.

Mulders, M., P. Rota, J. Icenogle, et al. 2016. Global measles and Rubella laboratory network support for elimination goals, 2010–2015. MMWR Morbidity and Mortality Weekly Report 65: 438–442. DOI: 10.15585/mmwr.mm6517a3.

National Geodetic Survey. 1986. *Geodetic Glossary*. Rockville, Maryland: National Geodetic Information Center.

Olson, J. 2006. Map projections and the visual detective: How to tell if a map is equal-area, conformal, or neither. *Journal of Geography* 105: 13–32. DOI: 10.1080/00221340608978655.

Robinson, A. 1985. Arno Peters and his new cartography. *The American Cartographer* 12(2): 103–111. DOI: 10.1559/152304085783915063.

Robinson, A. 1990. Rectangular world maps—no!. *Professional Geographer* 42(1): 101–104. DOI: 10.1111/j.0033-0124.1990.00101.x.

Saarinen, T. 1999. The Euro-centric nature of mental maps of the world. *Research in Geographic Education* 1(2): 136–78.

Saarinen, T., M. Parton, and R. Billberg. 1996. Relative size of continents on world sketch maps. *Cartographica* 33(2): 37–48. DOI: 10.3138/F981-783N-123M-446R.

Šavrič, B., B. Jenny, D. White, and D. Strebe. 2015. User preferences for world map projections. *Cartography and Geographic Information Science* 42(5): 398–409. DOI: 10.1080/15230406.2015.1014425.

Šavrič, B., B. Jenny, and H. Jenny. 2016. Projection wizard—An online map projection selection tool. *The Cartographic Journal* 53(2): 177–185. DOI: 10.1080/00087041.2015.1131938.

Slocum, T. A., R. M. McMaster, F. C. Kessler, H. H. Howard, and R. B. Mc Master. 2008. *Thematic Cartography and Geographic Visualization*. Upper Saddle River, NJ: Pearson.

Snyder, J. 1987. *Map Projections—A Working Manual*. Washington, D.C.: U.S. Geological Survey.

Snyder, J. 1993. *Flattening the Earth: Two-Thousand Years of Map Projections*. Chicago, Illinois: University of Chicago Press.

Thompson, A., and B. Hubbard. 2014. *A Comprehensive Population Dataset for Afghanistan Constructed Using GIS-Based Dasymetric Mapping Methods, Washington, D.C., United States Geological Survey Scientific Investigations Report 2013–5238*. 21, DOI: 10.3133/sir20135238.

Tobler, W. 1970. A computer movie simulating urban growth in the Detroit region. *Economic Geography* 46(Supplement): 234–240. DOI: 10.2307/143141.

United States Defense Mapping Agency and Topographic Center. 1973. *Glossary of Mapping, Charting, and Geodetic Terms*. 3rd edition. Washington, D.C.: U.S: Government Printing Office.

Vujakovic, P. 2002. Whatever happened to the 'New Cartography'?: The world map and development mis-education. *Journal of Geography in Higher Education* 26(3): 369–380. DOI: 10.1080/0309826022000019936.

Werner, R. 1993. A survey of preference among nine equator-centered map projections. *Cartography and Geographic Information Science* 20(1): 31–39. DOI: 10.1559/152304093782616733.

Wilkinson, L., and M. Friendly. 2009. The history of the cluster heat map. *The American Statistician* 63(2): 179–184. DOI: 10.1198/tas.2009.0033.

Index